International Series in Operations Research & Management Science

Volume 373

The book series **International Series in Operations Research and Management Science** encompasses the various areas of operations research and management science. Both theoretical and applied books are included. It describes current advances anywhere in the world that are at the cutting edge of the field. The series is aimed especially at researchers, advanced graduate students, and sophisticated practitioners.

The series features three types of books:

- Advanced expository books that extend and unify our understanding of particular areas.
- Research monographs that make substantial contributions to knowledge.
- Handbooks that define the new state of the art in particular areas. Each handbook will be edited by a leading authority in the area who will organize a team of experts on various aspects of the topic to write individual chapters. A handbook may emphasize expository surveys or completely new advances (either research or applications) or a combination of both.

The series emphasizes the following four areas:

Mathematical Programming: Including linear programming, integer programming, nonlinear programming, interior point methods, game theory, network optimization models, combinatorics, equilibrium programming, complementarity theory, multiobjective optimization, dynamic programming, stochastic programming, complexity theory, etc.

Applied Probability: Including queuing theory, simulation, renewal theory, Brownian motion and diffusion processes, decision analysis, Markov decision processes, reliability theory, forecasting, other stochastic processes motivated by applications, etc.

Production and Operations Management: Including inventory theory, production scheduling, capacity planning, facility location, supply chain management, distribution systems, materials requirements planning, just-in-time systems, flexible manufacturing systems, design of production lines, logistical planning, strategic issues, etc.

Applications of Operations Research and Management Science: Including telecommunications, health care, capital budgeting and finance, economics, marketing, public policy, military operations research, humanitarian relief and disaster mitigation, service operations, transportation systems, etc.

This book series is indexed in Scopus.

Yacob Khojasteh • S. M. Mousavi Jahan Abadi
Editors

Emerging Technologies in Supply Chains

Towards Digital Transformation

Springer

Editors
Yacob Khojasteh
Graduate School of Global Studies
Sophia University
Tokyo, Japan

S. M. Mousavi Jahan Abadi
Graduate School of Global Studies
Sophia University
Tokyo, Japan

ISSN 0884-8289 ISSN 2214-7934 (electronic)
International Series in Operations Research & Management Science
ISBN 978-3-032-01217-3 ISBN 978-3-032-01218-0 (eBook)
https://doi.org/10.1007/978-3-032-01218-0

This Springer imprint is published by the registered company Springer Nature Switzerland AG
The registered company address is: Gewerbestrasse 11, 6330 Cham, Switzerland

Preface

In an era of global uncertainty, supply chains have become vital to economic stability, social welfare, and even geopolitical power. The disruptions caused by pandemics, trade conflicts, climate-related disasters, and cybersecurity threats have highlighted the fragility of existing supply chain networks. As the foundation of global trade, supply chains must now navigate growing operational complexity while also responding to demands for greater transparency, sustainability, and resilience. Addressing these challenges requires not only incremental improvements but a systemic transformation of supply chain management practices.

Digital transformation stands at the center of this paradigm shift. By integrating advanced technologies into the structure of supply chain operations, organizations can build systems that are more agile, data-driven, and responsive to change. Digital transformation enables real-time visibility, predictive analytics, and automated decision-making—capabilities that are essential for navigating volatility and uncertainty. Beyond corporate performance, such transformation carries wider implications for economic resilience, public policy, and sustainable development.

Emerging technologies are the engines powering this digital revolution in supply chains. Data analytics and artificial intelligence (AI) enable organizations to process complex datasets, generate practical insights, and automate tasks with exceptional accuracy. Blockchain technologies are fostering trust and accountability through decentralized, immutable records. The Internet of Things (IoT) is creating digitally connected ecosystems where assets and products communicate their status continuously. Digital twin technology offers dynamic models for simulating and optimizing supply chain operations, while the metaverse introduces interactive environments for collaborative problem-solving, training, and remote management.

The potential impact of these technologies extends well beyond operational efficiency. AI and analytics can strengthen supply chain resilience by detecting risks early and optimizing mitigation strategies. Blockchain enhances compliance with regulations on sustainability, labor rights, and product verification. IoT devices enable real-time monitoring that supports proactive maintenance and risk mitigation. Digital twins allow policymakers and business leaders to simulate disruptions—whether from pandemics, natural disasters, or trade shifts—and develop coordinated response strategies. Metaverse applications provide new platforms for capacity building, stakeholder engagement, and decentralized governance in supply chains.

This book presents a comprehensive examination of these technological trends and their implications for supply chain practice, policy, and research. By bridging theory and application, the book provides a useful source for academics, industry practitioners, and policymakers engaged in shaping the future of global supply networks. It aspires to contribute to the broader agenda of building supply chains that are not only more efficient but also more resilient, sustainable, and equitable in an increasingly complex world.

The book is structured into four parts, each focusing on a critical aspect of this digital transformation: (1) Overview of Emerging Technologies and Their Applications, (2) Blockchain Applications for Enhancing Observability and Traceability, (3) Data-Driven Analytics for Resilient and Adaptive Supply Chains, and (4) Artificial Intelligence in Supply Chain Operations.

The first part of the book examines foundational digital technologies that are reshaping and transforming traditional supply chains into intelligent, responsive networks. The second part delves into the growing role of blockchain in enhancing traceability, trust, and transparency, while the third part focuses on data-driven analytics and advanced digital tools that strengthen resilience and adaptability in the face of disruption. The last part of the book explores the transformative power of artificial intelligence—including machine learning and generative AI—in optimizing operations, forecasting demand, and accelerating product development.

In the **first chapter** entitled "Enabling Supply Chain Transformation with Cutting-Edge Digital Tools," we examine how emerging technologies like AI, IoT, cloud computing, and digital twins are reshaping supply chains for greater resilience, efficiency, and performance. Drawing on case studies from industry leaders such as Heineken, Pfizer, and Amazon, the chapter illustrates the measurable impact of digital innovation and assesses transformation progress using a unique digital maturity scale. Emphasizing integrated implementation and human-centered change, it offers practical guidance for building agile, intelligent supply chains equipped to navigate growing complexity. In **Chap. 2,** Albana B. Qehaja presents a panoramic view of the evolution from traditional to digital supply chains, exploring key technologies and their impact on supply chain dynamics. It delves into strategic planning, the integration of digital technologies, and the creation of a digital ecosystem, illustrated by real-world case studies. The chapter emphasizes the importance of innovation, adaptability, and collaboration for success in this evolving field. In **Chap. 3**, by focusing on strategic decision-making in digital supply chains, Melisa Ozbiltekin-Pala presents a framework grounded in the Supply Chain Operations Reference (SCOR) Model Digital Standard. The framework details how digital tools enhance transparency, agility, and data-driven decision-making across the supply chain. It bridges theoretical models and practical applications, showing how organizations can align digital transformation efforts with broader strategic goals.

In **Chap. 4,** Markus Biehl explores the profound impact of Industry 4.0 technologies, collectively known as Transportation 4.0, on the transportation and logistics sectors. He discusses technologies like IoT, cyber-physical systems, big data, and AI, and their potential to improve supply chain visibility, fleet management, and automation. Readers will find a compelling case for how digital innovation can support triple bottom line goals—economic, environmental, and social—while reshaping mobility for a post-pandemic world. In **Chap. 5,** by delving into the complexities of change management in high-tech industries, Patrick Moder, Nour Ramzy, and Hans Ehm introduce a semantic web-based knowledge graph designed to streamline decision-making across the semiconductor supply chain. They explain how structured, machine-interpretable data can improve clarity, consistency, and automation in managing product changes. With practical validation using real-world data, the chapter showcases the potential of this approach to enhance operational accuracy and reduce delays. This is a forward-looking contribution that demonstrates how semantic technologies can improve change responsiveness in intricate global supply networks.

The **second part** of the book focuses on the transformative potential of blockchain technology in enhancing transparency and traceability across supply chains. Xuejian Gong, Mulang Song, Roger J. Jiao, and Linda Zhang begin this part with **Chap. 6** that presents a blockchain-based smart contracting system tailored for crowdsourced manufacturing. By integrating IoT-driven data streams and a three-layer system architecture, it enhances transparency, trust, and real-time decision-making in distributed manufacturing networks. A case study from the tank trailer industry illustrates the system's potential to improve coordination, performance, and operational efficiency in complex, decentralized supply chains. In **Chap. 7**, Elif Dursun and Yavuz Gunalay examine the application of blockchain technology to improve traceability in the food supply chain, using Kashar cheese production as a specific example. The chapter explores how blockchain can address food safety concerns, enhance transparency, and improve collaboration among stakeholders. It also discusses the challenges and future research directions in this area. In **Chap. 8,** Yanting Huang and Tianqin Shi investigate how blockchain can address key challenges in reverse supply chains—such as traceability, data gaps, and low consumer trust in remanufactured goods. They evaluate blockchain's role across three recycling modes and explore its interactions with advertising and information strategies. By highlighting blockchain's potential to increase transparency and efficiency, the chapter outlines a path toward more sustainable and data-driven circular supply chains.

The next four chapters presented in the **third part** of the book emphasize the crucial role of data analytics in creating supply chains capable of withstanding and adapting to disruptions. In the first chapter of this part (**Chap. 9**), Azadeh Dindarian explores the transformative potential of AI, big data, and IoT in enhancing supply chain resilience. Using real-world examples from companies like Amazon and Tesco, the chapter showcases how these technologies enable proactive risk management, improve visibility and agility, and optimize various aspects of supply chain operations. It also addresses the associated challenges and the need for responsible adoption. In **Chap. 10**, Luca Urciuoli and Mustafa Ç. Gurbuz investigate data-driven inventory policies to mitigate the challenges posed by demand and supply uncertainty in complex global networks. By simulating real-time visibility into inventory levels and shipment locations, the chapter demonstrates how data-driven policies can lead to improved service levels and more efficient inventory management.

In **Chap. 11,** Sirish K. Gouda and Prashanth Chandrasekharan examine the role of digital twins in visualizing, predicting, and managing supply chain disruptions. They present an architecture for integrating digital twins with organizational technology assets and explain how these models support agile, reliable, and responsive operations. Drawing on both primary and secondary data, they highlight how digital twins foster resilience through better forecasting, tracking, and decision-making. In **Chap. 12**, by combining the emerging capabilities of the metaverse and blockchain, Yu Cui and Prakash J. Singh propose a decentralized, Web 3.0-based model for enhancing supply chain resilience. Using Japanese case studies, they illustrate how Decentralized Autonomous Organization (DAO) mechanisms, immersive digital environments, and secure data infrastructures can support innovation, collaboration, and agility.

The **last part** in the book delves into the diverse applications of AI in optimizing various supply chain functions. The first chapter of this part introduces the concept of Causal Machine Learning (CML) and its potential to improve decision-making in supply chain management. By focusing on cause-and-effect relationships, CML addresses the limitations of traditional ML models that rely on correlations. In this chapter (**Chap. 13**), Alfredo Roa-Henriquez et al. provide a comprehensive overview of CML algorithms and their applications, highlighting their potential to revolutionize SCM research and practice.

Eric Lambourdière and Elsa Corbin in **Chap. 14** compare the effectiveness of ML-based forecasting methods with traditional statistical models. The chapter demonstrates how real-time data exchange, interoperability, and collaborative decision-making enhance forecasting accuracy. The findings support the growing value of ML in improving demand planning and advancing digital supply chain capabilities. In the last chapter of the book (**Chap. 15**), Tung N. Nguyen proposes a framework for applying generative AI tools—such as Google Colab with Gemini and Vertex AI—to streamline product image design. Focused on e-commerce and customization, the chapter shows how designers can quickly generate tailored, visually compelling product visuals. A case study on bicycle design is presented to illustrate the model's effectiveness in enhancing personalization and reducing development time while acknowledging the need for human oversight.

We would like to thank all the authors who contributed to this book. We appreciate their considerable efforts in preparing and revising their chapters, especially those who were asked to make multiple revisions.

We are also grateful to Camille C. Price, the former Series Editor of the *International Series in Operations Research & Management Science*, and Jialin Yan, Senior Editor, for their support throughout this project. Our thanks also go to the Springer team, including the project coordinator, Sneha Arunagiri, for their assistance and dedication.

Finally, we owe heartfelt thanks to our families for their unwavering support and understanding, which allowed us to devote the necessary time to complete this work. We lovingly dedicate this book to them.

Tokyo, Japan
June 2025

Yacob Khojasteh
S. M. Mousavi Jahan Abadi

Contents

About the Editors

Yacob Khojasteh is Professor of Operations Management at the Graduate School of Global Studies, Sophia University in Tokyo, Japan. He was Visiting Professor at the Department of Mechanical and Industrial Engineering, Toronto Metropolitan University (formerly Ryerson University), in Toronto, Canada, during 2021–2022. He received his Ph.D. in Systems and Information Engineering from the University of Tsukuba, Japan. He is the author or editor of several books, including *Supply Chain Risk Mitigation: Strategies, Methods, and Applications* (2022, Springer), *Supply Chain Risk Management* (2018, Springer), *Production Management* (2017, CRC Press), and *Production Control Systems* (2016, Springer). He has several years of professional experience in industry and consulting. His research interests include production and operations management, supply chain management, logistics and warehousing, systems modeling and optimization, digital transformation, and business analytics.

S. M. Mousavi Jahan Abadi is Senior Research Manager at FUJIFILM Software Co., Ltd. in Tokyo, Japan, where he has led research and development initiatives for over two decades. His work has focused on advancing technologies in digital transformation, cloud computing, data science, and artificial intelligence, contributing significantly to the innovation of numerous products and solutions. In addition to his corporate leadership, Dr. Mousavi Jahan Abadi possesses deep expertise in Japanese business practices and organizational culture. He holds a Ph.D. in Systems Engineering, and since 2021, he has served as Adjunct Professor at Sophia University in Tokyo, where he teaches undergraduate and graduate courses, including *Business Intelligence* and *Management Information Systems*. His combined experience in industry and academia reflects a strong commitment to bridging cutting-edge research with real-world business applications.

Part I
Overview of Emerging Technologies and Their Applications

Chapter 1
Enabling Supply Chain Transformation with Cutting-Edge Digital Tools

Yacob Khojasteh, S. M. Mousavi Jahan Abadi, and Chelsea Ng

Abstract The contemporary supply chain is navigating a significant digital shift, driving organizations to enhance systemic resilience, operational efficiency, and overall performance. This chapter explores how the emerging technologies such as Artificial Intelligence, Machine Learning, the Internet of Things, Cloud Computing, Advanced Analytics, and Digital Twins act as powerful catalysts in this essential transformation. Drawing on in-depth case studies of global leaders such as Heineken, Pfizer, Amazon, Procter & Gamble, and Unilever, it evaluates the strategic application and measurable impacts of these innovations on core supply chain objectives, comparatively assessing their Digital Transformation maturity using a novel Digital Transformation Scale. The analysis synthesizes key insights into practical principles for the effective integrated implementation of digital solutions. It comprehensively examines challenges, risks, and the critical importance of holistic transformation, evolving towards sophisticated "nerve center" capabilities, and emphasizes the human element in change management. This chapter guides organizations in leveraging these technologies to build intelligent, agile, and future-ready supply chains capable of succeeding amidst increasing complexity.

Keywords Digital supply chain · Emerging technologies · Artificial intelligence · Internet of things · Supply chain control tower

---

Y. Khojasteh
Graduate School of Global Studies, Sophia University, Tokyo, Japan

S. M. Mousavi Jahan Abadi (✉)
Graduate School of Global Studies, Sophia University, Tokyo, Japan

FUJIFILM Software Co., Ltd., Yokohama, Japan
e-mail: mousavi@sophia.ac.jp

C. Ng
Graduate Program in Global Studies (GPGS), Sophia University, Tokyo, Japan

Y. Khojasteh, S. M. M. Jahan Abadi (eds.), *Emerging Technologies in Supply Chains*, International Series in Operations Research & Management Science 373, https://doi.org/10.1007/978-3-032-01218-0_1

1.1 Introduction

In an environment characterized by unprecedented global interdependence and increased volatility, supply chains function as the essential framework of international commerce; however, they are perpetually confronted with a diverse array of challenges. These challenges encompass the necessity to ensure supply continuity, augment resilience against disruptions, and enhance profitability through improvements in cost structures and operational efficiency. Such prevailing pressures have substantially influenced the impetus towards the digitization of supply chains, as corporations endeavor to evolve in alignment with dynamic customer expectations, increase the efficiency of personnel activities through the automation of recurrent tasks—thereby enabling a focus on higher-level strategic activities—and achieve a comprehensive enhancement of business performance. Consequently, numerous organizations are strategically adopting digital transformation (DX) initiatives within their supply chains to elevate performance, diminish reliance on manual and paper-centric processes, effectively leverage data for reporting and analytical purposes, and facilitate data-driven decision-making.

The integration of emerging digital technologies into supply chain processes presents organizations with the potential to develop sophisticated digital operational systems, realize significant enhancements in the efficiency of their operations, and achieve reductions in operational expenditures (Ivanov, 2022). Moreover, these technologies are capable of fostering improvements in customer service, enabling more rapid and efficacious responses to supply chain disruptions, and ultimately contributing to enhanced supply chain resilience (Dieaconescu et al., 2022). Through the deliberate embedding of digital capabilities within products, services, and operational processes, organizations can more proficiently gather, analyze, and act upon data sourced from across their entire value chain, encompassing both suppliers and consumers, thereby unlocking novel avenues for establishing competitive advantage.

Despite the considerable promise offered by these technological advancements, the COVID-19 pandemic served as a powerful illustration of the vulnerabilities resident within many global supply chain architectures. Numerous organizations found themselves inadequately prepared for the scale and specific nature of the ensuing disruptions, which were frequently characterized by significant imbalances between supply and demand and were further intensified by governmental interventions, such as factory closures and shipping restrictions (Pellicelli, 2023a). Global supply chains with their substantial reliance on geographically diverse supplier networks, became exceptionally challenging to manage effectively during this period of crisis. A critical insight gained by many was the realization of an overemphasis on lean efficiency, often at the significant detriment of resilience, thereby catalyzing a strategic reorientation from predominantly "just-in-time" operational philosophies towards more robust "just-in-case" contingency approaches (Pellicelli, 2023b). A comprehensive understanding of how DX can substantively support this critical shift and address these fundamental objectives is of essential importance. This chapter, therefore, aims to thoroughly explore the application of emerging technologies within the context of supply chain DX, with a particular focus on their impact on performance, resilience, and efficiency, while also carefully considering the associated advantages and inherent risks. To guide this exploration, the chapter addresses the following initial key research questions:

- **Research Question 1**: In what ways does DX, through the leverage of emerging technologies, influence supply chain performance, resilience, and efficiency?
- **Research Question 2**: Which specific digital technologies make the most substantial contributions to the enhancement of supply chain performance, resilience, and efficiency?

Nevertheless, the path towards effective supply chain digitization is characterized by diverse challenges and inherent risks. Although the pandemic functioned as a catalyst for accelerated digitization efforts within some organizations seeking to strengthen their resilience, other entities experienced considerable delays or a notable lack of progress in their digital strategy progression. Current projections indicate a future wherein supply chains will increasingly depend on autonomous systems, with estimates suggesting that approximately 45% will leverage technologies such as robotics and drones by the year 2035 (Harapko, 2023). However, the attainment of a truly digital and autonomous supply chain paradigm extends significantly beyond the mere adoption of technology.

A common risk involves the development of a disconnected assortment technological solutions, typically characterized by constrained integration capabilities and minimal cross-functional collaboration across the entirety of the end-to-end supply chain (McArthur et al., 2021). Indeed, as will be demonstrated through the case studies in this chapter, leading organizations are actively working to overcome this disconnection by strategically implementing integrated digital platforms and ecosystems, moving towards the sophisticated 'nerve center' capabilities discussed later. Addressing this necessitates the seamless integration of interconnected systems that span planning, procurement, manufacturing, and logistics, crucially extending these connective frameworks across organizational boundaries to comprehensively include partners and suppliers.

This deficiency in profound, real-time, and bi-directional connectivity across the supply chain severely hinders its capacity to adapt dynamically to developing internal and external conditions throughout the broader value chain. Such disconnections almost consistently result in suboptimal performance levels, given that the intrinsic complexity of contemporary supply chains—with their numerous interdependent components—necessitates seamless horizontal and vertical integration to unlock their full operational potential. The criticality of this integration deficit becomes particularly acute during periods of systemic crisis, as was clearly demonstrated by the COVID-19 pandemic. Indeed, findings from a mid-2020 survey involving senior supply chain executives revealed that 85% were contending with challenges originating from inefficient digital technologies, and nearly half reported experiencing significant delays in their decision-making processes, a situation further compounded by the widespread transition to remote working arrangements

(McArthur et al., 2021). These pervasive implementation challenges underscore the urgent need for practical, actionable guidance, which leads to our final central line of investigation for this chapter:

- **Research Question 3**: How can organizations strategically and effectively implement these emerging digital technologies to maximize their realized benefits while concurrently mitigating the associated risks?

To address these critical areas of investigation and fulfill its stated objectives, this chapter is systematically structured as follows. Section 1.2 provides an overview of emerging technologies relevant to the DX of supply chains and deliberates upon the primary objectives for DX, specifically the enhancement of performance, resilience, and efficiency. Section 1.3 presents an analysis of the application of these technologies, drawing upon case studies of prominent global corporations, including Heineken, Pfizer, Amazon, Procter & Gamble, and Unilever. Subsequently, Sect. 1.4 presents the integrated results of this analysis, with a specific examination of how DX technologies impact the core objectives, alongside an outline of principles beneficial for their effective application. Finally, Sect. 1.5 concludes the chapter by providing a summary of the key findings and offering prospective insights into the future path of emerging technologies within supply chain management (SCM).

1.2 Emerging Technologies in DX of Supply Chains

1.2.1 Key Enabling Digital Technologies

The DX of supply chains is fundamentally driven by the strategic adoption and integration of a set of integrated emerging technologies. These technologies, when cohesively implemented, empower organizations to redefine their operational paradigms, enhance visibility, and unlock new efficiencies. Figure 1.1 provides a schematic overview of key digital technologies that are crucial in achieving substantial improvements in supply chain performance, resilience, and efficiency. Central to

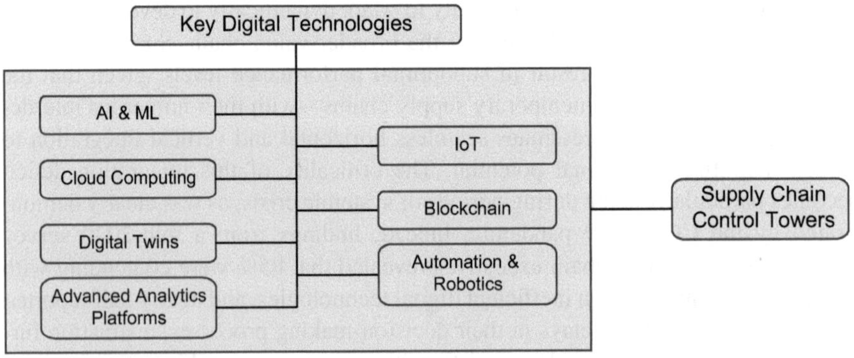

Fig. 1.1 Key digital technologies contributing to supply chain performance, resilience, and efficiency, and their integration into supply chain control towers

the effective utilization of these diverse technologies is the concept of integrated platforms, often manifested as Supply Chain Control Towers. These control towers serve as centralized hubs, ingesting, processing, and visualizing data from across the supply chain network. By consolidating information from multiple sources and leveraging the capabilities of the underlying technologies, these platforms provide end-to-end, real-time visibility, enabling data-driven decision-making and proactive management of potential disruptions. It is important to note that while not all organizations explicitly use the term "Supply Chain Control Tower," the underlying principle of achieving centralized, data-driven command and control functionality is key; the case studies presented later in this chapter will illustrate how leading companies implement such integrated capabilities, sometimes under different internal naming conventions such as "Pfizer's Digital Operations Centre (DOC)."

The key emerging technologies outlined in Fig. 1.1, and their roles in the DX of supply chains, are elaborated below:

- **Artificial Intelligence (AI) and Machine Learning (ML):** AI and ML represent a foundation of modern supply chain digitalization, offering capabilities that range from advanced forecasting to autonomous decision-making. AI algorithms can analyze vast datasets to identify patterns, predict demand fluctuations with greater accuracy, optimize inventory levels, and dynamically adjust logistics routes in response to real-time conditions (Baryannis et al., 2019). ML, a subset of AI, enables systems to learn from data and improve their performance over time without explicit programming, leading to continuous enhancements in areas such as predictive maintenance, quality control, and fraud detection (Nguyen et al., 2018). Their application significantly boosts operational efficiency, predictive capabilities for enhanced resilience, and overall performance through intelligent automation.
- **Internet of Things (IoT):** IoT encompasses a network of interconnected physical devices, vehicles, sensors, and other items embedded with software and connectivity, which enables these objects to collect and exchange data. In supply chains, IoT facilitates real-time tracking and tracing of goods, monitoring of environmental conditions (e.g., temperature for perishables), asset management, and enhanced warehouse operations through smart sensors (Rejeb et al., 2019). The granular and real-time data generated by IoT devices provides better visibility, enabling quicker responses to deviations, thereby improving efficiency and bolstering resilience against unforeseen events.
- **Cloud Computing:** Cloud computing provides on-demand access to a shared pool of configurable computing resources (e.g., networks, servers, storage, applications, and services) that can be rapidly provisioned and released with minimal management effort or service provider interaction (Mell & Grance, 2011). For supply chains, cloud platforms offer scalability, flexibility, and cost-efficiency for hosting SCM applications, storing vast amounts of data, and facilitating collaboration across geographically diverse partners. It serves as a foundational enabler for many other digital technologies, like AI and IoT, by providing the

necessary infrastructure and processing power, thus enhancing overall system performance and data accessibility.

- **Blockchain Technology:** Blockchain is a distributed, immutable ledger technology that facilitates secure and transparent recording of transactions and tracking of assets in a business network. In supply chains, blockchain can enhance traceability of products from origin to consumer, verify authenticity, improve security of transactions, and streamline processes through smart contracts (Kouhizadeh & Sarkis, 2018; Saberi et al., 2019). This increased transparency and trust among supply chain partners can improve efficiency by reducing disputes and paperwork, and significantly strengthen resilience by providing a verifiable audit trail and enhancing data integrity.
- **Digital Twins:** A digital twin is a virtual representation of a physical object, process, or system, which is dynamically updated with real-time data from its physical counterpart (Fuller et al., 2020). In the supply chain context, digital twins can be created for individual assets (like machinery or vehicles), warehouses, or even entire supply chain networks. They enable organizations to monitor performance, run simulations to test different scenarios (e.g., impact of disruptions, changes in network design), optimize processes, and perform predictive maintenance. This capability allows for proactive management, improved efficiency through optimization, and enhanced resilience by enabling stress-testing and contingency planning in a virtual environment.
- **Automation and Robotics:** This category includes both physical automation, such as autonomous mobile robots (AMRs) and automated storage and retrieval systems (AS/RS) in warehouses and manufacturing, and software-based automation like Robotic Process Automation (RPA) for administrative tasks (Fragapane et al., 2021). Robotics enhance the speed, accuracy, and efficiency of repetitive physical tasks like picking, packing, and sorting, while RPA can automate data entry, order processing, and communication, reducing errors and freeing up human resources for more strategic activities. This leads to significant efficiency gains, improved throughput, and can contribute to resilience by reducing reliance on manual labor in critical operations.
- **Advanced Analytics Platforms:** These platforms leverage sophisticated quantitative methods, including statistical analysis, optimization algorithms, and simulation modeling, to derive insights from large and complex supply chain datasets (Souza, 2014). Applications include demand sensing, inventory optimization, network design, risk assessment, and scenario planning. By transforming raw data into actionable intelligence, advanced analytics enable more informed strategic and tactical decisions, leading to improved operational performance, cost reductions, and enhanced ability to anticipate and mitigate risks, thereby contributing to resilience (Sodhi & Tang, 2019).

These technologies are often not deployed in isolation but are increasingly integrated to create complementary effects. The Supply Chain Control Tower, as shown in Fig. 1.1, illustrates such integration, acting as a command center that leverages these diverse digital capabilities to provide comprehensive oversight and control over the supply chain. By combining data streams from IoT devices, processing them with AI and advanced analytics on cloud platforms, and ensuring data integrity through mechanisms potentially supported by blockchain, control towers empower organizations to move towards more predictive, agile, and resilient supply chain operations.

1.2.2 Core Objectives of Supply Chain DX

The strategic impetus behind the DX of supply chains is multifaceted, with a primary focus on achieving significant improvements in three key areas: operational performance, systemic resilience, and resource efficiency. While Sect. 1.1 discussed the broad challenges and motivations for digitization, and Sect. 1.2.1 detailed the enabling emerging technologies, this section will specifically address these three core objectives that guide DX initiatives. Organizations undertake DX not solely to implement new technologies, but to fundamentally enhance their capacity to respond to a volatile global environment, meet evolving customer expectations, optimize resource utilization, and ultimately attain a sustainable competitive advantage.

The pursuit of enhanced **supply chain performance** through DX encompasses a range of improvements. This includes elevating customer service levels through faster order fulfillment, increased product availability, and greater transparency in tracking and delivery. Furthermore, performance improvements are realized through optimized cost structures, achieved by leveraging digital tools for better demand forecasting, inventory management, and logistics planning, thereby reducing waste and unnecessary expenditures. The automation of recurrent tasks, a key facet of DX, frees human capital to focus on strategic, value-added activities, further contributing to overall business performance and responsiveness (Dieaconescu et al., 2022). Data-driven decision-making, facilitated by advanced analytics and integrated platforms, underpins these performance gains, allowing for more precise and proactive management of supply chain operations (Ivanov, 2022).

Secondly, strengthening **supply chain resilience** has emerged as an essential objective, particularly highlighted by recent global disruptions. DX contributes to resilience by enhancing visibility across the entire supply network, enabling organizations to detect potential disruptions earlier and react more swiftly and effectively. Technologies such as IoT, AI, and digital twins allow for real-time monitoring, predictive risk assessment, and the simulation of various scenarios to develop robust contingency plans (Dieaconescu et al., 2022). This proactive approach, moving from reactive problem-solving to anticipatory management, is crucial for maintaining supply continuity and minimizing the impact of unforeseen events. The ability to quickly adapt sourcing, production, and distribution strategies in response to changing conditions is a characteristic of a digitally resilient supply chain.

Finally, the drive for greater **supply chain efficiency** is a constant in the competitive business environment, and DX offers significant avenues for its achievement.

This involves optimizing processes, reducing manual interventions and paperwork, and optimizing the utilization of assets and resources. Automation and robotics, for instance, can drastically improve the speed and accuracy of warehouse operations and manufacturing processes, while cloud computing and advanced analytics platforms enable more efficient data management and processing (Ivanov, 2022). By reducing errors, minimizing delays, and optimizing workflows from procurement to delivery, DX directly translates into lower operational costs and improved productivity. This focus on efficiency not only enhances profitability but also contributes to sustainability objectives by reducing waste and improving resource allocation.

In essence, these three objectives—performance, resilience, and efficiency—are interconnected and often mutually reinforcing. For instance, a more efficient supply chain is typically more performant, and enhanced visibility achieved for resilience purposes can also lead to efficiency gains through better coordination. The strategic application of the emerging technologies discussed in Sect. 1.2.1 is geared towards holistically advancing these objectives, enabling supply chains to become more agile, intelligent, and robust in the face of complex contemporary challenges.

In a recent study, Khojasteh et al. (2026) examined strategic approaches to digital transformation for building agile, high-performing, and resilient supply chains. Tracing the evolution toward Supply Chain 4.0, they highlighted the transformative role of AI, IoT, and big data across areas like transportation, warehousing, logistics, and procurement. Moreover, Khojasteh & Mousavi Jahan Abadi (2026) explored how organizations can harness digital transformation to achieve abovementioned supply chain goals.

1.3 Case Study Analysis: Emerging Technologies in Supply Chain Practice

1.3.1 Case Study Methodology and Analytical Framework

The methodological framework for the analysis presented in the following subsections, which examines the application of emerging technologies in real-world supply chains, begins with a review of existing academic literature. This foundational review focused on DX within SCM, serving to identify prevalent themes, key emerging technologies (as detailed in Sect. 1.2.1), and established best practices relevant to the digitalization of supply chain operations. Building upon this understanding, the core of the analysis employs a multiple case study approach. This approach was chosen to provide practical insights into how specific organizations are leveraging DX and the resultant impacts. The case studies concentrate on five prominent global corporations: Heineken, Pfizer, Amazon, Procter & Gamble, and Unilever.

These companies were selected based on several criteria: their significant global operational scale, their representation of diverse and complex industry sectors (including Fast-Moving Consumer Goods (FMCG)—beverages and consumer products, pharmaceuticals, and e-commerce/logistics), their recognized strategic engagement in substantive DX initiatives within their supply chains, and the availability of public domain information detailing these efforts.

Data for the case studies were systematically gathered from a range of publicly accessible sources. These sources included official company publications such as annual reports, sustainability reports, and investor briefings; dedicated company blog articles and press releases detailing specific DX projects and strategies; news articles and reports from reputable business, technology, and supply chain media outlets. While this approach provides a rich and diverse set of information, it is acknowledged that reliance on publicly available data inherently means that certain proprietary strategic details or internal operational nuances may not be fully captured.

For each case study company, the analysis focused on several key aspects. Firstly, the specific emerging digital technologies and platforms employed within their supply chain operations were identified, referencing the technological categories outlined in Sect. 1.2.1 (e.g., AI/ML, IoT, Cloud Computing, Automation/Robotics, Advanced Analytics, Blockchain, Digital Twins). Secondly, an assessment was made of the extent and nature of technology integration—specifically, examining whether these technologies operate in isolated segments of the supply chain or are integrated into a more cohesive, interconnected, and end-to-end system. Crucially, the analysis sought to evaluate the identifiable impact of these DX initiatives on the core strategic objectives of enhancing supply chain **performance**, strengthening **resilience**, and improving operational **efficiency**, as outlined in Sect. 1.2.2.

To provide a structured framework for evaluating and comparing the relative maturity of DX across the selected case study companies, a 'Digital Transformation Scale' (DX Scale) is introduced. This scale classifies the level of DX advancement based on a set of qualitative indicators, ranging from 'Low' (indicating minimal digital integration and reliance on offline processes) to 'Very High' (representing complete, cohesive integration of advanced technologies enabling real-time, data-driven, and adaptive supply chain operations). The specific indicators that define each level of the DX Scale are detailed in Table 1.1. This scale will be utilized in Sect. 1.3.7 to summarize the DX application levels observed across the case study companies.

1.3.2 DX in Heineken

Heineken, a global brewing company, has strategically adopted DX across its supply chain to enhance resilience, operational efficiency, and overall performance in a dynamic market. Their approach involves a complementary combination of investing in emerging technologies, improving organizational structures and processes, and developing a digitally adept organizational culture. This multi-faceted strategy aims to create an interconnected, data-driven supply chain capable of navigating volatility and capitalizing on new opportunities.

Table 1.1 DX scale indicators

DX scale	Indicators
Low	Minimal digital tools, primarily offline operations with limited data usage: basic digital tools for communication, minimal automation, primarily manual processes, and limited data use
Moderate	Basic digital tools in place, with data usage for reporting rather than decision-making: some automated systems (e.g., inventory management), digital workflows for internal processes, and a basic ERP (Enterprise Resource Planning) system
Intermediate	Integrated digital processes with data analytics: advanced ERP systems, beginning of data analytics for decision-making, partial supply chain visibility, and initial automation in operations
High	Widespread technology adoption, advanced data analytics, and integrated platforms but with potential gaps in integration: real-time data analytics, advanced automation, digital dashboards for monitoring, integration with supply chain partners, and use of predictive analytics
Very High	Complete, cohesive integration of technologies and platforms across all operations, enabling real-time data-driven decision-making and adaptive processes: predictive analytics, AI, real-time data sharing across the supply chain, automated decision-making processes, digital twin technology, and full supply chain visibility

A foundation of Heineken's DX is its Sales & Operations Planning (S&OP) evolution, notably through programs like S&OP FIT. This initiative leverages **Advanced Analytics Platforms**, enabling real-time, data-driven decision-making and sophisticated scenario analysis to effectively manage market volatility and uncertainty. For example, regional hubs, such as those in Africa, utilize simulations—a key capability of advanced analytics—to optimize resource allocation during disruptions, thereby improving both adaptability (resilience) and foresight (performance) (Kaporina et al., n.d.). This data-centric planning is critical for enhancing supply chain resilience against unforeseen events and improving forecasting performance.

Heineken has also focused on streamlining operations to reduce complexity and unlock efficiencies. The consolidation of 25 distinct supply and operations teams into a unified global structure has significantly enhanced decision-making agility and operational flexibility. This organizational shift was supported by substantial product and packaging simplification, including a 52% reduction in unique bottle types and a 50% decrease in secondary packaging variations (Ashcroft, 2023). Such rationalization directly contributes to efficiency by reducing changeover times and inventory holding costs, and to resilience by simplifying the supply chain network and mitigating potential points of failure. This is supported by robust ERP systems that provide a foundational data structure for such unified operations. Furthermore, Heineken's "heads, hearts, and hands" approach to change management empowers employees, fostering a culture capable of adapting to and leveraging digital tools, while the "tree mentality" emphasizes establishing strong foundational processes that support both productivity (efficiency) and agility (resilience) (Clark, 2024).

The application of **AI** and the **IoT** is central to Heineken's operational excellence. The "Connected Brewery" program is a prime example, utilizing IoT sensors to collect real-time operational data from equipment. This data then feeds into

AI-powered predictive maintenance models, allowing Heineken to anticipate potential equipment failures and schedule maintenance proactively, thus preventing costly disruptions and enhancing production efficiency and supply resilience (Drinkwater, 2023; Reynolds, 2023). Beyond manufacturing, AI-driven ordering systems are deployed to empower partners, such as bars, to monitor consumer demand more accurately. This improves responsiveness to market changes (performance) and optimizes inventory levels throughout the value chain (efficiency) (Reynolds, 2023). Specific AI tools, like AIDDA (AI-Driven Decision Automation), are employed to optimize inventory further, predict sales trends, and streamline overall sales processes. A notable outcome was a 5% uplift in sales efficiency in pioneering markets like Mexico, demonstrating a direct positive impact on performance (Reynolds, 2023).

To support these advanced digital capabilities, Heineken leverages **Cloud Computing** extensively. A comprehensive data integration model, reportedly involving **Digital Twins** for process visualization and analysis, centralizes data from operations across its 70 countries of operation onto a cloud-based infrastructure (Clark, 2024; Wolski, 2022). This centralized data repository, enabled by the scalability and accessibility of cloud platforms, facilitates consistent, data-informed decision-making across the global organization (performance) and mitigates the risks associated with isolated information systems, thereby improving overall operational efficiency.

Sustainability initiatives are also integrated into Heineken's operational strategy, with DX playing an enabling role. While achievements like a 24% reduction in carbon emissions and the adoption of low-complexity packaging are multi-faceted (Ashcroft, 2023), the data analytics and process optimization capabilities derived from digital tools contribute to more efficient resource utilization and waste reduction. Furthermore, the company's Knowledge & Insight Management (KIM) platform, which utilizes AI in the form of chatbots and automated processes, promotes the reuse of insights and accelerates decision-making. This not only leads to substantial time and cost savings (improving efficiency)—with research cost reductions reported in the hundreds of thousands of Euros annually—but also fosters a learning organization (CDOTrends editors, 2023).

Other specific platforms contributing to Heineken's DX include the Eazle sales platform, enhancing sales force productivity and order processing (**performance and efficiency**), and the Digifit platform, aimed at improving workforce productivity and collaboration in the digital realm (**efficiency**).

In summary, Heineken's DX strategy is characterized by the integration of advanced digital tools like AI, IoT, Cloud Computing, Digital Twins, and Advanced Analytics Platforms, coupled with significant organizational and process streamlining. This holistic approach has demonstrably enhanced their supply chain resilience, operational efficiency, and market adaptability and performance (Brans, 2022; Clark, 2024; Drinkwater, 2023). The combination of these technologies with purpose-built, integrated platforms such as Connected Brewery and S&OP FIT, and a strong focus on data integration across its global operations, analytics, and insights management, signifies an advanced stage of DX. This cohesive infrastructure, enabling real-time data utilization and predictive capabilities, aligns with the indicators for a 'High' classification on the DX Scale (as defined in Table 1.1), reflecting widespread technology adoption, advanced data analytics, and integrated platforms.

The key technologies, their advantages, and associated risks in the context of Heineken's DX are summarized in Table 1.2 (adapted from Ashcroft, 2023; Brans, 2022; CDOTrends editors, 2023; Clark, 2024; Drinkwater, 2023; Kaporina et al., n.d.; Reynolds, 2023; Wolski, 2022).

Table 1.2 Summary of Heineken's key DX technologies/platforms, advantages, and risks

Technology/ platform	Advantages	Risks
Core enabling technologies		
AI	Predictive maintenance (e.g., Connected Breweries), demand forecasting, inventory optimization, enhanced sales processes	High dependency on data quality and availability, potential biases in algorithms, complexity in model development and maintenance
ML	Enhanced scenario analysis (S&OP FIT), sales process optimization (e.g., 5% uplift in Mexico), powering predictive analytics	Requires substantial computational resources and skilled personnel, risk of inaccurate predictions if models are not well-tuned
Cloud Computing	Real-time collaboration and data access across 70 countries, centralized data for consistent decision-making, scalability, agility	Risks of data breaches or service outages, reliance on third-party providers, data governance and sovereignty concerns
IoT	Real-time issue reporting and monitoring (Connected Breweries), predictive maintenance enablement, improved operational responsiveness	Security vulnerabilities in IoT devices, data overload, integration challenges with legacy systems and across diverse protocols
Digital Twins	Enhanced process visualization, simulation for scenario planning (inferred), optimized data integration and modeling	High initial investment, complexity in development and maintenance, requires accurate real-time data feeds
Advanced Analytics Platforms	Streamlined operations, improved resource allocation, data-driven decision-making (S&OP FIT/ CORE), event-driven responsiveness	Complexity in implementation, requires specialized analytical skills, high initial costs for software and expertise
Foundational & supporting systems		
ERP Systems	Unified operations, reduced process complexity, improved coordination, foundational data management	High implementation and customization costs, need for extensive staff training, potential for rigidity if not configured well
Cybersecurity Measures	Protects sensitive data and interconnected digital systems, ensures operational continuity and regulatory compliance	Constant evolution of cyber threats requires continuous updates and investment, potential for insider threats

(continued)

Table 1.2 (continued)

Technology/ platform	Advantages	Risks
Specific Heineken platforms/programs		
S&OP CORE Program	Enhances strategic supply chain planning and alignment across regions, supports long-term forecasts	Requires substantial cross-functional data integration, complexity in coordinating global operations and stakeholder alignment
S&OP FIT Program	Real-time, data-driven decision-making, scenario analysis for disruptions, enhanced agility	Relies heavily on accurate and timely data input, can be resource-intensive for maintenance and updates
Eazle Sales Platform	Improves sales force productivity, optimises inventory visibility for sales, accelerates order processing	Potential difficulties in integration with existing legacy systems, user adoption challenges
Knowledge & Insight Management (KIM)	Centralizes insights, facilitates knowledge reuse, saves costs and time via AI chatbots and automated processes	Dependency on data consistency and quality of insights, requires regular updates and active user engagement to ensure relevance
Connected Brewery (IoT Platform)	Leverages IoT & AI for predictive maintenance, prevents disruptions, ensures real-time operational reporting	IoT device vulnerabilities, high implementation and maintenance costs for sensors and infrastructure
Digifit Platform	Enhances workforce productivity through digital tools, fosters collaboration, and streamlines digital adoption	Requires continuous updates to platform and training for users, ensuring relevance and effective utilization across workforce

1.3.3 DX in Pfizer

Pfizer, a global pharmaceutical leader, has strategically leveraged DX to significantly enhance its complex global supply chain and manufacturing operations, a necessity amplified by the unprecedented demands of the COVID-19 pandemic. Their approach integrates advanced digital technologies to improve predictive capabilities, operational flexibility, and overall responsiveness, targeting substantial gains in performance, resilience, and efficiency.

A critical component of Pfizer's DX is the application of Advanced Analytics Platforms and **AI/ML**. The Highly Orchestrated Supply Network (HOSuN) platform illustrates this, utilizing advanced predictive analytics to improve decision-making for products with long lead times, such as vaccines which can require up to 15 months for production. HOSuN enables optimal resource allocation by directing production activities to the most suitable facilities, thereby enhancing operational flexibility (performance) and ensuring efficient use of capacity (efficiency) (Pfizer Annual Review, 2016). This predictive capability was essential in managing the dynamic supply and demand for critical medicines.

Innovation in manufacturing processes is deeply intertwined with Pfizer's digital strategy. The adoption of Portable, Continuous, Miniature, and Modular (PCMM) manufacturing systems and Single-Use Bioreactor (SUB) technology represents a significant step in **Automation** and agile production. These systems reduce reliance on traditional, large-scale fixed facilities, allowing Pfizer to scale production with greater flexibility and rapidly deploy modular manufacturing units globally as needed. Notably, PCMM facilities require approximately 70% less physical space than conventional setups and facilitate quicker assembly and production initiation, directly strengthening manufacturing efficiency and systemic resilience by decentralizing production capabilities (Pfizer Annual Review, 2016).

To improve supply chain resilience and performance, Pfizer implemented a set of integrated digital tools, including the Digital Operations Centre (DOC). The DOC functions as a form of Supply Chain Control Tower, leveraging IoT devices for real-time tracking (e.g., maintaining stringent cold-chain conditions for vaccines) and AI-driven predictive models to monitor for potential disruptions and enable rapid, data-informed responses. During the COVID-19 pandemic, these capabilities were crucial. Real-time predictive models optimized clinical trial site selection, while virtual monitoring technologies allowed trials to continue with minimal interruption despite widespread access restrictions. This comprehensive digital ecosystem contributed significantly to accelerating vaccine development from a typical 10-year timeline to less than 1 year, culminating in the delivery of the COVID-19 vaccine just 9 months after development commenced (Martin, 2023; Moa, 2021). The use of Augmented Reality (AR) technology further supported operations by speeding up production training and equipment setup, contributing to efficiency and reducing ramp-up times (as noted in Table 1.3).

Logistics optimization, another key focus, benefited from **Automation** in alerts and **IoT-enabled real-time tracking**. These systems collectively reduced lead times and improved inventory control, achieving a remarkable 99.9% delivery success rate for critical products. Automated workflows further contributed to minimizing product waste, with real-time monitoring ensuring less than 0.1% loss for temperature-sensitive COVID-19 vaccines, a significant achievement in both efficiency and performance (Martin, 2023). The underlying **Cloud Computing** infrastructure (referred to as Cloud Analytics Platform in Table 1.3) provides the scalable data processing and storage capabilities essential for these real-time analytics and global coordination efforts.

Pfizer's DX initiatives have demonstrably enhanced operational efficiency across multiple domains. The company reported shortening time-to-market for new health innovations by over 50% and reducing cycle times for core manufacturing processes by up to 10%. Centralized digital systems helped eliminate redundant processes, while AI-driven real-time data analysis enabled faster decision-making and consolidation of clinical trials (Hall & Fonseca, 2024; Pfizer Annual Review, 2016). These improvements directly impact overall business performance and competitiveness.

Even in supporting functions like marketing and creative operations, DX has yielded significant efficiencies that indirectly benefit the supply chain by accelerating product launches. Pfizer employed AI tools like Adobe Firefly and integrated platforms such as Adobe Experience Manager (AEM) for content management and Workfront for project management. These technologies streamlined workflows,

Table 1.3 Summary of Pfizer's key DX technologies/platforms, advantages, and risks

Technology/platform	Advantages	Risks
Core enabling technologies		
AI & ML	Supports real-time data analysis, predictive modeling, improving decision-making and supply chain efficiency	Potential bias or errors in data training, model interpretability, dependency on data quality
IoT Devices	Enables real-time monitoring (e.g., cold-chain logistics), asset tracking, and predictive maintenance	Security vulnerabilities in IoT networks, data overload, integration challenges with legacy systems
Advanced Analytics Platforms (e.g., HOSuN)	Facilitates predictive analytics for long-lead-time products, optimal resource allocation, scenario planning	Requires continuous, high-quality data updates, complexity in model integration across global facilities
Cloud Computing (Cloud Analytics Platform)	Facilitates scalable, centralized data analysis, storage, and collaboration across the organization	Vulnerable to data breaches and service outages, reliance on third-party providers, data sovereignty
Automation (General & Process Specific)	Reduces waste, enhances inventory control, increases productivity, streamlines workflows	Requires significant capital expenditure for implementation, potential job displacement concerns
Specific Pfizer systems & applications		
Digital Operations Centre (DOC)	Provides real-time disruption monitoring, cold-chain condition maintenance, and rapid response capabilities	Vulnerable to cyberattacks due to real-time network dependencies, complexity of integrating diverse data feeds
PCMM System & SUB Platform	Enables scalable, flexible, modular production, reducing dependence on large facilities and setup time, enhances efficiency	High initial implementation costs, dependency on specialized modular facility coordination and supply chain
Virtual Monitoring	Ensures continuity of clinical trials during access restrictions, enables remote oversight	Dependent on robust digital infrastructure, may limit nuanced physical oversight or interaction
Predictive Models (for trials, supply)	Optimises trial site selection, accelerates product development, improves demand forecasting	Requires large, validated datasets and high computational resources, model accuracy dependent on assumptions
Augmented-Reality (AR) Technology	Speeds up production training, equipment setup, and potentially remote maintenance assistance	High costs for AR content development and hardware, potential user adaptation challenges
Supporting digital tools (marketing & operations)		

(continued)

Table 1.3 (continued)

Technology/platform	Advantages	Risks
Workfront	Streamlines project management and enhances cross-department collaboration	Potential challenges in user adoption and integration without proper training and change management
Adobe Creative Cloud & Adobe Firefly	Speeds up content creation, reduces time-to-market for materials, and increases creative efficiency	Subscription costs, integration challenges with other systems, ensuring brand compliance with AI content
Digital Asset Management (DAM)	Centralises marketing assets, reducing duplication and inefficiencies in content workflows	Requires diligent metadata management and regular updates, user adherence to processes
Content Management System (CMS)/Adobe Experience Manager (AEM)	Simplifies website and content creation, enabling rapid brand launches and ensuring brand consistency	Security vulnerabilities and potential downtime if not properly managed, high implementation/licensing costs

reportedly reducing content creation time by more than 50%, enabling the rapid launch of over 10 brand websites, and fostering improved collaboration between teams. This agility cuts costs and accelerates time-to-market for products, thereby enhancing overall organizational performance (Hall & Fonseca, 2024).

In summary, Pfizer's DX is characterized by the deep integration of Advanced Analytics Platforms (HOSuN), AI/ML, IoT, Automation (including PCMM/SUB and workflow automation), Cloud Computing, and elements suggestive of Digital Twin capabilities (DOC, virtual monitoring). These technologies are not isolated but are woven into cohesive systems like the DOC and HOSuN, facilitating comprehensive, real-time visibility and adaptive control across its supply chain and manufacturing operations. This level of integrated, data-driven decision-making, predictive capability, and operational adaptability clearly aligns Pfizer with the 'Very High' indicators on the DX Scale (as shown in Table 1.1). The company demonstrates "complete, cohesive integration of technologies and platforms across all operations, enabling real-time data-driven decision-making and adaptive processes," including extensive use of "predictive analytics, AI, real-time data sharing across the supply chain," and "full supply chain visibility."

The key technologies, their advantages, and associated risks in the context of Pfizer's DX are summarized in Table 1.3 (adapted from Hall & Fonseca, 2024; Martin, 2023; Moa, 2021; Pfizer Annual Review, 2016).

1.3.4 DX in Amazon

Amazon, a global e-commerce and cloud computing giant, has fundamentally architected its vast and complex supply chain around DX. The company continuously employs a sophisticated array of emerging technologies, many of which are developed and deployed through its Amazon Web Services (AWS) division, to achieve unparalleled levels of operational efficiency, customer-centric performance, and systemic resilience. This deep integration of digital capabilities is not merely an evolutionary step but a foundational element of Amazon's business model and competitive strategy.

Central to Amazon's supply chain DX is the widespread use of **AI and ML**. These technologies support critical functions, from demand forecasting to logistics optimization. Amazon's ML-driven demand planning systems, such as components within the AWS Supply Chain platform, provide real-time responsiveness to demand fluctuations, crucial for managing unexpected surges or disruptions. By continually refining algorithms to lower Mean Absolute Percentage Error (MAPE) values, Amazon significantly minimizes overstocking and stockouts, thereby enhancing inventory management efficiency and reducing associated costs (Balasubramanian, 2024). ML models also strengthen contingency planning by generating precise vendor lead-time estimates (e.g., through AWS Supply Chain Vendor Lead Time Insights), enabling the optimization of buffer stock levels and supporting strategies for supplier diversification, which directly contributes to supply chain resilience (Kane & Balasubramanian, 2024; Pantoja-Navajas, 2023; Selvaraj, 2024).

Specific AI applications, such as generative AI-powered onboarding agents and tools like Amazon Q (when applied to supply chain contexts), automate data ingestion and streamline supplier onboarding processes, reducing manual error rates and accelerating integration, which improves both efficiency and the speed at which resilience measures can be implemented (Metha, 2024; Shah, 2024).

The backbone of Amazon's digital supply chain is its extensive **Cloud Computing** infrastructure, primarily AWS. The AWS Supply Chain platform is a key example, offering a set of integrated of interconnected applications designed to unify data, provide real-time insights, and enhance decision-making. This platform centralizes data through services like the AWS Supply Chain Data Lake (SCDL) and utilizes AWS Simple Storage Service (S3) for scalable storage, forming a robust **Advanced Analytics Platform**. This allows for comprehensive visibility and coordination across the supply chain (Shah, 2024; Wishart, 2024). Capabilities like AWS Supply Chain N-Tier Visibility enhance communication with suppliers and allow for early identification of potential risks deep within the supply network, directly improving resilience (Kane & Balasubramanian, 2024). Automated data processes, supported by this cloud infrastructure and specific tools like a Data Orchestration Framework and Data Quality Engine, reduce manual interventions, cut error rates, and enable faster recovery from disruptions (Travers & Baeuml, 2024).

Extensive **Automation and Robotics** are characteristics of Amazon's fulfillment centers and logistics network. Warehouse Automation and Optimization (WAO) strategies, deploying fleets of robotic systems for picking, packing, and sorting, dramatically increase labor productivity, improve space utilization, and reduce fulfillment times. This high degree of automation enhances operational efficiency and contributes to resilience by reducing dependency on manual labor, especially during peak periods or labor shortages (Metha, 2024; Travers & Baeuml, 2024). **IoT** devices are widely deployed for real-time tracking of shipments and assets, monitoring environmental conditions, and enabling predictive maintenance on critical equipment. This granular, real-time data improves logistics performance, reduces lead times, and minimizes disruptions, further contributing to both efficiency and resilience (Kane & Balasubramanian, 2024).

Amazon also utilizes **Digital Twins** to simulate and optimize various aspects of its supply chain processes. By creating virtual replicas of fulfillment centers or logistics networks, fueled by real-time data, Amazon can test different scenarios, predict bottlenecks, and optimize layouts and workflows without disrupting physical operations. This capability is invaluable for risk management, capacity planning (potentially using tools like ProModel software), and enhancing overall operational efficiency and resilience (Shah, 2024). Technologies like LIDAR scanning contribute to creating accurate spatial data for these digital models and for optimizing robotic navigation within warehouses (Travers & Baeuml, 2024).

The cumulative impact of these DX initiatives is substantial. Real-time tracking and predictive analytics have demonstrably reduced lead times and improved warehouse operations (Kane & Balasubramanian, 2024). Amazon has reported significant outcomes such as a 30% reduction in excess inventory due to enhanced demand forecasting and marked improvements in fulfillment times (Shah, 2024; Wishart, 2024). Streamlined integration processes, for instance, with automated setups for SAP S/4HANA connections, have drastically cut supplier integration times to under 30 minutes, saving costs and enhancing operational agility (Metha, 2024). Furthermore, the company has cited a 40% increase in labor productivity and a 15% improvement in space utilization, reflecting its commitment to maximizing efficiency through these advanced technologies (Travers & Baeuml, 2024). The AWS Supply Chain Sustainability application also underscores a commitment to leveraging data for reducing waste and optimizing resource allocation (Shah, 2024).

In summary, Amazon's approach to DX in its supply chain is characterized by the deep, complementary integration of AI/ML, cloud computing (AWS), extensive automation and robotics, IoT, advanced analytics, and digital twin capabilities. These technologies are embedded within cohesive platforms like AWS Supply Chain, enabling real-time, data-driven decision-making, predictive insights, and highly automated processes. This holistic strategy directly enhances supply chain performance through speed and accuracy, builds profound resilience against disruptions via proactive risk management and adaptability, and drives significant operational efficiencies through optimized resource utilization and waste reduction. Amazon's operations exhibit "complete, cohesive integration of technologies and platforms across all operations, enabling real-time data-driven decision-making and adaptive processes," including "predictive analytics, AI, real-time data sharing across the supply chain, automated decision-making processes," and "full supply chain visibility," thereby categorizing its DX level as 'Very High' on the DX Scale (as outlined in Table 1.1).

The key technologies, their advantages, and associated risks in the context of Amazon's DX are summarized in Table 1.4 (adapted from Balasubramanian, 2024; Kane & Balasubramanian, 2024; Metha, 2024; Pantoja-Navajas, 2023; Selvaraj, 2024; Shah, 2024; Travers & Baeuml, 2024; Wishart, 2024).

Table 1.4 Summary of Amazon's key DX technologies/platforms, advantages, and risks

Technology/platform	Advantages	Risks
Core enabling technologies		
AI/ML	Improves demand forecasting, inventory optimization, contingency planning, automates data ingestion, and enhances decision-making accuracy	Risks of model bias or mispredictions if not properly trained/validated, dependency on vast high-quality data, complexity of models
Cloud Computing (leveraging AWS infrastructure)	Provides scalable, resilient, and secure infrastructure for data storage, processing, and application hosting; enables global collaboration	High dependency on cloud infrastructure availability, potential for data breaches or cyberattacks, ongoing operational costs
Automation & Robotics (incl. Warehouse Automation—WAO)	Increases labor productivity, improves space utilization, reduces fulfillment times, enhances operational consistency and throughput	High upfront capital investment, requires specialized maintenance and skills, potential job displacement concerns
IoT	Enables real-time tracking of goods and assets, predictive maintenance, monitoring of environmental conditions, improves logistics visibility	Security vulnerabilities in widespread IoT device networks, data overload, integration challenges with diverse systems
Advanced Analytics Platforms (often built on AWS services)	Facilitates sophisticated data analysis, predictive modeling, scenario simulation, and optimization for informed decision-making	Requires specialized analytical skills, complexity in developing and maintaining models, reliance on data quality
Digital Twins	Simulates supply chain processes for risk management, capacity planning, and process optimization without disrupting physical operations	High computational resources required, complexity in model development and maintenance, dependent on accurate real-time data feeds
Specific Amazon platforms & systems (largely AWS-based)		
AWS Supply Chain Platform (Overall Suite)	Centralizes supply chain data, provides end-to-end visibility, enables real-time decision-making, improves efficiency and coordination	High dependency on the AWS ecosystem, complexity of integrating diverse enterprise systems, subscription and usage costs

(continued)

Table 1.4 (continued)

Technology/platform	Advantages	Risks
AWS Supply Chain Demand Planning	Reduces stockouts and overstocking with ML-driven forecasting, improving inventory turnover and capital efficiency	Requires substantial, high-quality historical training data for accurate predictions; model adaptation to new patterns
AWS Supply Chain Vendor Lead Time Insights	Provides precise lead-time estimates for better buffer stock management and supplier diversification, enhancing resilience	Reliant on accurate and timely data from vendors; potential inaccuracies during widespread or novel disruptions
AWS Supply Chain N-Tier Visibility	Enhances multi-tier supplier communication, facilitates proactive risk identification and mitigation deep in the supply network	High complexity in integrating and mapping data across multiple supplier tiers, data sharing and trust concerns among partners
AWS Supply Chain Supply Planning	Optimizes inventory buffers, production plans, and resource allocation based on demand and supply constraints	Heavily reliant on accurate demand forecasts and real-time supply status; complexity in balancing competing objectives
AWS Supply Chain Sustainability	Supports sustainability goals by providing data and analytics to reduce waste, optimize resource allocation, and track carbon footprint	Initial implementation costs, reliability and completeness of sustainability data across the value chain
AWS Supply Chain Data Lake (SCDL)	Stores centralized supply chain data from disparate sources, enabling advanced analytics, ML model training, and comprehensive insights	Vulnerable to unauthorized access and data corruption if not properly secured, data governance and privacy compliance challenges
Generative AI Applications (e.g., for Onboarding, Amazon Q for SC tasks)	Automates tasks like supplier data ingestion, content generation for supply chain communications, accelerates issue resolution	Potential for errors or biases in AI-generated content if not adequately supervised, ethical considerations, data privacy
Amazon Bedrock	Facilitates development, deployment, and scaling of generative AI models, streamlining advanced analytics capabilities	High operational costs for model training and inference, need for specialized AI/ML engineering skills
Data Orchestration Framework & Data Quality Engine	Speeds up data ingestion from ERP systems (e.g., SAP S/4HANA), validates data for accuracy and consistency, improving decision quality	Requires robust integration with diverse existing systems, potential processing delays due to stringent quality checks
Other supporting technologies/tools		
LIDAR Scanning	Improves spatial accuracy for warehouse layout optimization, digital twin creation, and autonomous robotic navigation	High initial investment for hardware and software, requires specialized expertise for data processing and analysis
ProModel Software (or similar simulation tools)	Supports supply chain simulations for capacity planning, bottleneck analysis, and scenario analysis to test resilience strategies	Requires skilled operators, ongoing software licensing and update costs, model accuracy dependent on input data and assumptions

1.3.5 DX in Procter & Gamble

Procter & Gamble (P&G), a global leader in consumer goods, has strategically embraced DX across its extensive manufacturing operations and supply chain network to drive significant enhancements in efficiency, resilience, and overall performance. The company's approach focuses on integrating advanced digital technologies to augment its robust operational fundamentals, thereby creating scalable, data-driven systems that support its vast global enterprise (Malone et al., 2022).

At the core of P&G's DX strategy is the deployment of **Cloud Computing** and **Advanced Analytics Platforms**. A key enabler is the Manufacturing Data Cloud, often developed in collaboration with partners like GE Vernova (leveraging its Proficy suite including Proficy Plant Applications), which centralizes data collection from disparate sources across P&G's global manufacturing sites. This platform has reportedly reduced the physical infrastructure footprint by up to 80% and significantly streamlined operations (GE Digital, n.d.). By providing a unified view of operations and leveraging advanced analytics, the Manufacturing Data Cloud facilitates faster order fulfillment, reduces manual errors, and optimizes equipment utilization. This directly contributes to operational **efficiency** through cost reduction and process optimization, and to **performance** through improved responsiveness and accuracy. Furthermore, real-time insights derived from platforms like Proficy enable proactive risk mitigation, agile logistics adjustments, and robust contingency planning, thereby bolstering supply chain resilience against disruptions such as supply shortages or demand fluctuations (Chui & Fleming, 2011; GE Digital, n.d.).

AI and ML are pivotal in P&G's efforts to optimize manufacturing and supply chain processes. AI-driven tools are employed for predictive maintenance, detecting potential equipment issues and material quality problems early. This capability allows for faster recovery from issues and helps maintain uninterrupted production, leading to reduced downtime and improved Overall Equipment Effectiveness (OEE)—contributing to a reported $100 million reduction in manufacturing costs and a 10% improvement in OEE (Malone et al., 2022; Noyes, 2023). Within production lines, real-time quality control is enhanced by **Computer Vision** systems powered by AI, which identify product defects with high accuracy, thereby reducing waste (**efficiency**) and ensuring higher product quality (**performance**).

AI and **IoT** sensor data are also combined to optimize resource consumption, such as dynamically adjusting water and energy usage in manufacturing processes. This not only minimizes environmental impact and operational costs (**efficiency**) but also supports sustainability goals (Noyes, 2023). Beyond the factory floor, AI-powered analytics assist in strategic supplier diversification, enhancing resilience by reducing dependency on single sources (GE Digital, n.d.). Predictive analytics, fueled by ML models, enable P&G to anticipate potential disruptions, optimize warehouse operations, and minimize manual interventions, resulting in faster decision-making, a reported 20% improvement in delivery accuracy, and a 15% reduction in operational costs (Chui & Fleming, 2011). AI is also applied to reduce overpacking and accelerate new product development by facilitating integrated data flows between innovation centers and manufacturing plants (Noyes, 2023).

To enhance supply chain transparency and ensure supplier compliance, particularly in challenging conditions, P&G has incorporated **Blockchain Technology**. While specific widespread implementation details are less publicized, the strategic intent is to provide an immutable record of transactions and material provenance, which strengthens **resilience** by ensuring visibility and accountability within the supplier network (Chui & Fleming, 2011).

Automation extends beyond physical robotics to process automation and workforce enablement. Automated tracking systems are utilized in logistics and warehousing to improve accuracy and speed (**performance, efficiency**). While P&G utilizes robotics in its manufacturing lines for repetitive tasks, a significant aspect of its DX is the Integrated Work System (IWS). IWS is a foundational operational excellence program that standardizes processes and empowers employees across P&G's global sites. By digitizing elements of IWS and embedding digital tools within these standardized workflows, P&G has fostered a culture of continuous improvement and ownership, reportedly achieving high levels of employee engagement and significant reductions in production losses, thereby boosting overall productivity and **efficiency** (Malone, 2022). Digital adoption trackers ensure consistent application of these standards and provide visibility across sites, making it easier to scale innovations effectively.

The complementary application of these technologies has yielded substantial benefits. P&G's focus on leveraging digital tools to enhance its already strong operational fundamentals has resulted in significant cost reductions (e.g., a reported $175 million in costs and a 25% improvement in efficiency at one of its global brands) and improvements in agility and product quality (Malone et al., 2022). These digital solutions transform operational strengths into scalable systems that also support sustainability objectives, for instance, through AI-driven adjustments in manufacturing that reduce material waste (Noyes, 2023).

In summary, Procter & Gamble has made significant progress in its DX, particularly within its manufacturing and operational domains. The company has widely adopted advanced digital technologies like AI/ML, IoT, Cloud Computing, Advanced Analytics, and process automation, integrated through platforms such as the Manufacturing Data Cloud and the overarching Integrated Work System. This has led to demonstrable improvements in operational efficiency, product quality (performance), and the ability to manage disruptions (resilience). As noted by Malone et al. (2022), P&G is "more than halfway through a DX of its manufacturing operations" and is "now looking to extend its DX beyond manufacturing to its entire supply chain." This ongoing journey and the current emphasis, particularly the developmental stage of achieving a fully interconnected, end-to-end digital supply chain that functions as a holistic, adaptive system, affirm a classification of 'High' on the DX Scale. The extension to achieve the "complete, cohesive integration... across all operations" characteristic of a 'Very High' level is clearly an active part of their ongoing strategy.

The key technologies, their advantages, and associated risks in the context of Procter & Gamble's DX are summarized in Table 1.5 (adapted from Chui & Fleming, 2011; GE Digital, n.d.; Malone, 2022; Malone et al., 2022; Noyes, 2023).

Table 1.5 Summary of Procter & Gamble's key DX technologies/platforms, advantages, and risks

Technology/platform	Advantages	Risks
Core enabling technologies		
Cloud Computing (incl. Manufacturing Data Cloud, potentially with partners like GE Vernova)	Centralizes data collection, reduces infrastructure footprint (e.g., by 80%), enhances scalability, real-time data access and insights across global operations, improves order fulfilment	High dependency on cloud provider stability and security; significant initial integration efforts; ongoing operational & data storage costs
AI	Enables predictive maintenance, predictive analytics for demand/ disruptions, optimizes resource use (water, energy), reduces waste (e.g., overpacking), enhances product consistency and quality control, accelerates product development	Risk of errors or biases in algorithm predictions if data or models are flawed; requires substantial high-quality data; ethical considerations
ML	Powers predictive analytics, improves resource allocation, optimizes manufacturing processes, enables sophisticated data-driven decision-making	Requires substantial and representative training data; potential for model bias; complexity in model development and maintenance
IoT Sensors	Enables real-time monitoring of equipment and processes, optimises resource usage by providing granular data, reduces downtime by early issue detection	Potential network vulnerabilities and cybersecurity threats; data overload; requires robust connectivity and integration infrastructure
Advanced Analytics Platforms (e.g., leveraging Proficy)	Provides actionable insights for predictive maintenance, supply chain logistics, process optimisation, risk mitigation, and contingency planning	Requires high-quality, integrated data; necessitates specialized analytical skills for interpretation and model building
Blockchain Technology	Enhances supply chain transparency, helps ensure supplier compliance, facilitates secure tracking of goods, and supports risk mitigation through verified data	Implementation complexity and cost; scalability challenges; interoperability with legacy systems; energy consumption of some protocols
Automation & Robotics	Improves efficiency and speed in production lines, reduces manual interventions and errors, enables automated tracking systems in logistics and warehousing	High upfront capital investment and ongoing maintenance costs; potential job displacement concerns requiring workforce reskilling
Computer Vision	Automates defect detection on production lines, ensures consistent quality control, provides data for process improvement	Requires high computational power and extensive, accurately labeled training datasets; performance can be affected by environmental factors

(continued)

Table 1.5 (continued)

Technology/platform	Advantages	Risks
Edge Computing	Processes data locally (e.g., on factory floor) for faster, real-time insights and decision-making, reduces network latency and dependency on centralized cloud resources	Hardware constraints at the edge; potential scalability challenges; managing distributed edge devices and security
P&G specific systems/approaches		
Integrated Work System (IWS)—Digitally Enhanced	Standardizes processes globally, increases employee engagement and ownership, reduces production losses, drives operational excellence and productivity	May face resistance from employees during adoption if change management is inadequate; requires consistent global implementation
Proficy Plant Applications (from GE Vernova)	Tracks and analyzes production data, provides real-time operational insights, supports OEE improvement, and facilitates standardized reporting	Complexity in integration with diverse legacy systems and equipment; requires skilled personnel for configuration and maintenance

1.3.6 DX in Unilever

Unilever, a global leader in the Fast-Moving Consumer Goods (FMCG) sector, has strategically embedded DX within its extensive supply chain and operational frameworks. The company's approach aims to enhance agility, streamline complex operations, improve resource utilization, and ultimately deliver superior consumer experiences and sustainable outcomes by leveraging advanced digital technologies, including AI, ML, IoT, Cloud Computing, and Advanced Analytics Platforms (Law, 2024; Unilever PLC, 2023).

A foundation of Unilever's DX strategy is the widespread application of **AI and ML** across its value chain. To strengthen supply chain resilience, AI is utilized to rapidly identify and test alternative ingredients during disruptions, enabling the company to maintain product availability and quality without significant delays. For instance, AI-driven virtual simulations can drastically reduce the time required to assess millions of potential ingredient combinations, accelerating new product development and reformulation cycles, which is crucial for adapting to volatile raw material markets (Unilever PLC, 2023). This enhances resilience by providing flexibility in sourcing and contributes to performance by speeding time-to-market. Furthermore, AI-powered demand forecasting models have significantly reduced manual forecasting efforts by a reported 30%, improving accuracy and allowing for more efficient inventory planning (Law, 2024; Unilever PLC, 2024).

Unilever's Customer Connectivity Model, a sophisticated **AI-powered Advanced Analytics Platform**, illustrates its data-driven approach. This model performs an estimated 13 billion computations daily to synchronize real-time consumer purchase data with supply chain operations. By merging customer and supplier data streams into a unified ecosystem, Unilever minimizes disruptions, reduces lead times, and optimizes inventory levels across its network (Unilever PLC, 2024). A pilot program with Walmart Mexico, leveraging this connectivity, achieved a 98% on-shelf availability, showcasing a direct positive impact on performance and customer satisfaction (Unilever PLC, 2024). This real-time responsiveness also contributes to resilience by allowing swift adjustments to demand shifts.

The adoption of the Kinaxis Supply Chain Planning (SCP) platform, another key **Advanced Analytics solution**, has fundamentally transformed Unilever's planning capabilities. Covering approximately 80% of the company's turnover and managing over 130,000 Stock Keeping Units (SKUs) across six regional hubs, this integrated platform has yielded substantial quantitative benefits. These include a 20-basis-point reduction in operational costs (efficiency), a 15% improvement in planner productivity (efficiency), and a 5% decrease in working capital requirements (efficiency). Moreover, the Kinaxis platform has enabled Unilever to reduce days of inventory outstanding while simultaneously improving service levels for constrained inventory from 70% to over 90% (performance) (Banker, 2023). Such advanced planning capabilities also enhance resilience by enabling more effective scenario planning and response to supply constraints.

Cloud Computing serves as a critical enabler for Unilever's digital ambitions, with the company's migration to Microsoft Azure providing operational flexibility, scalability for its data-intensive AI and analytics applications, and enhanced resilience against global or localized disruptions (Law, 2024). This cloud infrastructure supports the vast data processing needs of systems like the Customer Connectivity Model and facilitates seamless data integration across diverse parts of the organization, improving overall decision-making speed and quality (performance and efficiency).

Automation and the **IoT** are significantly improving Unilever's operational efficiency and responsiveness. IoT-enabled smart cabinets for B2B customers, for example, automate inventory monitoring and restocking processes, helping to prevent stockouts, improve order fulfillment accuracy (performance), and reduce waste from expired or overstocked products (efficiency) (Unilever PLC, 2023). Within its manufacturing and logistics operations, automated workflows and robotics play a crucial role. Automated warehouse operations have reportedly cut human errors by 30% and improved order accuracy (performance, efficiency). This optimization also contributes to sustainability goals by enabling more efficient logistics, resulting in fewer trucks on the road, reduced fuel consumption, and lower carbon emissions (Unilever PLC, 2023, 2024). The co-creation of solutions with AI and robotics, particularly within Unilever's AI Horizon3 Lab, has also led to disruptive product launches and numerous patents, indicating an innovation pipeline that can feed into more agile and efficient production systems (Law, 2024).

Unilever is also exploring **Digital Twin** technology through virtual simulations, particularly in product development for testing ingredient combinations (Unilever PLC, 2023). Broader applications of Digital Twins could further optimize manufacturing processes and supply chain network design by allowing for scenario modeling and proactive issue resolution, enhancing both efficiency and resilience. While the "Be My Eyes" app, which integrates AI (ChatGPT-4) to assist visually impaired consumers with product information, is primarily a consumer-facing initiative, the real-time insights and feedback gathered can indirectly inform demand sensing and product design, contributing to a more responsive supply chain (Law, 2024).

Sustainability is a key driver and outcome of Unilever's DX initiatives. Digital tools are instrumental in lowering material costs through optimized formulations, reducing overproduction via improved forecasting and demand sensing, and optimizing supply chain logistics to minimize environmental impact. AI-driven adjustments in distribution and manufacturing processes have demonstrably decreased waste and accelerated product-to-market timelines, aligning efficiency gains with sustainability objectives (Law, 2024; Unilever PLC, 2023).

In summary, Unilever's DX journey is characterized by a comprehensive and deeply integrated application of advanced technologies—including AI/ML, IoT, Cloud Computing, Advanced Analytics Platforms, and Automation—across its supply chain, manufacturing, and consumer engagement channels. The organization demonstrates "complete, cohesive integration of technologies and platforms across all operations, enabling real-time data-driven decision-making and adaptive processes." This is evidenced by its interconnected systems like the Customer Connectivity Model and the Kinaxis SCP platform, its widespread use of "predictive analytics, AI, real-time data sharing across the supply chain," and its achievement of "full supply chain visibility" in critical areas. These capabilities, coupled with significant quantifiable improvements in performance, resilience, and efficiency, firmly place Unilever at the 'Very High' level on the DX Scale (as defined in Table 1.1).

The key technologies, their advantages, and associated risks in the context of Unilever's DX are summarized in Table 1.6 (adapted from Banker, 2023; Law, 2024; Unilever PLC, 2023, 2024).

1.3.7 Comparative Summary of DX Applications

The detailed analyses of Heineken, Pfizer, Amazon, Procter & Gamble, and Unilever reveal a consistent strategic imperative among these leading global corporations: the leveraging of emerging digital technologies to drive substantial improvements in supply chain performance, resilience, and efficiency. Key technologies, as outlined in Sect. 1.2.1 and Fig. 1.1—most notably AI and ML, Cloud Computing, the IoT, Advanced Analytics Platforms, Automation/Robotics, and increasingly Digital Twins and Blockchain—are central to their transformation efforts. While the specific technological focus and the architecture of their digital platforms vary according to distinct industry demands, company scale, and overarching strategic priorities, all five organizations demonstrate a profound commitment to DX as a fundamental enabler of operational excellence and competitive advantage.

Table 1.6 Summary of Unilever's key DX technologies/platforms, advantages, and risks

Technology/platform	Advantages	Risks
Core enabling technologies		
AI/ML	Accelerates product/ingredient testing, enables precise demand forecasting (30% manual effort reduction), real-time operational decision-making, powers Customer Connectivity Model	High dependency on data quality and volume, potential for model biases, complexity in model validation
IoT (e.g., Smart Cabinets)	Automates inventory restocking, reduces stockouts and waste, improves order fulfillment accuracy, enables real-time asset monitoring	Security vulnerabilities in connected devices, data overload, network dependency, integration complexity
Cloud Computing (Microsoft Azure)	Provides operational flexibility and resilience, scalable infrastructure for AI/ analytics, facilitates global data integration and collaboration	Data security and privacy concerns, reliance on third-party provider, potential vendor lock-in, costs
Advanced Analytics Platforms (e.g., Kinaxis SCP, Customer Connectivity Model)	Optimizes supply chain planning (130k SKUs), reduces operational costs (20 BPS), improves planner productivity (15%), working capital (5%), enhances service levels (to >90%), enables real-time demand-supply synchronization	Requires high-quality, consistent data inputs; complexity in model development and maintenance; skilled personnel
Automation & Robotics (Workflows, Warehouse Operations)	Streamlines production, reduces manual errors (30% in warehouses), improves order accuracy, frees resources for innovation, enhances logistics efficiency	High initial investment, workforce skill gaps and resistance to change, maintenance requirements
Digital Twins (via Virtual Simulations for ingredients/ formulations)	Drastically reduces time for testing ingredient combinations, accelerates product development, potential for broader process/network optimization	High computational demands, complexity in model development and validation, requires accurate real-time data
Specific Unilever initiatives/platforms		
Customer Connectivity Model	Enables real-time synchronisation of consumer data with supply chain ops (13B daily computations), boosts agility, 98% on-shelf availability in pilots	High dependency on continuous, accurate data streams; integration complexity across diverse systems
Kinaxis Supply Chain Planning (SCP) Platform	Manages ~80% of turnover, significant cost, productivity, working capital, and service level improvements	Dependency on platform provider; requires robust data governance and user training for effective utilization

(continued)

Table 1.6 (continued)

Technology/platform	Advantages	Risks
AI Horizon3 Lab	Drives AI and robotics innovation for disruptive product/packaging launches that can influence supply chain agility and material use	Long R&D cycles for tangible SC impact, translating research into scalable operational solutions
"Be My Eyes" Integration (leveraging AI)	Enhances real-time consumer engagement and provides insights into product usability/accessibility, indirectly informing demand signals	Primary focus is consumer support, indirect SC impact; reliance on AI model reliability and data privacy
Blockchain Technology (Exploratory/ Potential)	Potential for improved supply chain transparency, traceability, and enhanced supplier compliance	Implementation complexity, scalability challenges, interoperability with existing systems, industry adoption

A significant commonality observed across these diverse companies is the strategic development of integrated data ecosystems. These ecosystems, frequently supported by scalable Cloud Computing infrastructure, are designed to enable end-to-end visibility, facilitate sophisticated data analytics, and support agile, data-driven decision-making. This architectural approach aligns closely with the concept of Supply Chain Control Towers (discussed in Sect. 1.2.1), which function as centralized hubs that consolidate data and analytical capabilities from multiple underlying technologies to provide comprehensive oversight and intelligent control. Explicit examples include Pfizer's DOC and Amazon's AWS Supply Chain platform. Similar integrated systems are seen in Heineken's advanced S&OP systems developed on centralized data, Procter & Gamble's Manufacturing Data Cloud for operational intelligence, and Unilever's complementary Customer Connectivity Model and Kinaxis Supply Chain Planning platform. These examples illustrate varying breadths of integration but share the common goal of unifying data for enhanced SCM.

Applying the DX Scale and its indicators (as detailed in Table 1.1), the case study companies largely exhibit advanced stages of DX maturity. Heineken and Procter & Gamble are assessed at a 'High' level of DX. Heineken demonstrates this through widespread technology adoption in its S&OP FIT and Connected Brewery initiatives, significantly enhancing planning agility and operational monitoring. Procter & Gamble showcases robust DX within its manufacturing domain through an extensive Manufacturing Data Cloud and AI-driven process optimizations, with ongoing efforts to expand these digital capabilities throughout its entire supply chain.

Pfizer, Amazon, and Unilever are classified at a 'Very High' level. Pfizer's exemplary DX is evidenced by its agile response capabilities, such as the HOSuN platform and the DOC, which were pivotal during the COVID-19 pandemic. Amazon, as a digital-native entity, has architected its entire operational framework upon deeply embedded, AI-powered, and highly automated systems, prominently featuring its AWS Supply Chain platform. Unilever displays comprehensive DX through sophisticated, interconnected platforms like its Customer Connectivity Model and the Kinaxis SCP, which have yielded substantial, quantifiable improvements across its global supply chain operations.

Table 1.7 provides a comparative summary, detailing the key digital technologies applied, the principal platforms leveraged, the assessed DX Scale, and, crucially, the differentiated impacts of these DX initiatives on the core objectives of supply chain performance, resilience, and efficiency for each case study company.

This comparative summary underscores the comprehensive approaches these leading corporations are adopting in their DX endeavors. While the deployment of advanced technology is a crucial facilitator, the case studies also highlight, implicitly and explicitly, the indispensable roles of strategic alignment with business objectives, effective organizational change management, the development of digital skills, and fostering a culture of continuous innovation and data-driven decision-making. These non-technological factors are essential in fully realizing the targeted benefits of these significant digital investments. The next section will further analyze how these DX technologies specifically impact the core objectives of performance, resilience, and efficiency, and will explain key principles beneficial for their effective and strategic application within supply chains.

1.4 Synthesized Findings and Discussion

This section synthesizes the primary findings derived from the comprehensive case study analyses of Heineken, Pfizer, Amazon, Procter & Gamble, and Unilever, as detailed in Sect. 1.3. The analysis investigated how these leading global corporations are leveraging emerging digital technologies to transform their supply chains, with a specific focus on the impacts on performance, resilience, and efficiency. A key observation is the advanced state of DX maturity across these organizations. As summarized in Table 1.7, three of the five companies (Pfizer, Amazon, Unilever) are assessed at a 'Very High' level on the DX Scale (Table 1.1), while the remaining two (Heineken, Procter & Gamble) are at a 'High' level. This indicates a significant and widespread commitment to, and successful implementation of, DX initiatives at the forefront of diverse industries.

A consistent theme emerging from the case studies is the strategic adoption of emerging technologies, including AI and ML, Cloud Computing, the IoT, Advanced Analytics Platforms, Automation and Robotics, and, to a growing extent, Digital Twins and Blockchain. Crucially, the DX journey within these organizations has often evolved from implementing technologies in siloed functional areas towards creating holistic, integrated digital ecosystems that span the entire supply chain. This shift is particularly evident in companies achieving 'Very High' DX maturity, where interconnected platforms enable end-to-end visibility and data-driven decision-making.

Table 1.7 Comparative summary of DX applications, key technologies, platforms, and impacts in case study companies

Company	DX scale	Key technologies applied	Key platforms/systems deployed	Illustrative impact on performance	Illustrative impact on resilience	Illustrative impact on efficiency
Heineken	High	• AI/ML (predictive analytics, advanced algorithms for S&OP) • Cloud Computing (global cloud infrastructure) • IoT (sensor integration for real-time monitoring) • Advanced Analytics Platforms • ERP integration • Cybersecurity measures	• S&OP CORE Program • S&OP FIT program (supported by FuturMaster BLOOM) • Cloud-based Data Platforms • Eazle Sales Platform, Knowledge & Insight Management (KIM) Platform • Connected Brewery IoT Platform • Digifit Platform	• Enhanced sales process efficiency (e.g., 5% uplift via AIDDA) • Real-time data-driven decision-making and improved forecasting via S&OP FIT • Increased responsiveness to market dynamics through integrated planning and operational platforms	• Improved capacity to manage disruptions via S&OP FIT scenario analysis and simulations • Reduced equipment downtime from predictive maintenance (Connected Brewery) • Enhanced adaptability through event-driven architecture responding to demand spikes	• Streamlined global operations via unified organizational structure and ERP • Optimized inventory levels and reduced operational complexity through AI and process simplification • Significant cost and time savings via knowledge management platforms (KIM)
Pfizer	Very High	• AI/ML (predictive models, data analytics for decision support) • IoT (real-time tracking devices) • Advanced Analytics Platforms, Automation (PCMM/SUB, workflow automation) • Cloud Computing • Virtual Monitoring • Augmented Reality • Collaboration Software (e.g., Workfront)	• PCMM System • Highly Orchestrated Supply Network (HOSuN) • SUB Platform • Digital Operations Center (DOC) • Cloud Analytics Platform • Adobe Experience Manager (AEM) for content	• Accelerated product development and delivery (e.g., COVID-19 vaccine in 9 months) • Significantly shortened time-to-market for innovations (>50%) • 99.9% delivery success rate for critical products via real-time tracking & logistics optimization	• Enhanced real-time disruption monitoring and rapid response via DOC • Optimized resource allocation across global facilities using HOSuN • Increased manufacturing flexibility, scalability, and decentralized production capability through PCMM/SUB modular systems	• Reduced cycle times in core manufacturing (up to 10%) • Minimized product waste (e.g., <0.1% for COVID-19 vaccines) through precise monitoring and control • Eliminated redundant processes and enabled faster decision-making via centralized digital systems and automated workflows

Amazon	Very High	• AI/ML (demand forecasting, vendor lead time insights, generative AI for onboarding, Amazon Q) • Cloud Computing (AWS infrastructure) • Automation & Robotics (WAO) • IoT (real-time tracking) • Advanced Analytics Platforms (AWS SC suite) • Digital Twins • LIDAR Scanning • Data Orchestration	• AWS Supply Chain Platform (incl. Demand Planning, Vendor Lead Time Insights, N-Tier Visibility, Supply Planning, Sustainability modules) • AWS Supply Chain Data Lake (SCDL) • Warehouse Automation and Optimization (WAO) • AWS S3 • Amazon Bedrock • ProModel (or similar for simulation)	• Reduced order fulfillment times and improved delivery accuracy via extensive automation and advanced logistics • Real-time responsiveness to demand fluctuations via ML-driven dynamic forecasting • Improved inventory availability and customer service	• Proactive risk identification and mitigation deep in the supply network (N-Tier Visibility) • Dynamic contingency planning with precise vendor lead-time estimates and optimized buffer stocks • Support for supplier diversification strategies	• Significant reduction in excess inventory (30%) via ML demand planning • Increased labor productivity (40%) and improved space utilization (15%) in fulfillment centers via WAO • Streamlined supplier onboarding processes reducing manual effort and time using AI and data frameworks
Procter & Gamble	High	• Cloud Computing (Manufacturing Data Cloud) • AI/ML (predictive maintenance, computer vision for QC) • IoT (sensors for resource optimization) • Advanced Analytics (Big Data Analytics) • Automation & Robotics • Blockchain (for transparency) • Edge Computing	• GE Vernova's Manufacturing Data Cloud (leveraging Proficy suite e.g., Proficy Plant Applications, Proficy Lite) • Integrated Work System (IWS)—digitally enhanced	• Faster order fulfillment and improved responsiveness via Manufacturing Data Cloud • Enhanced product quality and consistency through AI-powered Computer Vision • 20% improvement in delivery accuracy through data-driven logistics	• Strengthened risk mitigation and agile logistics adjustments through real-time insights from analytics platforms • AI-supported supplier diversification to reduce dependency • More robust contingency planning capabilities and faster issue recovery	• Substantial manufacturing cost reductions (e.g., $100M) and OEE improvement (10%) via AI and advanced analytics • 15% reduction in operational costs • Optimized resource utilization (water, energy) in manufacturing • Streamlined global processes and reduced production losses via digitized IWS

(continued)

Table 1.7 (continued)

Company	DX scale	Key technologies applied	Key platforms/systems deployed	Illustrative impact on performance	Illustrative impact on resilience	Illustrative impact on efficiency
Unilever	Very High	• AI/ML (ingredient testing, demand forecasting, customer data processing) • IoT (Smart Cabinets) • Cloud Computing (Microsoft Azure) • Advanced Analytics Platforms (Big Data Analytics) • Automation (Automated Workflows, Robotics) • Digital Twins (Virtual Simulations) • Blockchain (exploratory) • ERP	• Kinaxis Supply Chain Planning (SCP) Platform (RapidResponse) • Customer Connectivity Model, AI Horizon3 Lab • "Be My Eyes" Integration (with ChatGPT-4) • Microsoft Azure (Cloud platform)	• Improved on-shelf availability (e.g., 98% via Customer Connectivity Model) • Enhanced service levels (>90% for constrained inventory via Kinaxis) • Accelerated new product development (NPD) and reformulation cycles via AI simulation	• Increased flexibility in sourcing through AI-driven rapid testing of alternative ingredients • Real-time demand-supply synchronization minimizing disruptions via Customer Connectivity Model • Enhanced scenario planning and response to supply constraints via Kinaxis SCP	• Notable reduction in operational costs (20 BPS) and working capital (5%) via Kinaxis SCP • Improved planner productivity (15%) • 30% reduction in manual forecasting effort via AI • 30% reduction in warehouse human errors through automation

The next sub-section will delve deeper into dissecting how these specific digital technologies, as applied by the case study companies, directly influence the core objectives of supply chain performance, resilience, and efficiency. This exploration aims to address the central inquiries of this chapter, particularly Question 1: "In what ways does DX, through the leverage of emerging technologies, influence supply chain performance, resilience, and efficiency?" and Question 2: "Which specific digital technologies make the most substantial contributions to the enhancement of supply chain performance, resilience, and efficiency?".

1.4.1 Impact of DX Technologies on Performance, Resilience, and Efficiency

DX, through the strategic integration of emerging technologies, fundamentally enhances supply chain capabilities, driving significant improvements in operational performance, systemic resilience, and resource efficiency. These technologies empower organizations by augmenting visibility, enabling sophisticated data analytics, fostering intelligent automation, and supporting proactive, data-driven decision-making to navigate complex challenges and optimize multi-faceted operations. The case studies illustrate how various technologies contribute to these objectives:

AI and ML AI and ML are pivotal in elevating supply chain capabilities.

- **Performance**: These technologies significantly improve demand forecasting accuracy, as seen with Amazon's ML-driven systems ensuring product availability and efficient customer delivery, and Unilever's 30% reduction in manual forecasting effort (Law, 2024). Pfizer's use of AI to rapidly scale production for its COVID-19 vaccine highlights its role in responsive performance during critical demand surges (Moa, 2021). AI also drives performance through process optimization, such as Heineken's AI-Driven Decision Automation (AIDDA) improving sales efficiency (Reynolds, 2023).
- **Resilience**: Predictive capabilities enable better anticipation of disruptions and optimized contingency planning. Amazon uses ML for precise vendor lead-time estimates to optimize buffer stocks (Pantoja-Navajas, 2023), while Unilever leverages AI to quickly test alternative ingredients during shortages (Unilever PLC, 2023). P&G applies AI for strategic supplier diversification (GE Digital, n.d.).
- **Efficiency**: AI/ML automates decision-making processes, optimizes inventory levels, and streamlines workflows. Heineken employs AI to streamline inventory management (Drinkwater, 2023; Reynolds, 2023), Amazon minimizes overstocking, and P&G reported a $100 million reduction in manufacturing costs due to AI-driven optimizations (Malone et al., 2022).

IoT IoT networks of connected devices provide granular, real-time data, fostering enhanced visibility and control.

- **Performance**: Real-time tracking improves logistics accuracy and customer service, as demonstrated by Unilever's smart cabinets streamlining last-mile deliveries and improving on-shelf availability (Unilever PLC, 2023).
- **Resilience**: Continuous monitoring of assets and environmental conditions allows for early warnings and proactive interventions. Heineken's "Connected Brewery" program uses IoT for predictive maintenance, minimizing equipment downtime and ensuring supply continuity (Drinkwater, 2023; Reynolds, 2023). Pfizer's real-time tracking of cold-chain conditions for vaccines was crucial for maintaining product integrity during distribution (Martin, 2023).
- **Efficiency**: Predictive maintenance reduces operational disruptions and costs. Automated data capture from IoT devices reduces manual labor and errors, optimizing resource allocation and asset utilization across the supply chain.

Cloud Computing Cloud platforms provide the scalable and flexible infrastructure necessary for modern digital supply chains.

- **Performance**: Centralized data access facilitates faster, more informed decision-making across global operations, improving responsiveness. Companies like Heineken (Clark, 2024; Wolski, 2022) and Unilever (Law, 2024) leverage cloud for global data integration and real-time analytics.
- **Resilience**: Cloud infrastructure offers operational scalability and business continuity, ensuring system availability and agility during demand peaks or disruptions, as demonstrated by Pfizer's vaccine distribution management.
- **Efficiency**: Reduces the need for extensive on-premise IT infrastructure, lowers operational costs, and enables efficient data storage, processing, and sharing, leading to streamlined processes. P&G, for example, reduced its physical infrastructure footprint by up to 80% via its Manufacturing Data Cloud (GE Digital, n.d.).

Digital Twins Virtual replicas of physical assets, processes, or entire supply chain networks, dynamically updated with real-time data, enable simulation and optimization.

- **Performance**: Allows for testing and optimizing processes or new product introductions in a virtual environment before physical implementation, speeding time-to-market as seen in Unilever's use of virtual simulations for product testing (Unilever PLC, 2023).
- **Resilience**: Enables organizations to simulate disruption scenarios, assess potential impacts, and develop robust contingency plans. Heineken utilizes scenario analysis (inferred from use of Digital Twins for process visualization and S&OP FIT capabilities) to refine operations and prepare for disruptions (Kaporina et al., n.d.; Clark, 2024; Wolski, 2022). Amazon uses digital twins for optimizing fulfillment center layouts and workflows (Shah, 2024).

- **Efficiency**: Optimizes resource allocation, identifies bottlenecks, and streamlines processes by virtually modeling and testing changes, leading to cost savings and improved throughput.

Blockchain Technology Distributed ledger technology enhances transparency, security, and traceability.

- **Performance**: Improves trust and reduces disputes among supply chain partners, leading to smoother transactions and potentially better quality assurance, as Procter & Gamble aims by tracking supplier compliance (Chui & Fleming, 2011).
- **Resilience**: Provides an immutable record of transactions and product provenance, which can be crucial for verifying authenticity, managing recalls, and ensuring supply chain integrity, especially in complex, multi-tier networks. Unilever's exploration for verifying sustainability claims supports ethical sourcing and brand resilience (Unilever PLC, 2023).
- **Efficiency**: Streamlines processes by reducing paperwork, automating verification through smart contracts, and enhancing data accuracy among participants.

Automation and Robotics Includes both physical automation (robots in warehouses/manufacturing) and software automation (RPA for administrative tasks).

- **Performance**: Increases speed, accuracy, and throughput in repetitive tasks. Unilever's automated warehouses improved order accuracy, leading to better customer service (Unilever PLC, 2023). Pfizer's PCMM systems enable rapid production scaling (Pfizer Annual Review, 2016).
- **Resilience**: Reduces reliance on manual labor in critical operations, which can be vital during labor shortages or disruptions. Amazon's extensive warehouse automation (Travers & Baeuml, 2024) ensures operational continuity.
- **Efficiency**: Drives significant productivity gains, reduces labor costs, and minimizes errors. Amazon reported a 40% increase in labor productivity through warehouse automation (Travers & Baeuml, 2024). Pfizer's use of AI models to streamline workflows also exemplifies efficiency gains through automation (Hall & Fonseca, 2024).

Furthermore, many companies are strategically deploying these technologies to advance sustainability objectives, which are increasingly intertwined with overall performance and resilience. Heineken's adoption of low-complexity packaging (Ashcroft, 2023) and P&G's use of AI to optimize resource consumption (Noyes, 2023) illustrate how DX supports environmental goals, often through enhanced efficiency. Event-driven systems, as exemplified by Amazon's real-time risk alerts, directly bolster resilience by enabling dynamic and rapid responses to unexpected operational changes and disruptions.

A critical aspect amplifying the impact of these individual technologies is their integration into cohesive platforms, often functioning like Supply Chain Control Towers. As discussed in Sect. 1.2.1, these centralized hubs (or similar integrated data ecosystems) provide end-to-end, real-time visibility and analytical capabilities by consolidating data from diverse sources. Although not always explicitly labeled as "control towers" by all companies, the functional capabilities observed in Pfizer's DOC, Amazon's AWS Supply Chain platform, and Unilever's Customer Connectivity Model align strongly with this concept. Such integration is key to unlocking holistic improvements in performance, resilience, and efficiency.

In essence, the effective application of these emerging digital technologies, particularly when integrated into cohesive data-driven platforms, provides organizations with the essential tools to not only navigate the inherent uncertainties of modern supply chains but also to unlock new levels of operational excellence. The synthesis of these technologies directly supports the strategic objectives of enhancing performance, bolstering resilience, and improving efficiency. As conceptually illustrated in Fig. 1.1, key digital technologies form the building blocks that, when integrated, often through platforms similar to Supply Chain Control Towers, enable comprehensive visibility, advanced analytics, and intelligent automation, thereby transforming supply chain capabilities.

1.4.2 Guiding Principles for Effective DX Technology Application

Building upon the understanding of how various DX technologies contribute to supply chain performance, resilience, and efficiency (as elaborated in Sect. 1.4.1), this section addresses the principles underpinning the strategic and effective application of these technologies. The successful DX of supply chains extends beyond mere technological adoption, necessitating a holistic approach that integrates technology, processes, and people to navigate an increasingly complex and dynamic global landscape. Historically, while digitalization efforts focused on enhancing operational efficiency and reducing costs, some organizations have encountered challenges in fully embracing and integrating Industry 4.0 technologies. A significant contributing factor to these challenges has been an underemphasis on "softer" elements, such as human resource management and organizational change management, particularly when the pace of technological application outstrips the capacity for employee skill development (Liboni et al., 2019). The case studies of Heineken, Procter & Gamble, and Unilever notably illustrate a conscious focus on the "people" dimension as a vital aspect of their supply chain DX journeys.

Heineken's change management strategy revolved around strong leadership, comprehensive capability building, and effective, targeted communication. Their "heads, hearts, and hands" approach not only articulated the rationale behind the transformation but also focused on engaging employees emotionally and practically. Heineken clearly prioritized people over mere technology deployment, emphasizing proactive leadership engagement, the enhancement of employee skills, and the alignment of communication strategies with local operational realities to ensure a smoother transformation process (Ashcroft, 2023, 2024). Furthermore, the digital changes implemented provided real-time insights that benefited employees, suppliers, and customers alike, thereby improving operational decision-making and enhancing the overall user experience (Wolski, 2022).

Similarly, Procter & Gamble's IWS program underscores the central role of people in successful transformation. This initiative extends beyond the implementation of new technologies to encompass the fostering of a culture characterized by employee ownership and continuous improvement. The company actively focuses on cultivating an environment encouraging to proactive problem-solving and sustained employee engagement, thereby moving beyond a reactive, crisis-management operational mode (Malone et al., 2022). Unilever's DX also places a strong emphasis on its people, identifying culture change, robust capacity building, and the strategic development of essential digital capabilities as key pillars of its transformation agenda. The company aims to leverage technology and automation not to displace human capital, but to transition employees towards more value-added tasks by eliminating non-productive activities, such as those associated with bureaucratic systems (Infosys Limited, n.d.). Likewise, Amazon, a company at the forefront of automation and AI deployment, has made significant public commitments to upskilling its workforce. Initiatives such as its large-scale investment in training programs aim to equip employees with new skills necessary to thrive in increasingly automated and technologically advanced operational environments, thereby managing the human impact of rapid technological integration and fostering workforce adaptability.

These examples collectively demonstrate that successful DX is contingent not only on adopting new technologies but, crucially, on reshaping the organizational culture and augmenting the capabilities of the people who drive and sustain the change.

As supply chains continue to develop concurrently with technological advancements, the progressive digitalization of these chains is fundamentally reshaping operational paradigms, with activities becoming increasingly reliant on sophisticated technologies. However, a fragmented approach to digitalizing discrete parts of the supply chain may not lead to sustainable or transformative change. An effective strategy, therefore, involves a paradigm shift from isolated technology transformation initiatives towards a more comprehensive, technology-enabled, business-wide transformation. It is by adopting such a holistic approach—characterized by end-to-end integration of both business processes and technological infrastructure—that organizations can substantially improve the resilience of their supply chains. This integrated approach also enhances their ability to respond swiftly and effectively to unexpected external disturbances (Bauer et al., 2024), primarily because this mode of transformation interconnects disparate parts of the supply chain and provides widespread data visibility throughout its entirety.

Several case study companies, particularly those achieving a 'Very High' DX Scale rating, such as Pfizer, Amazon, and Unilever, exemplify this advanced stage of transformation. These organizations have progressively transformed their supply chains by establishing what can be conceptualized as "supply chain nerve centers"—digital, end-to-end control towers that go beyond traditional functional

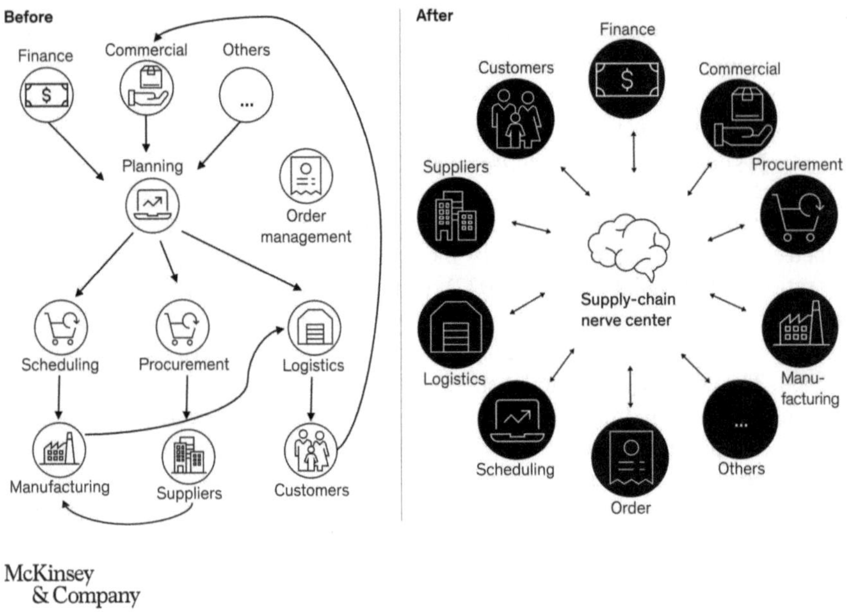

Fig. 1.2 Supply-chain nerve centers could simplify internal structures (McArthur et al., 2021)

silos. These nerve centers extend comprehensive end-to-end planning capabilities by interconnecting previously isolated functions. They leverage digital technologies, including AI and IoT, to sense impending shocks or disturbances earlier and more deeply across the supply chain. Furthermore, they facilitate data-driven decision-making at multiple operational and strategic levels and often involve a reimagining of the operating model, grounded in the understanding that technology alone is insufficient to deliver sustainable improvements in supply chain performance (McArthur et al., 2021). This integrated operational model can also lead to a simplification of internal organizational structures, as illustrated in Fig. 1.2.

For instance, Pfizer has employed a wide array of technologies, including AI, data analytics, ML, IoT devices, AR, and predictive modeling, enabling sophisticated insights, automation, and data-driven decisions. Pfizer's integrated platforms, such as the Digital Operations Center (DOC), PCMM system, HOSuN, and SUB platform, constitute a mature infrastructure designed for real-time, agile, and high-quality manufacturing and supply chain performance. The DOC, specifically, is an AI-powered system for monitoring and optimizing end-to-end supply chain operations, functioning as a true nerve center (LogiPharma, 2022; Moa, 2021). The deep integration of predictive analytics, IoT, and real-time monitoring within platforms like DOC and HOSuN facilitates comprehensive visibility and adaptive control across its supply chain and manufacturing processes.

Amazon, a digital-native entity, extensively utilizes AI, ML, IoT, robotics, automation, digital twins, and LIDAR scanning to optimize operations, predict outcomes, and enhance supply chain efficiency. Its ML capabilities, such as those embedded within AWS Supply Chain for Vendor Lead Time Insights and Demand Planning, empower Amazon to make precise data-driven decisions for inventory management and demand forecasting (Selvaraj, 2024). The AWS Supply Chain Platform, AWS Supply Chain Data Lake (SCDL), and AWS Simple Storage Service (S3) provide the backbone for real-time data access, storage, and sophisticated analytics, enabling the centralization and orchestration of data across global operations, thus ensuring consistency and responsiveness (Shah, 2024; Wishart, 2024). Specialized modules like AWS Supply Chain N-Tier Visibility enhance multi-tier supplier communication and risk identification (Kane & Balasubramanian, 2024). Amazon's systems operate with a high degree of interconnectedness, characteristic of a nerve center, facilitating real-time data processing and pervasive visibility across complex supply chain networks, thereby enhancing resilience, efficiency, and performance.

Unilever leverages AI for predictive analytics, prominently demonstrated by its AI-Powered Customer Connectivity Model, which processes over 13 billion computations daily and integrates real-time consumer purchase data to optimize inventory and demand forecasting (Unilever PLC, 2024). The company integrates IoT devices, cloud platforms like Microsoft Azure, and advanced planning tools such as the Kinaxis SCP platform, fostering seamless data sharing and adaptive operations (Banker, 2023; Law, 2024). These integrated systems provide extensive supply chain visibility and have yielded significant quantifiable benefits, including reduced lead times, substantially improved on-shelf availability (e.g., 98% in the Walmart Mexico pilot), and a 30% streamlining of forecasting efforts (Unilever PLC, 2023, 2024). The AI-Powered Customer Connectivity Model and Kinaxis SCP platform function as unifying technologies, synchronizing operations across functions and contributing to significant improvements in cost, inventory management, and service levels (Banker, 2023). This comprehensive integration across functions connects the supply chain end-to-end, enhancing its resilience, efficiency, and agility.

In contrast, organizations such as Heineken and Procter & Gamble, assessed at a 'High' level on the DX Scale, demonstrate substantial digital advancements but exhibit different evolutionary paths or strategic emphases compared to the comprehensive end-to-end integration embodied by a fully developed "nerve center." Heineken utilizes a set of advanced technologies including AI, ML, IoT, predictive analytics, and ERP systems, supported by cloud computing and cybersecurity measures. Platforms like the S&OP FIT program, Connected Brewery, and KIM Platform represent a robust ecosystem for data-driven decisions and streamlined operations (Ashcroft, 2023; Drinkwater, 2023; Reynolds, 2023). However, their journey towards completely seamless, real-time, end-to-end visibility and harmonized operations across all global segments, particularly integrating external partners into a unified digital ecosystem, appears to be an ongoing evolution. While significant data integration has been achieved (Clark, 2024; Wolski, 2022), complexities in harmonizing diverse technological frameworks or legacy systems across a vast global footprint may present challenges to achieving the full depth of integration seen in 'Very High' DX companies.

Procter & Gamble has made significant progress in incorporating advanced technologies like IoT, AI, ML, robotics, and computer vision, particularly within its manufacturing domain, leveraging platforms such as GE Vernova's Proficy suite and its digitally-enhanced IWS to drive operational efficiency, quality control, and process innovation (GE Digital, n.d.; Malone et al., 2022). Their current strategic emphasis, as noted in Sect. 1.3.5, has been on perfecting these operational and manufacturing systems. While these efforts are crucial and yield substantial benefits, the extension towards a fully interconnected, end-to-end digital supply chain that seamlessly integrates all functions from suppliers to customers is an area of continued development, distinguishing its 'High' rating from a 'Very High' one. The IWS, while fostering operational excellence and employee capability, does not inherently create the pervasive, real-time cross-functional data flows characteristic of a fully realized supply chain nerve center.

These observations align with their 'High' DX Scale rating, indicating advanced digital capabilities that are continually evolving towards more comprehensive, end-to-end integration, similar to the 'nerve center' models seen in 'Very High' DX companies. The strategic pursuit of such interconnected systems, like supply chain nerve centers, which unify previously isolated supply chain functions, presents both new opportunities and considerable challenges (McArthur et al., 2021).

In summary, the effective application of DX technologies in supply chains, as demonstrated by leading organizations and guided by principles such as those for designing nerve centers, goes beyond just technological implementation. It requires a strategic, holistic commitment that encompasses integrating technology with business processes, fostering a supportive organizational culture, empowering personnel through skill development, and establishing agile governance frameworks. By thoughtfully addressing these multifaceted aspects, organizations can strategically leverage emerging technologies to achieve sustained improvements in supply chain performance, resilience, and efficiency, thereby securing a competitive advantage in an ever-evolving global landscape.

1.5 Conclusions

DX has clearly evolved beyond being just an enhancement—it has become a fundamental necessity for organizations striving to thrive in today's increasingly competitive, volatile, and uncertain global environment. As this chapter has explained, while many companies embark on their digital supply chain journey with a primary focus on increasing operational efficiency and reducing costs, the ultimate value and transformative potential are realized through the achievement of comprehensive, end-to-end integration. This integration is essential, enabling seamless connectivity, widespread real-time visibility, and agile, data-driven decision-making across the entirety of supply chain functions, from initial procurement and strategic sourcing through to final distribution and customer engagement.

This chapter's exploration, particularly through the case studies of Heineken, Pfizer, Amazon, Procter & Gamble, and Unilever, has revealed that emerging technologies—notably AI and ML, the IoT, Cloud Computing, Advanced Analytics Platforms, Automation and Robotics, and increasingly Digital Twins—are pivotal in profoundly enhancing supply chain performance, bolstering resilience against disruptions, and driving significant operational efficiencies. These organizations, assessed at 'High' to 'Very High' levels on the DX Scale, exemplify the strategic pursuit of integrated digital ecosystems. Initiatives such as Pfizer's Digital Operations Centre, Amazon's AWS Supply Chain platform, Unilever's Customer Connectivity Model, Heineken's Connected Brewery, and P&G's Manufacturing Data Cloud, while differing in specific architectures, all strive towards creating cohesive, intelligent systems, akin to supply chain control towers or "nerve centers," which centralize data and analytics to optimize planning and execution. The realized benefits are substantial, ranging from accelerated product development and enhanced delivery accuracy, as seen at Pfizer, to significant inventory reduction and productivity gains at Amazon, and improved on-shelf availability and cost optimization at Unilever.

However, the successful application of these advanced technologies and the realization of their full benefits are contingent upon more than technological expertise alone. A critical recurring theme, consistently underscored across the analyzed successful transformations, is the indispensable emphasis on the human element and organizational factors. As noted in the context of lean practices, a stronger focus on "soft practices" can constitute a significant competitive advantage. Effective change management strategies, such as Heineken's "heads, hearts, and hands" approach, and Unilever's strategic focus on capacity building, digital capability development, and fostering an adaptive culture, highlight that it is the people within an organization who ultimately drive meaningful and sustainable change. Indeed, the "softer" elements of implementation are critical, particularly as employees must adapt to and master the rapidly advancing technological landscape. Therefore, building essential digital skills, cultivating a culture of ownership and continuous improvement, and securing proactive leadership engagement are essential to ensure that digital tools are not merely implemented but are leveraged effectively to innovate processes, adapt to disruptions, and maintain resilience.

Furthermore, effective DX necessitates a strategic shift from isolated technology deployments or partial digitalization efforts towards a holistic, technology-enabled, organization-wide transformation. By embracing an approach that cohesively integrates business processes and technological infrastructure across the entire supply chain, companies can substantially strengthen their systemic resilience and enhance their capacity to respond swiftly and effectively to unforeseen external disturbances. This integrated transformation fosters greater connectivity between disparate supply chain components and ensures data visibility throughout the system, thereby enabling quicker, more informed, and strategically sound decision-making, particularly in disruptive scenarios. Successfully architecting such integrated systems, often evolving into sophisticated "supply chain nerve centers," involves making deliberate design choices.

While the advantages of DX—enhanced visibility, improved agility leading to better performance, proactive risk mitigation bolstering resilience, and optimized resource utilization driving efficiency—are compelling, organizations must also diligently manage the associated risks. These include ensuring robust data security and privacy, managing the complexity of integrating new technologies with legacy systems, mitigating technological dependencies, addressing potential skill gaps within the workforce, and managing the substantial initial investment and ongoing operational costs.

Ultimately, the pathway to a truly transformed supply chain lies in achieving a complementary balance between advanced technologies and human ingenuity. By purposefully integrating sophisticated digital tools with a skilled and empowered workforce capable of maximizing their potential, organizations can unlock not only immediate operational efficiencies and cost savings but also sustained agility, enhanced resilience, and a decisive competitive advantage.

References

Ashcroft, S. (2023, April 26). "Ninja" approach key to Heineken supply chain transformation. *Supply Chain Digital Magazine*. Retrieved April 25, 2025, from https://supplychaindigital.com/digital-supply-chain/heineken-europe-ninja-key-to-supply-chain-transformation.

Ashcroft, S. (2024, February 29). Heineken toasts success of supply chain transformation. *Supply Chain Digital Magazine*. Retrieved April 25, 2025, from https://supplychaindigital.com/articles/heineken-toasts-success-of-supply-chain-transformation.

Balasubramanian, V. (2024, August 15). A prescriptive approach for AWS Supply Chain proof of values (PoVs). *Amazon Supply Chain and Logistics*. Retrieved April 25, 2025, from https://aws.amazon.com/blogs/supply-chain/a-prescriptive-approach-to-scoping-aws-supply-chain-proof-of-values-povs/

Banker, S. (2023, July 11). Unilever's relentless commitment to S-curve transformations. *Forbes*. Retrieved April 25, 2025, from https://www.forbes.com/sites/stevebanker/2023/07/11/unilevers-relentless-commitment-to-s-curve-transformations/

Baryannis, G., Validi, S., Dani, S., & Antoniou, G. (2019). Supply chain risk management and artificial intelligence: state of the art and future research directions. *International Journal of Production Research, 57*(7), 2179–2202.

Bauer, F., Dilda, V., Gong, L., Groger, C., Jensen, B., Spuntrup, F. S., & Vedpathak, K. (2024, January 29). Tech-enabled transformations: Three supply chain success stories. *McKinsey & Company*. Retrieved April 25, 2025, from https://www.mckinsey.com/capabilities/operations/our-insights/tech-enabled-transformations-three-supply-chain-success-stories.

Brans, P. (2022, October 3). Digital transformation is brewing at Heineken: Computer Weekly. *ComputerWeekly.com*. Retrieved April 25, 2025, from https://www.computerweekly.com/news/252525617/Digital-transformation-is-brewing-at-Heineken

CDOTrends editors. (2023). Heineken brews the future beer with data. Retrieved April 25, 2025, from https://www.cdotrends.com/story/18218/heineken-brews-future-beer-data

Chui, M., & Fleming, T. (2011, November 1). Inside P&G's digital revolution. *McKinsey & Company*. Retrieved April 25, 2025, from https://www.mckinsey.com/industries/consumer-packaged-goods/our-insights/inside-p-and-ampgs-digital-revolution.

Clark, D. (2024, March 15). The Heineken digital transformation journey. *HICX.com*. Retrieved April 25, 2025, from https://www.hicx.com/blog/heineken-digital-transformation-journey/.

Dieaconescu, R. I., Belu, M. G., Paraschiv, D. M., & Joldes, C. S. (2022). Supply chain digital transformation. In *BASIQ 2022 international conference on new trends in sustainable business and consumption*. Retrieved April 25, 2025, from https://doi.org/10.24818/basiq/2022/08/079.

Drinkwater, D. (2023, February 21). What Heineken's CIO is brewing for better connectivity. *CIO*. Retrieved April 25, 2025, from https://www.cio.com/article/455951/what-heinekens-cio-is-brewing-for-better-connectivity.html.

Fragapane, G., de Koster, R., Sgarbossa, F., & Strandhagen, J. O. (2021). Planning and control of autonomous mobile robots for intra-logistics: Literature review and research agenda. *European Journal of Operational Research, 294*(2), 405–426.

Fuller, A., Fan, Z., Day, C., & Barlow, C. (2020). Digital twin: Enabling technologies, challenges and open research. *IEEE Access, 8*, 108952–108971.

GE Digital. (n.d.). Procter & Gamble improves manufacturing performance with digitization. Retrieved April 25, 2025, from https://www.ge.com/digital/customers/digitization-step-change-procter-gamble-improves-performance.

Hall, E., & Fonseca, L. (2024, March 26). How Pfizer is transforming its content supply chain to bring scientific breakthroughs to market at lightspeed. Retrieved April 25, 2025, from https://business.adobe.com/blog/the-latest/how-pfizer-is-transforming-content-supply-chain.

Harapko, S. (2023, January 6). How covid-19 impacted supply chains and what comes next. *EY*. Retrieved April 25, 2025, from https://www.ey.com/en_pl/insights/supply-chain/how-covid-19-impacted-supply-chains-and-what-comes-next.

Infosys Limited. (n.d.). Success stories: Unilever digital transformation case study: A people-led digital transformation. *Infosys*. Retrieved April 25, 2025, from https://www.infosys.com/industries/consumer-packaged-goods/case-studies/people-led-digital-transformation.html.

Ivanov, D. (2022). Viable supply chain model: Integrating agility, resilience and Sustainability Perspectives—lessons from and thinking beyond the COVID-19 pandemic. *Annals of Operations Research, 319*(1), 1411–1431.

Ivanov, D., Dolgui, A., & Sokolov, B. (2022). Cloud supply chain: Integrating Industry 4.0 and digital platforms in the "supply chain-as-a-service". *Transportation Research Part E: Logistics and Transportation Review, 160*, 102676.

Kane, D., & Balasubramanian, V. (2024, July 26). Mitigating operational risks with AWS supply chain work order insights. *Amazon Supply Chain and Logistics*. Retrieved April 25, 2025, from https://aws.amazon.com/blogs/supply-chain/mitigating-operational-risks-with-aws-supply-chain-work-order-insights/

Kaporina, E., Bonomi, G., & Sewell, M. (n.d.). Heineken hubs: A planning transformation to leverage cross border resources. Heineken Hubs: A planning transformation to leverage cross border resources. Retrieved April 25, 2025, from https://www.futurmaster.com/case-studies/heineken-webinar#webinar.

Khojasteh, Y., & Mousavi Jahan Abadi, S. M. (2026). *Digital transformation of supply chains: Business objectives and applications*. Springer. https://link.springer.com/book/9783032012050

Khojasteh, Y., Mousavi Jahan Abadi, S.M., & Ng, C. (2026). Leveraging digital transformation in supply chains: Strategies for agility, performance, and resilience. In Y. Khojasteh & S.M. Mousavi Jahan Abadi (Eds.), *Digital transformation of supply chains: Business objectives and applications*. Springer. DOI: 10.1007/978-3-032-01206-7_1

Kouhizadeh, M., & Sarkis, J. (2018). Blockchain practices, potentials, and perspectives in greening supply chains. *Sustainability, 10*(10), 3652.

Law, M. (2024, January 22). How Unilever uses AI & digital solutions in its operations. *Technology Magazine*. Retrieved April 25, 2025, from https://technologymagazine.com/articles/how-unilever-is-using-ai-and-digital-solutions.

Liboni, L. B., Cezarino, L. O., Jabbour, C. J., Oliveira, B. G., & Stefanelli, N. O. (2019). Smart industry and the pathways to HRM 4.0: Implications for SCM. *Supply Chain Management: An International Journal, 24*(1), 124–146.

LogiPharma. (2022, February 15). Pfizer accelerates digital transformation with the launch of a Digital Operations Centre. Retrieved April 25, 2025, from https://logipharmaeu.wbresearch.com/blog/pfizer-accelerates-digital-transformation-with-the-launch-of-a-digital-operations-centre.

Malone, M. (2022, October 7). Transform your processes before going digital. *EY*. Retrieved April 25, 2025, from https://www.ey.com/en_dk/alliances/transform-your-processes-before-going-digital.

Malone, M., Henderson, C., Hakkennes, B., & Gregory, C. (2022). Digital transformation, EY's Alliance with P&G shows why you need to go back to basics first. Retrieved April 25, 2025, from https://www.ey.com/content/dam/ey-unified-site/ey-com/en-gl/alliances/documents/ey-iws-digital-transformation-pov-eyg-no-008159-22gbl.pdf.

Martin, J. (2023, July 3). The supply chain innovations behind the world's largest vaccine programme. *Pharmaceutical Technology*. Retrieved April 25, 2025, from https://www.pharmaceutical-technology.com/sponsored/the-supply-chain-innovations-behind-the-worlds-largest-vaccine-programme/.

McArthur, S., Sankur, A., & Shah, K. (2021, January 22). Building a digital bridge across the supply chain with nerve centers. *McKinsey & Company*. Retrieved April 25, 2025, from https://www.mckinsey.com/capabilities/operations/our-insights/building-a-digital-bridge-across-the-supply-chain-with-nerve-centers

Mell, P., & Grance, T. (2011). *The NIST definition of cloud computing*. National Institute of Standards and Technology, Special Publication 800-145.

Metha, A. (2024, April 17). Accelerating data ingestion and on-boarding with new AWS supply chain capabilities. *Amazon Supply Chain and Logistics*. Retrieved April 25,2025, from https://aws.amazon.com/blogs/supply-chain/accelerating-data-ingestion-and-on-boarding-with-new-aws-supply-chain-capabilities/.

Moa, A. (2021, July 22). What Pfizer's digital transformation strategy can teach us about making pharma more agile and effective. Retrieved April 25, 2025, from https://www.tribecaknowledge.com/blog/what-pfizers-digital-transformation-strategy-can-teach-us-about-making-pharma-more-agile-and-effective.

Nguyen, T., Zhou, L., Spiegler, V., Ieromonachou, P., & Lin, Y. (2018). Big data analytics in supply chain management: A state-of-the-art literature review. *Computers & Operations Research, 98*, 254-264.

Noyes, K. (2023, January 28). Procter & Gamble CIO: 'everybody must own the digital story'. *WSJ. Wall Street Journal*. Retrieved April 25, 2025, from https://deloitte.wsj.com/cio/procter-gamble-cio-everybody-must-own-the-digital-story-01674840235

Pantoja-Navajas, D. (2023, November 30). AWS supply chain announces new upstream capabilities. *Amazon Supply Chain and Logistics*. Retrieved April 25, 2025, from https://aws.amazon.com/blogs/supply-chain/aws-supply-chain-announces-new-upstream-capabilities/

Pellicelli, M. (2023a). Never so much attention to the digital transformation of supply chains. *The Digital Transformation of Supply Chain Management*, 199–211.

Pellicelli, M. (2023b). The need for a different approach. *The Digital Transformation of Supply Chain Management*, 85–100.

Pfizer Annual Review. (2016). Transforming delivery of high quality products. Retrieved September 30, 2024, from https://www.pfizer.com/sites/default/files/investors/financial_reports/annual_reports/2016/transforming-delivery-of-high-quality-products/index.html.

Rejeb, A., Keogh, J. G., & Treiblmaier, H. (2019). Leveraging the internet of things and blockchain technology in supply chain management. *Future Internet, 11*(7), 161.

Reynolds, C. (2023, August 14). Balancing optimisation and innovation: Q&A with Heineken digital chiefs. *Verdict*. Retrieved April 25, 2025, from https://www.verdict.co.uk/balancing-optimisation-and-innovation-qa-with-heineken-digital-chiefs/.

Saberi, S., Kouhizadeh, M., Sarkis, J., & Shen, L. (2019). Blockchain technology and its relationships to sustainable supply chain management. *International Journal of Production Research, 57*(7), 2117–2135.

Selvaraj, S. V. (2024, August 29). Enhancing supply chain agility with AWS supply chain vendor lead time insights. *Amazon Supply Chain and Logistics*. Retrieved April 25, 2025, from https://aws.amazon.com/blogs/supply-chain/enhancing-supply-chain-agility-with-aws-supply-chain-lead-time-insights/.

Shah, A. (2024, July 12). Addressing industry-specific supply chain challenges with AWS Supply Chain. *Amazon Supply Chain and Logistics*. Retrieved April 25, 2025, from https://aws.amazon.com/blogs/supply-chain/addressing-industry-specific-supply-chain-challenges-with-aws-supply-chain/.

Sodhi, M. S., & Tang, C. S. (2019). Research opportunities in supply chain transparency. *Production and Operations Management, 28*(12), 2946–2959.

Souza, G. C. (2014). Supply chain analytics. *Business Horizons, 57*(5), 595–605.

Travers, E., & Baeuml, M. (2024, May 3). AWS simulation and digital twin to increase warehouse productivity. *Amazon Supply Chain and Logistics*. Retrieved April 25, 2025, from https://aws.amazon.com/blogs/supply-chain/aws-simulation-and-digital-twin-to-increase-warehouse-productivity/.

Unilever PLC. (2023, April 4). How AI and digital help us innovate faster and smarter. *Unilever*. Retrieved April 25, 2025, from https://www.unilever.com/news/news-search/2023/how-ai-and-digital-help-us-innovate-faster-and-smarter/.

Unilever PLC. (2024, July 31). Utilising AI to redefine the future of customer connectivity. *Unilever*. Retrieved April 25, 2025, from https://www.unilever.com/news/news-search/2024/utilising-ai-to-redefine-the-future-of-customer-connectivity/.

Wishart, M. (2024, May 8). Digital and physical transformation for supply chains resiliency. *Amazon Supply Chain and Logistics*. Retrieved April 25, 2025, from https://aws.amazon.com/blogs/supply-chain/digital-and-physical-transformation-for-supply-chains-resiliency/.

Wolski, C. (2022, June 1). How Heineken is digitally transforming. *Digital CxO*. Retrieved April 25, 2025, from https://digitalcxo.com/article/how-heineken-is-digitally-transforming/

Chapter 2
Transforming Supply Chains in the Digital Age: Strategies, Technologies, and Best Practices

Albana Berisha Qehaja

Abstract Digital technologies are revolutionizing supply chain management in today's fast-paced business environment. This chapter examines supply chain transformation in the digital age, focusing on strategies and best practices. It begins with an overview of the evolution from traditional to digital supply chains and highlights the impact of this shift. The chapter discusses the role of technology in contemporary supply chains, emphasizing key technologies driving digital transformation and their effects on supply chain dynamics. It also explores strategic planning, integrating digital technologies in operations, and creating a digital ecosystem. Case studies illustrate best practices and provide key insights and takeaways. Future trends emphasizing sustainability and resilience are also outlined. This chapter utilizes a mixed-methods approach, integrating a literature review and case studies to thoroughly understand supply chain management in the digital era. Supply chain leaders are urged to foster innovation, adaptability, and collaboration to overcome challenges and develop resilient, customer-focused supply chains effectively. Keeping up-to-date with emerging technologies and industry trends is essential for long-term success in this continually evolving field.

Keywords Digital supply chain · Strategies · Best practices · Technology adoption · Sustainability

A. B. Qehaja (✉)
Faculty of Economy, University of Prishtina, Prishtina, Republic of Kosovo
e-mail: albana.berisha@uni-pr.edu

Y. Khojasteh, S. M. M. Jahan Abadi (eds.), *Emerging Technologies in Supply Chains*, International Series in Operations Research & Management Science 373, https://doi.org/10.1007/978-3-032-01218-0_2

2.1 Introduction

There has been a recent increase in academic papers on business, management, and information systems, highlighting the impact of digital technologies (Schneider & Kokshagina, 2021). In today's business environment, digital technologies have transformed various aspects of organizational operations, including Supply Chain Management (SCM) (MacCarthy & Ivanov, 2022; Yang et al., 2021). The COVID-19 pandemic has notably accelerated the pace of digital transformation across diverse industries (Bigliardi et al., 2022; Kraus et al., 2022; Li, 2020). Despite recent advancements, digital transformation (DX) studies are still in their early stages (Queiroz et al., 2021).

The Digital Supply Chain (DSC) concept is still in its early stages in the academic world. Previous studies have mainly focused on Industry 4.0 and the technological drivers of DSC, including the Internet of Things (IoT), Big Data (BD), Cloud Computing (CC), and Unmanned Aerial Vehicles (UAV), among others (Ageron et al., 2020). From the perspective of Industry 4.0, emerging technologies are impacting traditional supply chains and accelerating the shift to DSC (Queiroz et al., 2021). Industry 4.0 technologies disrupt traditional supply chain models, requiring companies to rethink their strategies (McKinsey & Company, 2016).

The traditional paradigms of SCM have been disrupted, paving the way for a new era of DSCs (Ivanov et al., 2021; MacCarthy & Ivanov, 2022). Transitioning from a traditional supply chain to a digital one is seen as a competitive advantage, creating lasting value for organizations (Bonnet & Westerman, 2020; Makris et al., 2019). This transformation has brought unprecedented challenges and opportunities, necessitating a strategic rethinking of SCM practices (Faridi & Malik, 2020; Schneider & Kokshagina, 2021).

Integrating digital technologies into SCM has brought about significant change. It has transformed supply chain operations from linear, sequential processes to interconnected, synchronized, and flexible DSC (Oh et al., 2023). This change has improved operational efficiency and enabled real-time visibility, predictive analytics, and customer-centric services (Pinto et al., 2023; Tay & Loh, 2022). Supply chain management involves structuring and administrating components such as products and information throughout the supply chain (Khojasteh & Abdi, 2016). According to Ivanov et al. (2021, p. 5), "Supply chain management is a cross-department and cross-enterprise integration and coordination of material, information, and financial flows to transform and use the supply chain resources in the most rational way along the entire value chain, from raw material suppliers to customers". There is a broad consensus in both scholarly and professional writings that utilizing digital technologies can enhance and automate numerous aspects of supply chain management within organizations and throughout the supply chain (MacCarthy & Ivanov, 2022).

This chapter aims to delve deeply into the evolution of SCM in the context of digital transformation. It clarifies the strategies and best practices related to DSC, providing supply chain leaders with the knowledge and insights necessary to navigate this digital revolution.

This chapter is organized to provide a comprehensive understanding of digital SCM. It starts with an overview of SCM and its development, emphasizing the influence of digital transformation. Subsequent sections delve into the digital age's paradigm shift, technology's role in modern supply chains, and the strategies for digital SCM. The chapter also presents best practices through illustrative case studies. Finally, it concludes with a look into future trends in digital SCM.

This chapter combines theoretical insights with practical observations to offer a comprehensive perspective on SCM in the digital era. It emphasizes the importance of innovation, adaptability, and collaboration in this age and stresses the need for supply chain leaders to stay updated on emerging technologies and industry trends (Gezgin et al., 2017; Rasool et al., 2022).

In the digital age, supply chain leaders must embrace innovation and adaptability. They must be willing to disrupt traditional business models and processes and leverage digital technologies to create value. Furthermore, collaboration is key in the DSC, enabling organizations to share resources, information, and capabilities, enhancing overall supply chain performance.

Remaining current with the latest technological advancements and industry trends is key to achieving long-term success in the ever-shifting supply chain management landscape. Therefore, this chapter serves as a valuable resource for supply chain leaders seeking to understand and navigate the complexities of digital transformation in SCM.

2.2 Evolution of Supply Chain Management

2.2.1 Traditional Supply Chain Management

The roots of SCM can be traced back to initial studies on postponement, system dynamics, and the bullwhip effect (Forrester, 1961), collaboration (Bowersox, 1969), multi-tier inventory management (Geoffrion & Graves, 1974), Just-In-Time (JIT), and lean production. The term "SCM" was first employed in the article titled "*SCM: Logistics Catches up with Strategy*" authored by Oliver and Webber in 1982 (Ivanov et al., 2021). Harvard Professor Michael Porter's seminal work, *Competitive Advantage*, illustrated how a company could become more profitable by strategically analyzing the five primary processes on which its value chain framework is built: inbound logistics, operations, outbound logistics, sales and marketing, and service (Porter, 1985). Despite the gap of over two decades between their respective works, both Forrester and Porter identified that a vertical strategy—characterized by the segregation of each department and group into isolated silos—was detrimental to the sustained growth and vitality of a company (Blanchard, 2021).

Historically, supply chain management has played a crucial role in business operations (Ageron et al., 2020; Ivanov et al., 2021; MacCarthy & Ivanov, 2022). It is characterized by linear, sequential processes involving various entities such as suppliers, manufacturers, distributors, and retailers, all working in a coordinated manner to deliver products and services to the end customer (Khojasteh & Abdi, 2016). The primary focus of traditional SCM has been on optimizing individual elements of the supply chain, often at the expense of overall supply chain performance (Wang & Li, 2020). While effective in a stable and predictable business environment, this approach has limitations in today's dynamic and uncertain world (Zekhnini et al., 2021).

In the traditional SCM model, information flow across the supply chain was often slow and disjointed, leading to inefficiencies such as excess inventory, stockouts, and missed delivery deadlines (Bigliardi et al., 2022; Gezgin et al., 2017). Furthermore, the lack of transparency and visibility across the supply chain made it difficult to respond quickly to changes in demand or supply disruptions (Khojasteh et al., 2022). Despite its recognition almost three decades ago, a small percentage of companies' full adoption of supply chain management remains limited (Blanchard, 2021).

2.2.2 *Impact of Digital Transformation on Supply Chain Management*

The adoption and reliance on digital technologies to perform business activities are referred to as "digital transformation" or "digitalization" (Rasool et al., 2022). Digitalizing the supply chain represents a strategic theme within the academic and professional communities involved in logistics and supply chain management (Makris et al., 2019). The rapid advancement of digital technology has brought about significant changes in various industries in recent decades (Endres et al., 2020). New companies are challenging established ones using digital tools to create platform ecosystems and drive industry-wide transformations (He et al., 2023; Sinitsyna & Nekrasov, 2024). The ongoing industrial revolution, often called Industry 4.0, is fundamentally based on granting full autonomy to existing automated industrial systems (Rana & Daultani, 2023).

Digital technologies have profoundly impacted SCM, disrupting traditional paradigms and introducing new possibilities (Ageron et al., 2020; Kraus et al., 2022). Digital transformation in SCM involves using digital technologies to improve supply chain operations' efficiency, effectiveness, and flexibility (MacCarthy & Ivanov, 2022; Yang et al., 2021). It encompasses many technologies, including CC, BDA, AI, machine learning, IoT, and blockchain (Bigliardi et al., 2022; Dutta et al., 2020).

While there is a global emphasis on studying and understanding digital transformation, and despite numerous attempts to define it precisely, there currently exists no universally accepted definition of digital transformation. The boundaries that could potentially define it remain unclear (Kraus et al., 2021). According to Vial (2019, p. 118), digital transformation is the "process that aspires to augment an entity by initiating significant modifications to its properties via the amalgamation of various technologies". Kraus et al. (2021) state that DX is associated with digital strategy, supply chain management, leadership, value creation, and entrepreneurship. It has also allowed for greater collaboration among supply chain partners, improving efficiency and effectiveness (Aamer et al., 2023; Pinto et al., 2023; Preindl et al., 2020). It has also allowed for greater collaboration among supply chain partners, leading to improved efficiency and effectiveness (Li & Zhao, 2024). For instance, companies can share real-time data with their supply chain partners through cloud-based platforms. This enables them to coordinate their activities more effectively and respond quickly to demand or supply disruptions (Gezgin et al., 2017).

Digital transformation involves organizational structure and culture changes (Rasool et al., 2022). This includes transitioning from management-led to data-oriented decision-making, promoting interdisciplinary collaboration instead of segregated silos, and embracing agile, experimental, and spontaneous methods over risk-averse strategies (Gezgin et al., 2017; Levallet & Chan, 2018).

2.2.3 Transition from Traditional to Digital: Challenges and Opportunities

The structure of the DSC is not a traditional linear supply chain, but rather a digital supply network that is more dynamic. This system utilizes information technology to integrate supply chain activities for seamless material flows. The configuration of the DSC is contingent upon the goals and strategies of the organization (Makris et al., 2019). The digitalization of the supply chain offers the potential for new digital structures, enhanced abilities, and more efficient information systems and information technology for the integration, planning, administration, and regulation of the supply chain (MacCarthy & Ivanov, 2022).

The transition from traditional to digital SCM is not without its challenges. These include significant investment in technology and infrastructure, new skills and competencies, and the need to manage change effectively (Faridi & Malik, 2020). For example, adopting digital technologies demands a substantial initial investment, which might be prohibitive for small and medium-sized enterprises (SMEs). Furthermore, employees need to be trained to use these technologies effectively, which requires additional investment in training and development (Cichosz et al., 2020). Although DSCs offer several advantages over traditional supply chains, adopting digital technologies is neither easy nor inexpensive. This adoption impacts firm performance and short and long-term competitiveness (Rasool et al., 2022).

Arunachalam et al. (2018) identified two main challenges in adopting big data analytics in supply chain management: organizational and technical difficulties. Organizational challenges include time-consuming processes, behavioral issues, insufficient resources, security and privacy concerns, investment challenges, and skills shortages. Technical challenges include data scalability, data quality, and a lack of techniques and procedures.

However, the opportunities presented by digital SCM far outweigh the challenges. Digital SCM offers the potential for improved operational efficiency, increased agility, and enhanced customer satisfaction (Kashem et al., 2024; Li & Zhao, 2024). It also opens up new avenues for innovation and competitive advantage (Bonnet & Westerman, 2020; Kraus et al., 2022; Oh et al., 2023). For instance, through the use of big data analytics (BDA), companies can gain insights into customer behavior and preferences, enabling them to tailor their products and services to meet customer needs more effectively (Maheshwari et al., 2021; Pinto et al., 2023). IoT technologies allow companies to monitor their products' location and status in real-time, leading to more efficient inventory management and reducing stockouts and excess inventory (Gezgin et al., 2017). Abbu and Gopalakrishna (2022) studied how big data analytics is employed by leading grocery corporations like Walmart, Kroger, Albertsons, Target, and Amazon Whole Foods for digital transformation. The research underscores the pivotal role of big data in multiple facets of DX.

According to Li (2020), three new approaches are currently being adopted by several leading global organizations for leading digital transformation:

1. *Innovating by Experimenting*: Employing experiments as a means to assess, adjust strategy, and direct execution (Alibaba, Google, Amazon, VMWare)
2. *Radical transformation through incremental approaches*: The strategy is frequently formulated, assessed, and adjusted throughout the execution process (Amazon, Slack, JD, Didi) and
3. *Dynamic sustainable advantages via temporary advantages*: Utilizing temporary benefits to dynamically and cumulatively establish sustainable advantages (Baidu; Uber; VMWare, Slack)

In conclusion, the evolution of SCM from traditional to digital has transformed how businesses operate, offering new opportunities for efficiency, agility, and customer satisfaction. However, it also presents new challenges that need to be managed effectively. As we move forward, it is clear that digital SCM will play a critical role in shaping the future of business operations.

2.3 The Digital Age: A Paradigm Shift in Supply Chain Management

2.3.1 Role of Technology in Modern Supply Chains

While the significance of ensuring a reliable supply has been recognized throughout history, the phrase 'supply chain management' did not come into use until the 1980s (Ellram & Cooper, 2014). The field of supply chain management has developed in tandem with Information Technology (IT) over the past three decades (MacCarthy & Ivanov, 2022).

Technology is a key driver in the operation of supply chains in the current digital landscape. It serves as the backbone of digital supply chain management, enabling real-time visibility, data-driven decision-making, and enhanced operational efficiency (Kashem et al., 2024; Preindl et al., 2020). The decision-making process within supply chains is progressively more dependent on the exchange of information. Consequently, many decisions are based on available data, making them data-driven (Arunachalam et al., 2018).

Technology facilitates seamless integration and collaboration among supply chain partners, improving coordination and responsiveness (Gezgin et al., 2017). Furthermore, adopting digital technologies has been found to significantly impact the sustainability of supply chains, contributing to their environmental and social performance (Junge & Straube, 2020; Maheshwari et al., 2021). Jahanabadi et al. (2022) identified sixteen determinants that influence the sustainability of the chicken meat supply chain through a comprehensive review of literature and expert consultations.

A digital twin is a virtual system representation used to understand performance, improve processes, and increase value-added activities (Park et al., 2019). It is not an isolated device or entity. Instead, it is a comprehensive system of interconnected elements that require specialized technical knowledge from various fields, such as sensor technology and data analytics (Sarker, 2023). Kamble et al. (2022) extensively analyzed 98 studies on digital twins. The review underscores the significance of digital twins in developing sustainable and intelligent manufacturing systems. It also emphasizes the growing potential of digital twin applications in supply chains, attributing this to technological advancements such as IoT, CC, and blockchain.

2.3.2 Key Technologies Driving the Digital Transformation

Several key technologies are driving the digital transformation in supply chain management. These include:

Artificial Intelligence (AI) Artificial Intelligence is currently under comprehensive examination, particularly emphasizing its potential benefits to SCM by implementing AI-driven business strategies (Rana & Daultani, 2023). It enables intelligent automation, predictive analytics, and advanced decision-making in supply chain operations (Dutta et al., 2020; MacCarthy & Ivanov, 2022). It allows for the automation of routine tasks, freeing human resources for more strategic activities (Kraus et al., 2021).

Internet of Things IoT allows real-time tracking and monitoring of goods, enhancing transparency and efficiency in the supply chain (Sarkis et al., 2021). It enables the collection and analysis of vast amounts of data from various points in the supply chain, leading to improved decision-making and forecasting (Bigliardi et al., 2022; Maheshwari et al., 2021). IoT devices can efficiently virtualize, trace, and verify products within the supply chain. Connected physical IoT devices can scrutinize and evaluate the origin of materials and products to ascertain if they are derived from authenticated renewable resources (Sarkis et al., 2021).

Blockchain Blockchain technology serves as a platform for resource distribution, allowing suppliers to share surplus resources through a peer-to-peer network without the need for mutual trust. This decentralized system fosters trust and transparency, eliminating the need for transaction intermediaries (Sarkis et al., 2021). Blockchain ensures secure and transparent transactions, fostering trust and collaboration among supply chain partners (Korpela et al., 2017). It provides a decentralized and immutable record of transactions, reducing the risk of fraud and enhancing traceability (Kamble et al., 2022).

Big Data Analytics BDA provide valuable insights into customer behavior and market trends, enabling proactive and customer-centric supply chain strategies (Pinto et al., 2023; Yang et al., 2021). They allow for the analysis of large volumes of structured and unstructured data, leading to improved forecasting and demand planning (Van Veldhoven & Vanthienen, 2022). Big data is widely anticipated to positively impact SCM by enabling organizations to make more data-driven, strategic, and informed decisions (Addo-Tenkorang & Helo, 2016). Also, BDA is a powerful tool that can efficiently address questions related to operations and supply chain risk (Maheshwari et al., 2021).

Arunachalam et al. (2018, p. 419) describe big data analytics in supply chain management as "the ability of organisations to collect and organise supply chain data from heterogeneous systems distributed across organisational boundaries, analyse it either batch-wise or real-time or near real-time and visualise it intuitively to create proactive supply chain system and support decision making".

2.3.3 Impact and Implications of Technologies on Supply Chain Dynamics

The integration of these technologies has profound implications for supply chain dynamics. They enhance operational efficiency and transform how supply chain partners interact and collaborate (Gezgin et al., 2017). For instance, AI and BDA enable companies to predict market trends and make informed decisions, thereby reducing uncertainties and improving supply chain performance (Arunachalam et al., 2018; MacCarthy & Ivanov, 2022; Preindl et al., 2020). IoT enhances real-time visibility across the supply chain, enabling companies to respond promptly to changes or disruptions (Sarkis et al., 2021). Blockchain fosters trust and transparency among supply chain partners, facilitating secure and efficient transactions (Korpela et al., 2017).

Moreover, adopting these technologies can significantly improve supply chain sustainability (Khojasteh & Abdi, 2016; MacCarthy & Ivanov, 2022; Yang et al., 2021). For instance, IoT can enable better tracking of products throughout their lifecycle, contributing to waste reduction and improved resource efficiency (Rehman Khan et al., 2022). Similarly, AI and BDA can help companies optimize their operations and reduce their environmental footprint.

2.4 Strategies for Digital Supply Chain Management

In the digital transformation era, businesses leverage various strategies to optimize their supply chains, improve customer satisfaction, lower operational costs, and mitigate risks associated with supply chain disruptions. These strategies offer businesses a structured approach to navigating shifting market landscapes and capitalize on emerging DSC business models. Organizations must develop fundamental capabilities to achieve significant supply chain performance, including leveraging digitalization and integrating technologies with employees, customers, and suppliers throughout the supply chain (Queiroz et al., 2021). A successful DSC requires a blend of digital tools, strategies, and methods that enhance interactions between external customers, suppliers, and internal employees. To maximize the benefits of the DSC, it is crucial to utilize innovative strategies, including digital transformation with technologies (Nasiri et al., 2020).

Based on McKinsey's experiences, companies can achieve greater benefits by developing a comprehensive vision for their future supply chains, systematically assessing their current performance, and creating a strategic roadmap for long-term transformation (Angevine et al., 2021; Gezgin et al., 2017). Preindl et al. (2020) identified three key transformation strategies for the supply chain. Firstly, classical multiyear strategies with long-term visions are common, often divided into phases from situation analysis to an optimal future state. Secondly, some companies rapidly implement Industry 4.0 technologies to stay competitive, requiring fast decision-making and a mix of bottom-up and top-down projects. Thirdly, some focus on achieving key performance indicators using Industry 4.0 technologies without a specific strategy.

Here are some other key strategies, ordered from planning to execution:

Strategic Planning for Digital Transformation The digital transformation of supply chain management begins with strategic planning. This process requires setting clear objectives, identifying Key Performance Indicators (KPIs), and developing a roadmap for integrating digital technologies. It necessitates a thorough understanding of the current state of the supply chain, the capabilities of available technologies, and the potential challenges and opportunities that digital transformation may bring (Chawla & Goyal, 2022; Li, 2020; Schneider & Kokshagina, 2021).

Effective company transformation depends on a forward-thinking and innovative concept for the future supply chain. It requires understanding the company's position among competitive pressures and trends and adapting to customers' evolving needs. The vision for the supply chain should align with the company's strategic goals (Gezgin et al., 2017).

Integration of Digital Technologies in Supply Chain Operations Once the strategic plan is in place, the next step is the integration of digital technologies in supply chain operations. This complex process involves selecting the appropriate technologies, adapting business processes, and training staff to use the new tools effectively. Consequently, there is a growing trend of employees receiving extensive training in digital competencies (Van Veldhoven & Vanthienen, 2022). Also, a strong IT infrastructure is required to support the new technologies and ensure seamless data flow across the supply chain (Cichosz et al., 2020). The root of failure is frequently a misalignment between the strategy's formulation and implementation (Yang et al., 2021).

Supply Chain 4.0 The integration of advanced digital technologies such as big data, AI, Cloud Computing, and blockchain within Supply Chain Management is characterized in the literature as Intelligent Supply Chains or Supply Chain 4.0 (Rana & Daultani, 2023). As part of the integration process, businesses are moving towards Supply Chain 4.0, which refers to applying advanced technologies like the IoT, robotics, and BDA in supply chain management. These technologies enable companies to address new customer requirements, supply-side challenges, and efficiency improvement expectations (Frederico et al., 2020; Kashem et al., 2024). Data analytics can revolutionize retail supply chain management by optimizing supply chains, improving inventory management, and enhancing logistics operations (Gezgin et al., 2017; Leão & da Silva, 2021; Pinto et al., 2023). Makris et al. (2019) have also studied how multinational corporations from five distinct sectors can adjust to embrace Supply Chain 4.0, thereby gaining a competitive advantage from this shift.

Forecasting and Demand Planning Businesses can enhance their forecasting and demand planning capabilities by integrating digital technologies. Effective demand forecasting helps businesses minimize storage costs and avoid stockouts and overstock. This strategy leverages past sales data, market analysis, and economic indicators to forecast future demand (Junge & Straube, 2020; Simchi-Levi & Timmermans, 2021).

Building a Digital Supply Chain Ecosystem The final step in the digital transformation process is building a DSC ecosystem. This involves creating a network of interconnected entities, including suppliers, manufacturers, distributors, retailers, and customers, all linked through digital technologies. This ecosystem enables real-time visibility, collaboration, and coordination across the supply chain, improving efficiency and responsiveness (Al Mashalah et al., 2022; Gezgin et al., 2017; Li & Zhao, 2024).

By following these strategies, businesses can effectively navigate their digital transformation journey, enhancing their competitiveness in today's complex global business environment. Practitioners are advised to consider a digital supply chain-specific strategy to address the challenges of digital transformation and Industry 4.0 for lasting success (Preindl et al., 2020).

2.5 Best Practices in Digital Supply Chain Management

2.5.1 Case Studies of Successful Digital Supply Chain Transformations

2.5.1.1 Toyota

Founded in 1937, Toyota has established itself as a global leader in the automotive industry, renowned for its innovative Toyota Production System (TPS) (AbiEdu, 2024). It is not merely a factory operating system but also an organizational culture. Its core philosophy emphasizes empowering and developing individuals to swiftly identify and resolve issues, fostering a mindset of continuous improvement. This philosophy extends beyond Toyota's internal operations to include its suppliers (Shih, 2022).

Toyota has optimized its production processes and inventory management by integrating IoT, AI, and advanced analytics, significantly enhancing operational efficiency. The company's lean manufacturing approach, which emphasizes minimizing waste and maximizing value, combined with its strategic use of digital technologies, has enabled Toyota to maintain high standards of quality and efficiency. Additionally, Toyota's commitment to sustainability and resilience is evident through its efforts to diversify suppliers, enhance transparency, and implement robust risk management strategies (AbiEdu, 2024).

Digital Transformation Initiatives

IoT integration: Toyota integrates IoT sensors to collect real-time data from its manufacturing facilities and supply chain, enabling accurate tracking of components and materials. This integration helps minimize delays and enhances efficiency (Shih, 2022).

Big data and advanced analytics: Toyota can use big data analytics to forecast demand more accurately and optimize inventory levels. This approach has minimized excess stock and reduced waste, aligning with lean manufacturing principles (INEAK, 2023).

Artificial intelligence: AI has been integrated into Toyota's production processes to optimize operations. This includes robot-assisted production for precision and consistency and machine learning algorithms for quality inspection and defect rectification (INEAK, 2023; Toyota, 2023).

Blockchain technology: Toyota has adopted blockchain technology to improve transparency and traceability within its supply chain. This implementation helps verify product authenticity and mitigates the risk of counterfeiting (van der Meulen et al., 2022).

Cloud computing: Toyota's shift to cloud computing has facilitated real-time collaboration among manufacturing units and teams. Cloud solutions' centralized data storage and access have provided scalable computational resources for analytics and simulations, improving supply chain efficiency (INEAK, 2023).

Resilience and adaptability: Toyota's digital transformation has significantly enhanced its supply chain resilience, effectively managing disruptions and maintaining operational continuity during crises, like the COVID-19 pandemic.

These initiatives have led to several notable outcomes, including:

- Real-time monitoring and data analytics have significantly improved production efficiency and reduced downtime.
- Advanced analytics and AI have streamlined inventory management, resulting in substantial cost savings.
- Blockchain technology has enhanced transparency and trust across the supply chain, ensuring product authenticity and reducing the risk of counterfeiting.
- Enhanced resilience and adaptability have enabled Toyota to manage disruptions effectively and maintain operational continuity during crises.

Toyota's digital supply chain transformation is a powerful example of how established companies can leverage modern technologies to stay ahead in a competitive market. By embracing IoT, AI, BDA, blockchain, and CC, Toyota has streamlined its operations and enhanced its ability to respond to disruptions and maintain continuity. The company's focus on sustainability, transparency, and innovation underscores its commitment to creating value for its customers and stakeholders. As Toyota continues to evolve, its strategic use of digital technologies will undoubtedly play a crucial role in sustaining its leadership in the automotive industry.

2.5.1.2 Amazon

Founded in 1994, Amazon has grown from an online bookstore to a global e-commerce giant renowned for its innovative use of digital technologies to optimize its supply chain. Amazon's commitment to customer satisfaction, operational efficiency, and continuous innovation has positioned it as a leader in digital supply chain transformation (Hybrow Labs, 2023).

Amazon's digital supply chain transformation is a benchmark for the industry. The company has extensively used robotics, automation, and AI to streamline its supply chain operations (Damtew & Goshu, 2024).

Digital Transformation Initiatives

IoT integration: Amazon utilizes IoT technology in its warehouses to manage and monitor inventory in real-time. These devices track product movements, ensuring accurate stock levels and reducing the risk of stockouts or overstock situations.

Big data and advanced analytics: Amazon employs big data analytics to predict demand, optimize inventory management, and enhance customer experiences. By processing large volumes of data, Amazon can identify purchasing trends and adjust its supply chain operations to meet anticipated needs (Hybrow Labs, 2023).

Artificial intelligence: Amazon employs AI to forecast demand, manage inventory, and optimize delivery routes. This technology ensures timely deliveries and helps reduce operational costs (Pantoja-Navajas, 2024b).

Robotics and automation: Amazon's warehouses utilize sophisticated robotics to manage tasks such as sorting, packing, and shipping. This automation has greatly enhanced operational efficiency and lowered labor costs and human errors (Pantoja-Navajas, 2024a).

Blockchain technology: Amazon has investigated implementing blockchain technology in its supply chain to improve transparency and traceability. This approach aids in verifying product authenticity and securing transactions (Hybrow Labs, 2023).

Cloud Computing: Amazon leverages its Amazon Web Services (AWS) platform to integrate various supply chain functions, providing real-time visibility and enhancing decision-making capabilities (Pantoja-Navajas, 2024b). The development of innovative AWS models on the cloud platform has positively impacted supply chain resilience, performance, and inclusiveness. These advancements offer valuable insights for researchers, policymakers, and industry leaders (Damtew & Goshu, 2024).

These initiatives have led to several notable outcomes, including:

- Integrating robotics and automation has greatly enhanced operational efficiency, leading to reduced labor costs and fewer human errors. This improvement has enabled Amazon to process orders with increased speed and accuracy.
- AI and machine learning technologies have enhanced Amazon's ability to forecast demand and optimize delivery routes. This has resulted in faster delivery times, higher customer satisfaction, and more efficient resource use.
- AWS supply chain solutions have given Amazon the scalability to adapt quickly to changing market demands. This flexibility has supported the company's rapid growth and ability to handle peak periods effectively.
- Exploiting blockchain technology has improved transparency and traceability within Amazon's supply chain. This has helped verify product authenticity, secure transactions, and build trust with customers and partners.
- Amazon's digital transformation has significantly enhanced its ability to manage disruptions and maintain operational continuity. This resilience was particularly evident during crises such as the COVID-19 pandemic, where Amazon met customer needs despite global supply chain challenges.
- The use of big data and advanced analytics has empowered Amazon to make informed, data-driven decisions. This has optimized inventory management,

reduced waste, and aligned supply chain operations with customer demand patterns.

Amazon's supply chain is renowned for its speed and efficiency, setting a high standard for the industry. The company's dedication to rapid product delivery compels other retail giants to rethink their operational strategies. However, relying solely on technology is insufficient for maintaining competitiveness. Many companies that try to replicate Amazon's success often fall short because they overlook the need to transform their management systems and corporate cultures. Amazon is not just participating in the new digital economy; it is actively shaping it (Calvo, 2019).

The next phase of supply chain management will incorporate generative AI to automate manual tasks, enhance information accuracy, and boost resource productivity. Amazon Q, an interactive generative AI assistant in AWS Supply Chain set to launch later this year, will analyze data to identify business risks and opportunities, offer explainability, and assist customers in evaluating trade-offs. It will proactively detect significant exceptions, recommend actions to swiftly mitigate issues, and respond to customer inquiries, thus replacing the traditional method of manually gathering and analyzing data from various systems (Wishart, 2024).

2.5.2 Lessons Learned and Key Takeaways

Integration of advanced technologies:

- IoT, AI, and advanced analytics are crucial for optimizing supply chain operations. Toyota employs IoT for real-time data collection, AI to optimize production processes, and advanced analytics to manage inventory effectively. Amazon employs IoT in warehouses, AI for demand forecasting, and machine learning for delivery route optimization.
- Investing in these technologies enhances operational efficiency, reduces costs, and improves supply chain performance.

Importance of real-time data:

- Real-time monitoring and data analytics are essential for informed decision-making and improving production efficiency. Toyota's real-time data reduces downtime and enhances efficiency, while Amazon's real-time information flows and big data analytics optimize inventory management and customer experiences.

Scalability and flexibility:

- Rapidly scaling operations in response to market demands is vital for competitiveness. Toyota uses cloud computing for scalable analytics and simulations, and Amazon's AWS solutions provide scalability to adapt to market changes.

Enhancing transparency and trust:

- Transparency in the supply chain fosters trust and ensures product authenticity. Blockchain enhances transparency and traceability for both Toyota and Amazon.

Customer-centric approach:

- Building loyalty through timely deliveries and reliable service requires prioritizing customer satisfaction. Amazon's AI and machine learning optimize delivery routes and forecast demand, leading to faster deliveries and higher satisfaction.

Sustainability and resilience:

- Sustainable practices and robust risk management are crucial for long-term success. Toyota diversifies suppliers, enhances transparency, and implements risk management strategies, while Amazon's digital transformation enhances disruption management and operational continuity.

These lessons from Toyota and Amazon's digital supply chain transformations provide valuable insights for businesses aiming to enhance their supply chain operations through advanced technologies, real-time data, scalability, transparency, customer-centric approaches, and sustainability.

2.6 The Future of Supply Chain Management in the Digital Age

The future of supply chain management in the digital age is set to be shaped by several key trends:

1. **Generative AI in Operations**: Generative AI (GenAI), a specialized branch of AI, holds the potential to transform supply chain management, logistics, and procurement. GenAI-powered software engines can handle significantly larger datasets than traditional machine learning models and evaluate an almost limitless array of variables. Additionally, GenAI can self-learn the intricacies of a company's supply chain ecosystem, continuously refining and enhancing its analytical capabilities.
2. **Digital Supply Chains**: The 'smart' supply chain, enabled by a raft of technological developments, is well on its way to becoming the new normal (Enrique et al., 2022). This new approach to supply chain management enables organizations to respond more swiftly to daily demands, proactively tackle problems, and minimize errors and inefficiencies. Additionally, it can boost visibility, transparency, and traceability. Farahani et al. (2017) proposed six dimensions for DSCs in the automotive supply industry: digital performance measurement, digital IT and technology, digital suppliers, digital production systems, digital logistics and inventory, and digital customers.
3. **Sustainability and Social Responsibility**: As we move towards a future where autonomous, self-learning machines efficiently handle the entire supply chain process, organizations must prioritize extracting relevant, clean, and well-governed data to maximize their technology investments. As organizations face increasing pressure to fulfill evolving Environmental, Social, and Governance (ESG) as well as Scope 3 commitments, the importance of data becomes para-

mount. Furthermore, studies indicate that integrating digital technologies within supply chain operations significantly influences the sustainability of these supply chains, enhancing their overall environmental and social performance (Junge & Straube, 2020; MacCarthy & Ivanov, 2022).

4. **Supply Chain 4.0**: In the current landscape, some scholars propose that organizations can create Supply Chain 4.0, often called Smart or Intelligent Supply Chains, depending on the context. Supply Chain 4.0 involves the application of advanced digital technologies such as IoT, AI, robotics, and big data analytics to improve information flows and operational activities (Enrique et al., 2022). The adoption of IoT, advanced robotics, and big data analytics in supply chain management is anticipated to enhance performance and customer satisfaction substantially. These technologies enable real-time tracking and monitoring of goods, enhancing transparency and efficiency in the supply chain (Pinto et al., 2023; Preindl et al., 2020; Sarkis et al., 2021).
5. **Emerging Technologies**: Mastering emerging technologies, including AI, distributed ledger technologies, low-code/no-code platforms, and fleet electrification, is vital for organizations aiming to transition smoothly to a new digital architecture and implement it seamlessly. These technologies offer new opportunities for innovation and competitive advantage (Bonnet & Westerman, 2020; Makris et al., 2019; Schneider & Kokshagina, 2021).

These trends highlight the increasing focus on DSCs, BDA, AI, and the importance of circular and sustainable networks. Organizations must grasp these trends and take immediate steps to initiate their transformation to stay competitive.

2.7 Conclusions

In this chapter, we embarked on a comprehensive exploration of the transformation of supply chain management in the digital age. We delved into the paradigm shift brought about by the digital age, discussing the role of technology in modern supply chains and the key technologies driving digital transformation. We also explored digital supply chain management strategies, including strategic planning for digital transformation, integrating digital technologies in supply chain operations, and establishing a DSC ecosystem.

Furthermore, we presented the best digital supply chain management practices, featuring case studies of successful DSC transformations and key takeaways.

Several key trends are anticipated to shape the future of supply chain management in the digital age, including the rise of generative AI, the creation of 'smart' supply chains, and a heightened emphasis on sustainability and social responsibility.

The digital transformation of supply chains goes beyond merely adopting new technologies. It is about rethinking supply chain management from a digital perspective, embracing new working methods, and creating value in an interconnected world. As such, supply chain leaders, practitioners, and policymakers are more important than ever. They are the ones who will steer their organizations and communities through this transformation, tackle the challenges, and grasp the opportunities presented by the digital age.

We also examined Toyota and Amazon's DSC transformations. Toyota's integration of IoT, AI, and advanced analytics has optimized its production processes and inventory management, enhancing operational efficiency and sustainability. Amazon's use of robotics, automation, and AI has streamlined its supply chain operations, improving customer satisfaction and resilience. These case studies illustrate how established companies can leverage modern technologies to stay competitive and responsive in a rapidly evolving market.

For practitioners, this chapter underscores the importance of strategic planning, effective integration of digital technologies, and creating a DSC ecosystem. It also highlights the need for a focus on social responsibility and sustainability in digital supply chain management.

For policymakers, the findings of this chapter suggest the need for supportive policies and regulations that facilitate the digital transformation of supply chains. This could include policies encouraging investment in digital technologies, training and education in digital skills, and promoting sustainable practices in digital supply chain management.

Future research could focus on emerging technologies such as generative AI and their potential impact on supply chain management. Additionally, more case studies of successful digital transformations could provide valuable insights and lessons learned.

In summary, this chapter provides a comprehensive perspective on the digital transformation of supply chains, highlighting the importance of strategic planning, the role of technology, and the need for sustainable practices. It serves as a valuable resource for supply chain leaders, practitioners, and policymakers seeking to understand and navigate the complexities of digital transformation in supply chain management.

References

Aamer, A., Sahara, C. R., & Al-Awlaqi, M. A. (2023). Digitalization of the supply chain: transformation factors. *Journal of Science and Technology Policy Management, 14*(4), 713–733. https://doi.org/10.1108/jstpm-01-2021-0001

Abbu, H., & Gopalakrishna, P. (2022). Digital transformation powered by big data analytics: The case of retail grocery business. In *Proceedings of the 55th Hawaii international conference on system sciences*. https://doi.org/10.24251/hicss.2022.231

AbiEdu. (2024, June 9). *Case study: Supply chain management at Toyota*. Retrieved from https://abiedu.com/case-study-supply-chain-management-at-toyota/

Addo-Tenkorang, R., & Helo, P. T. (2016). Big data applications in operations/supply-chain management: A literature review. *Computers & Industrial Engineering, 101*, 528–543. https://doi.org/10.1016/j.cie.2016.09.023

Ageron, B., Bentahar, O., & Gunasekaran, A. (2020). Digital supply chain: challenges and future directions. *Supply Chain Forum: An International Journal, 21*(3), 133–138. https://doi.org/10.1080/16258312.2020.1816361

Al Mashalah, H., Hassini, E., Gunasekaran, A., & Bhatt, D. (2022). The impact of digital transformation on supply chains through e-commerce: Literature review and a conceptual framework. *Transportation Research Part E: Logistics and Transportation Review, 165*, 102837. https://doi.org/10.1016/j.tre.2022.102837

Angevine, C., Keomany, J., Thomsen, J., & Zemmel, R. (2021). *Implementing a digital transformation at industrial companies*. McKinsey.

Arunachalam, D., Kumar, N., & Kawalek, J. P. (2018). Understanding big data analytics capabilities in supply chain management: Unravelling the issues, challenges and implications for practice. *Transportation Research Part E: Logistics and Transportation Review, 114*, 416–436. https://doi.org/10.1016/j.tre.2017.04.001

Bigliardi, B., Filippelli, S., Petroni, A., & Tagliente, L. (2022). The digitalization of supply chain: a review. *Procedia Computer Science, 200*, 1806–1815. https://doi.org/10.1016/j.procs.2022.01.381

Blanchard, D. (2021). *Supply chain management best practices*. Wiley. https://doi.org/10.1002/9781119738275

Bonnet, D., & Westerman, G. (2020). The new elements of digital transformation. *MIT Sloan Management Review, 62*(2).

Bowersox, D. J. (1969). Physical distribution development, current status, and potential. *The Journal of Marketing, 33*(1), 63–70.

Calvo, J. (2019, July 26). The miracle of Amazon's digital supply chain. *GLOBIS Insights*. Retrieved from https://globisinsights.com/future-of-work/ai/the-miracle-of-amazons-digital-supply-chain/

Chawla, R. N., & Goyal, P. (2022). Emerging trends in digital transformation: a bibliometric analysis. *Benchmarking: An International Journal, 29*(4), 1069–1112. https://doi.org/10.1108/bij-01-2021-0009

Cichosz, M., Wallenburg, C. M., & Knemeyer, A. M. (2020). Digital transformation at logistics service providers: barriers, success factors and leading practices. *The International Journal of Logistics Management, 31*(2), 209–238. https://doi.org/10.1108/ijlm-08-2019-0229

Damtew, A. W., & Goshu, Y. Y. (2024). The impacts of AWS from digital visions to action for supply chain resilience, performances, and inclusiveness. *The International Journal of Advanced Manufacturing Technology, 130*(9), 4821–4834. https://doi.org/10.1007/s00170-023-12919-4

Dutta, P., Choi, T. M., Somani, S., & Butala, R. (2020). Blockchain technology in supply chain operations: Applications, challenges and research opportunities. *Transportation Research Part e: Logistics and Transportation Review, 142*, 102067. https://doi.org/10.1016/j.tre.2020.102067

Ellram, L. M., & Cooper, M. C. (2014). Supply chain management: It's all about the journey, not the destination. *Journal of Supply Chain Management, 50*(1), 8–20. https://doi.org/10.1111/jscm.12043

Endres, H., Stoiber, K., & Wenzl, N. M. (2020). Managing digital transformation through hybrid business models. *Journal of Business Strategy, 41*(6), 49–56. https://doi.org/10.1108/jbs-07-2019-0142

Enrique, D. V., Lerman, L. V., de Sousa, P. R., Benitez, G. B., Santos, F. M. B. C., & Frank, A. G. (2022). Being digital and flexible to navigate the storm: How digital transformation enhances supply chain flexibility in turbulent environments. *International Journal of Production Economics, 250*, 108668. https://doi.org/10.1016/j.ijpe.2022.108668

Farahani, P., Meier, C., & Wilke, J. (2017). Digital supply chain management agenda for the automotive supplier industry. In G. Oswald & M. Kleinemeier (Eds.), *Shaping the digital enterprise: Trends and use cases in digital innovation and transformation* (pp. 157–172). Springer. https://doi.org/10.1007/978-3-319-40967-2_8

Faridi, M. R., & Malik, A. (2020). Digital transformation in supply chain, challenges and opportunities in SMEs: a case study of Al-Rumman Pharma. *Emerald Emerging Markets Case Studies, 10*(1), 1–16. https://doi.org/10.1108/eemcs-05-2019-0122

Forrester, J. W. (1961). *Industrial dynamics*. MIT Press.
Frederico, G. F., Garza-Reyes, J. A., Anosike, A., & Kumar, V. (2020). Supply Chain 4.0: concepts, maturity and research agenda. *Supply Chain Management: An International Journal, 25*(2), 262–282. https://doi.org/10.1108/scm-09-2018-0339
Geoffrion, A. M., & Graves, G. W. (1974). Multicommodity distribution system design by Benders decomposition. *Management Science, 20*(5), 822–844.
Gezgin, E., Huang, X., Samal, P., & Silva, I. (2017). *Digital transformation: Raising supply-chain performance to new levels* (pp. 1–10). McKinsey.
He, T., Liu, W., Shao, X., & Tian, R. G. (2023). Exploring the digital innovation process and outcome in retail platform ecosystems: disruptive transformation or incremental change. *Electronic Commerce Research*, 1–30. https://doi.org/10.1007/s10660-023-09699-0
Hybrow Labs. (2023). *5 lessons for digital transformations from a case study on Amazon*. Retrieved from https://hybrowlabs.com/blog/5-lessons-for-digital-transformations-from-a-case-study-on-amazon
INEAK. (2023). *Toyota's digital transformation: Embracing the future*. Retrieved from https://www.ineak.com/toyotas-digital-transformation-embracing-the-future/
Ivanov, D., Tsipoulanidis, A., Schönberger, J., Ivanov, D., Tsipoulanidis, A., & Schönberger, J. (2021). Basics of supply chain and operations management. In D. Ivanov, A. Tsipoulanidis, & J. Schönberger (Eds.), *Global supply chain and operations management: A decision-oriented introduction to the creation of value* (pp. 3–19). Springer. https://doi.org/10.1007/978-3-030-72331-6_1
Jahanabadi, E. A., Mousavi, S. N., Moosavihaghighi, M. H., & Eslami, M. R. (2022). Analysis of the causal relationship between factors affecting the management of a sustainable supply chain of chicken meat. *Iranian Journal of Biosystems Engineering, 53*(3), 231–248. https://doi.org/10.22059/IJBSE.2022.335980.665465
Junge, A. L., & Straube, F. (2020). Sustainable supply chains–digital transformation technologies' impact on the social and environmental dimension. *Procedia Manufacturing, 43*, 736–742. https://doi.org/10.1016/j.promfg.2020.02.110
Kamble, S. S., Gunasekaran, A., Parekh, H., Mani, V., Belhadi, A., & Sharma, R. (2022). Digital twin for sustainable manufacturing supply chains: Current trends, future perspectives, and an implementation framework. *Technological Forecasting and Social Change, 176*, 121448. https://doi.org/10.1016/j.techfore.2021.121448
Kashem, M. A., Shamsuddoha, M., & Nasir, T. (2024). Digital-era resilience: Navigating logistics and supply chain operations after COVID-19. *Businesses, 4*(1), 1–17. https://doi.org/10.3390/businesses4010001
Khojasteh, Y., & Abdi, M. R. (2016). Japanese supply chain management. In P. Haghirian (Ed.), *Routledge handbook of Japanese business and management* (pp. 307–316). Routledge.
Khojasteh, Y., Xu, H., & Zolfaghari, S. (2022). *Supply chain risk mitigation*. Springer. https://doi.org/10.1007/978-3-031-09183-4
Korpela, K., Hallikas, J., & Dahlberg, T. (2017). Digital supply chain transformation toward blockchain integration. In *Proceedings of the annual Hawaii international conference on system sciences*. https://doi.org/10.24251/hicss.2017.506.
Kraus, S., Durst, S., Ferreira, J. J., Veiga, P., Kailer, N., & Weinmann, A. (2022). Digital transformation in business and management research: An overview of the current status quo. *International Journal of Information Management, 63*, 102466. https://doi.org/10.1016/j.ijinfomgt.2021.102466
Kraus, S., Jones, P., Kailer, N., Weinmann, A., Chaparro-Banegas, N., & Roig-Tierno, N. (2021). Digital transformation: An overview of the current state of the art of research. *Sage Open, 11*(3), 21582440211047576. https://doi.org/10.1177/21582440211047576
Leão, P., & da Silva, M. M. (2021). Impacts of digital transformation on firms' competitive advantages: A systematic literature review. *Strategic Change, 30*(5), 421–441. https://doi.org/10.1002/jsc.2459

Levallet, N., & Chan, Y. E. (2018). Role of digital capabilities in unleashing the power of managerial improvisation. *MIS Quarterly Executive, 17*(1), 1–21.

Li, F. (2020). Leading digital transformation: three emerging approaches for managing the transition. *International Journal of Operations & Production Management, 40*(6), 809–817. https://doi.org/10.1108/ijopm-04-2020-0202

Li, P., & Zhao, X. (2024). The impact of digital transformation on corporate supply chain management: Evidence from listed companies. *Finance Research Letters, 60*, 104890. https://doi.org/10.1016/j.frl.2023.104890

MacCarthy, B. L., & Ivanov, D. (2022). The digital supply chain—Emergence, concepts, definitions, and technologies. In *The digital supply chain* (pp. 3–24). Elsevier. https://doi.org/10.1016/b978-0-323-91614-1.00001-0

Maheshwari, S., Gautam, P., & Jaggi, C. K. (2021). Role of Big Data Analytics in supply chain management: current trends and future perspectives. *International Journal of Production Research, 59*(6), 1875–1900. https://doi.org/10.1080/00207543.2020.1793011

Makris, D., Hansen, Z. N. L., & Khan, O. (2019, April). Adapting to supply chain 4.0: an explorative study of multinational companies. *Supply Chain Forum: An International Journal, 20*(2), 116–131. Taylor & Francis. https://doi.org/10.1080/16258312.2019.1577114.

McKinsey & Company. (2016). *Supply Chain 4.0: The next-generation digital supply chain.* McKinsey & Company. Retrieved from https://www.mckinsey.com/capabilities/operations/our-insights/supply-chain-40–the-next-generation-digital-supply-chain

Nasiri, M., Ukko, J., Saunila, M., & Rantala, T. (2020). Managing the digital supply chain: The role of smart technologies. *Technovation, 96*, 102121. https://doi.org/10.1016/j.technovation.2020.102121

Oh, K., Yoo, H., & Jeong, E. (2023). Research trends in digital transformation in supply chain based on bibliometric and network analysis. *Journal of Business Economics and Management, 24*(6), 1042–1058. https://doi.org/10.3846/jbem.2023.20649

Pantoja-Navajas, D. (2024a, January 12). NRF 2024: Strengthening digital commerce with AWS Supply Chain. *AWS Supply Chain Blog*. Retrieved from https://aws.amazon.com/blogs/supply-chain/nrf-2024-strengthening-digital-commerce-with-aws-supply-chain/

Pantoja-Navajas, D. (2024b, January 9). 2024 supply chain management predictions: Navigating the future with advanced technologies. *AWS Supply Chain Blog*. Retrieved from https://aws.amazon.com/blogs/supply-chain/2024-supply-chain-management-predictions-navigating-the-future-with-advanced-technologies/

Park, K. T., Nam, Y. W., Lee, H. S., Im, S. J., Noh, S. D., Son, J. Y., & Kim, H. (2019). Design and implementation of a digital twin application for a connected micro smart factory. *International Journal of Computer Integrated Manufacturing, 32*(6), 596–614. https://doi.org/10.1080/0951192x.2019.1599439

Pinto, M. R., Salume, P. K., Barbosa, M. W., & de Sousa, P. R. (2023). The path to digital maturity: A cluster analysis of the retail industry in an emerging economy. *Technology in Society, 72*, 102191. https://doi.org/10.1016/j.techsoc.2022.102191

Porter, M. (1985). *Competitive advantage: Creating and sustaining superior performance* (pp. 39–43). Free Press.

Preindl, R., Nikolopoulos, K., & Litsiou, K. (2020). Transformation strategies for the supply chain: the impact of industry 4.0 and digital transformation. *Supply Chain Forum: An International Journal, 21*(1), 26–34. https://doi.org/10.1080/16258312.2020.1716633

Queiroz, M. M., Pereira, S. C. F., Telles, R., & Machado, M. C. (2021). Industry 4.0 and digital supply chain capabilities: A framework for understanding digitalisation challenges and opportunities. *Benchmarking: An International Journal, 28*(5), 1761–1782. https://doi.org/10.1108/bij-12-2018-0435

Rana, J., & Daultani, Y. (2023). Mapping the role and impact of artificial intelligence and machine learning applications in supply chain digital transformation: a bibliometric analysis. *Operations Management Research, 16*(4), 1641-11666. https://doi.org/10.1007/s12063-022-00335-y

Rasool, F., Greco, M., & Grimaldi, M. (2022). Digital supply chain performance metrics: a literature review. *Measuring Business Excellence, 26*(1), 23–38. https://doi.org/10.1108/mbe-11-2020-0147

Rehman Khan, S. A., Ahmad, Z., Sheikh, A. A., & Yu, Z. (2022). Digital transformation, smart technologies, and eco-innovation are paving the way toward sustainable supply chain performance. *Science Progress, 105*(4), 00368504221145648. https://doi.org/10.1177/00368504221145648

Sarker, D. (2023). *Achieving digital transformation success: Interplay of performance, business models and digital twins in supply chain management.* Master's thesis, NTNU. Retrieved from https://ntnuopen.ntnu.no/ntnu-xmlui/bitstream/handle/11250/3101222/no.ntnu%3Ainspera%3A146715325%3A91423017.pdf?sequence=1

Sarkis, J., Kouhizadeh, M., & Zhu, Q. S. (2021). Digitalization and the greening of supply chains. *Industrial Management & Data Systems, 121*(1), 65–85. https://doi.org/10.1108/imds-08-2020-0450

Schneider, S., & Kokshagina, O. (2021). Digital transformation: What we have learned (thus far) and what is next. *Creativity and Innovation Management, 30*(2), 384–411. https://doi.org/10.1111/caim.12414

Shih, C. W. (2022). What really makes Toyota's production system resilient. *Harvard Business Review.* Retrieved from https://hbr.org/2022/11/what-really-makes-toyotas-production-system-resilient

Simchi-Levi, D., & Timmermans, K. (2021). A simpler way to modernize your supply chain. *Harvard Business Review, 99*(5), 133–141. Retrieved from https://hbr.org/2021/09/a-simpler-way-to-modernize-your-supply-chain

Sinitsyna, A., & Nekrasov, A. (2024). Digital transformation tools for sustainable supply chains in the life cycle processes. In M. Sari & A. Kulachinskaya (Eds.), *Digital transformation: What are the smart cities today?* (pp. 209–219). Springer Nature. https://doi.org/10.1007/978-3-031-49390-4_15

Tay, H. L., & Loh, H. S. (2022). Digital transformations and supply chain management: a Lean Six Sigma perspective. *Journal of Asia Business Studies, 16*(2), 340–353. https://doi.org/10.1108/jabs-10-2020-0415

Toyota. (2023). *Toyota and generative AI: It's here, and this is how we're using it.* Retrieved from https://pressroom.toyota.com/toyota-and-generative-ai-its-here-and-this-is-how-were-using-it/

van der Meulen, N., Mooney, J. G., & Beath, C. M. (2022). *Digital in the driver's seat: Accelerating Toyota's transformation to mobility services* (Working Paper No. 454). MIT Center for Information Systems Research. Published April 4, 2022. Retrieved from https://cisr.mit.edu/publication/MIT_CISRwp454_ToyotaMobilityServices_VanderMeulenMooneyBeath

Van Veldhoven, Z., & Vanthienen, J. (2022). Digital transformation as an interaction-driven perspective between business, society, and technology. *Electronic Markets, 32*(2), 629–644. https://doi.org/10.1007/s12525-021-00464-5

Vial, G. (2019). Understanding digital transformation: A review and a research agenda. *Journal of Strategic Information Systems, 28*(2), 118–144. https://doi.org/10.1016/j.jsis.2019.01.003

Wang, Q., & Li, J. J. (2020). Managing and repairing supply chains upon disruptions. In T. Y. Choi, J. J. Li, D. S. Rogers, T. Schoenherr, & S. M. Wagner (Eds.), *The Oxford handbook of supply chain management* (pp. 334–356). Oxford University Press. https://doi.org/10.1093/oxfordhb/9780190066727.013.20

Wishart, M. (2024, May 8). *Digital and physical transformation for supply chains resiliency.* Amazon Web Services. Retrieved from https://aws.amazon.com/blogs/supply-chain/digital-and-physical-transformation-for-supply-chains-resiliency/

Yang, M., Fu, M., & Zhang, Z. (2021). The adoption of digital technologies in supply chains: Drivers, process and impact. *Technological Forecasting and Social Change, 169*, 120795. https://doi.org/10.1016/j.techfore.2021.120795

Zekhnini, K., Cherrafi, A., Bouhaddou, I., Benghabrit, Y., & Garza-Reyes, J. A. (2021). Supply chain management 4.0: a literature review and research framework. *Benchmarking: An International Journal, 28*(2), 465–501. https://doi.org/10.1108/bij-04-2020-0156

Chapter 3
A SCOR-Based Framework for Applying Digital Transformation Technologies in Supply Chain Strategic Decisions

Melisa Ozbiltekin-Pala

Abstract The transition from traditional supply chain management to digital supply chains has created radical changes in accelerating data flows, automating decision-making processes, and using analytical tools. Digital transformation has provided flexibility and efficiency in supply chain management, making decision-making processes more dynamic and data-oriented. This chapter provides a framework for integrating strategic decisions and digital-based tools and techniques in digital supply chains based on the SCOR (Supply Chain Operations Reference) Model Digital Standard (Version 14, 2022). The chapter examines the effects of digital transformation on supply chains and discusses in detail the integration of digital tools and techniques into decision-making processes. The structure provided by the SCOR Model shows how strategic decisions can be supported with digital data to make supply chain management more transparent and effective.

Keywords Digital transformation · Decision-making · SCOR digital standard (version 14, 2022) model · Supply chain management

3.1 Introduction

Digital transformation and the digital supply chain have grown in importance recently. Due to the quick growth of digital technology, which also offers businesses a ton of options, digital supply chains are becoming more complex (Mubarik & Khan, 2024). In the rapidly evolving global context, decision-making processes are becoming increasingly crucial in these complex digital supply chain structures, particularly digital transformation and digital supply chain connections (Akbari et al., 2024).

M. Ozbiltekin-Pala (✉)
Department of Logistics Management, Faculty of Business, Yaşar University, İzmir, Turkey
e-mail: melisa.ozbiltekin@yasar.edu.tr

Y. Khojasteh, S. M. M. Jahan Abadi (eds.), *Emerging Technologies in Supply Chains*, International Series in Operations Research & Management Science 373, https://doi.org/10.1007/978-3-032-01218-0_3

Due to the swift advancement of digital technologies, businesses can improve their efficiency and competitiveness in digital supply chain management (Brintrup et al., 2024). At this point, enhancing strategic decision-making procedures becomes essential to digital transformation (Dixit et al., 2024). Businesses may efficiently analyze data inside their digital supply chains to make more accurate decisions by implementing better data management and analytics. Adopting cutting-edge business models and technologies also helps organizations become faster and more adaptive, encouraging them to approach strategic decision-making processes dynamically (Ramesh et al., 2024). Therefore, businesses can gain a competitive edge by combining digital transformation with strategic decision-making processes in digital supply chain management. This approach enables a comprehensive analysis of digital supply chains, emphasizing the necessity of including all digital supply chain components in the digital transformation process to improve overall business performance.

Understanding the decision-making processes in digital supply chains and their digital tools and techniques is essential for companies to ensure their sustainability in a competitive environment (Li et al., 2024). Since it is a subject that needs to be studied and integrated, in this context, the dimensions of the Supply Chain Operations Reference (SCOR) model constitute a critical theoretical infrastructure to provides a practical road map and handle all processes (Guo & Mantravadi, 2024). The SCOR Digital Standard (Version 14, 2022) Model is a framework developed by the Supply Chain Council (now part of APICS/ASCM) to help organizations analyze, improve, and optimize their supply chain operations. Version 14 integrates digital capabilities, emphasizing end-to-end supply chain visibility, automation, artificial intelligence, and analytics. It maintains the Plan, Source, Make, Deliver, Return, and Enable structure while incorporating emerging technologies and modern performance metrics. This version enables businesses to align their supply chain strategies with digital transformation efforts, improving efficiency, resilience, and customer satisfaction in an increasingly dynamic and interconnected global market. The SCOR Model covers the digital supply chain's dimensions, processes, and relationships. This model is used as a framework to evaluate and improve strategic decision-making processes in digital supply chains by guiding the processes of determining, implementing, and improving supply chain strategies (Ricárdez-Estrada et al., 2024). Strategic decision-making processes, which will be evaluated within the scope of the five principal dimensions of the SCOR model (plan, source, make, deliver, and return), must be carried out with digital tools and techniques in the digitalizing environment (Raja et al., 2024).

Therefore, this chapter is dedicated to evolving decision-making procedures in digital supply chains, a topic of significant relevance in digital transformation in changing environments. The focus is on integrating digital-based tools and techniques with strategic decision-making processes in digital supply chains based on the SCOR Model. The proposed framework aims to comprehensively understand how digital supply chains evolve in dynamic contexts, how strategic decision-making processes operate, and how these processes are harmoniously integrated with digital tools and approaches. Moreover, the primary aim of this chapter is to present a framework for integrating strategic decisions and digital-based tools and techniques in decision-making in digital supply chains based on the SCOR Model. As previously discussed, the transformation of the digital supply chain can be complex and dynamic. However, this proposed framework offers solutions that can significantly enhance digital supply chain efficiency, pricing strategies, customer satisfaction, and competitive advantage. Digital tools and techniques have revolutionized and enhanced these decision-making processes in recent years.

In summary, the objectives of this chapter can be summarized as follows: to understand strategic decision-making processes in digital supply chain management and to identify critical points in these processes, to investigate the usage areas and advantages of digital tools and techniques in digital supply chain management, to evaluate how digital tools can be integrated with strategic decision-making processes and the impact of this integration on digital supply chain performance in the context of dimensions of SCOR model, to contribute to businesses making more effective and efficient decisions in digital supply chain management through the proposed framework in the context of dimensions of SCOR model, to contribute the study's results to the academic literature and offer suggestions for application in the industry.

This chapter is organized as follows: Sect. 3.2 explains strategic decisions in digital supply chains based on the SCOR Model. Section 3.3 covers a proposed framework for integrating strategic decisions and digital-based tools and techniques in decision-making in digital supply chains based on the SCOR model. Lastly, Sect. 3.4 gives discussions and conclusions.

The following section explains decision-making in the digital supply chain through digital transformation.

3.2 Strategic Decisions in Digital Supply Chains Based on the SCOR Model

Recently, decision-making mechanisms have radically changed with the rapid development of digital technologies and the evolution of digital supply chains (AlMulhim, 2021). The changing environment of decision-making mechanisms in digital supply chains is closely related to the innovations brought by digital transformation (Zhao & Wang, 2024). While traditional supply chain management is based on physical data flows and manual processes, digital transformation has radically changed this structure (Brau et al., 2024). Digitalization has made decision-making processes more efficient by increasing access to information and analysis capabilities. These technological advances create a competitive advantage for large-scale organizations and small businesses (Sadeghi et al., 2024).

This digital transformation in supply chains has made decision-making processes not only faster but also more dynamic and harmonious (Zhao & Wang, 2024).

With their real-time data collection and analysis capabilities, digital technologies provide decision-makers instant and predictable information, allowing them to react quickly to environmental changes (Gardas & Narwane, 2024). This change makes it possible to strategically shape supply chain strategies, considering various factors such as sustainability, risk management, and customer demands. As a result, digital transformation makes decision-making mechanisms in supply chains more flexible and efficient.

The SCOR Model provides a comprehensive reference framework for effectively managing supply chains (Ayyildiz & Taskin Gumus, 2021). The model identifies processes: plan, order, source, transform, fulfill, return, but more digital-focused. In other words, in order to enhance supply chain performance, it incorporates cutting-edge digital technologies like artificial intelligence (AI), the Internet of Things (IoT), blockchain, automation, and cloud computing, business intelligence (BI) while maintaining the fundamental SCOR structure— plan, order, source, transform, fulfill, return. In order to maintain supply chains' competitiveness in a world that is becoming more and more data-driven, the revised model adds new performance criteria that emphasize digital efficiency, sustainability, and agility. These processes cover all supply chain stages, enabling organizations to balance demand and supply capacity, procure materials, produce products, make customer deliveries, and manage returns (Ozbiltekin-Pala et al., 2023). This structure reflects the supply chain's cyclical, continuous, and connected nature, allowing processes to be more synchronized and flexible. SCOR has also been modernized to include sustainability and digital know-how. These updates support retail, omnichannel strategies, and strategic sourcing process changes. It also includes durability, economic, and sustainability metrics and benchmarks (APICS, 2022).

Adding the People and Practices aspects in SCOR Version 14 is a significant improvement, as it now highlights digital skills, analytical capabilities, and automation expertise as critical qualifications for supply chain workers. Additionally, the framework promotes circular economy tactics and green supply chains by incorporating sustainability concepts. By embedding digital tools and innovative technologies into every process, SCOR 14 provides organizations with a more agile, efficient, and resilient supply chain model that aligns with Industry 4.0 and the future of supply chain management. As shown in Fig. 3.1, This model modernizes the traditional SCOR's Plan, Source, Make, Deliver, and Return processes and introduces the concept of Orchestrate, a new high-level management process (APICS, 2022). Orchestrate is the central management layer that connects supply chain processes, data-driven and integrated. SCOR aims to manage the entire process in an end-to-end synchronous manner instead of optimizing supply chain processes only at certain stages. Orchestrate provides real-time decision-making, optimization, and automation using digital tools throughout the supply chain (APICS, 2022).

From a digital supply chain perspective, digital transformation adds significant richness to the framework offered by the SCOR Model and creates radical changes at every stage of the supply chain (Pal, 2024). While technological innovations enable the core processes of the SCOR Model to be managed more efficiently and flexibly, they also reduce uncertainties in the digital supply chain, strengthen risk

Strategic Decisions in Digital Supply Chains based on SCOR Digital Standard Model	Plan	Demand Forecasting and Planning
		Resource Planning
		Strategic Network Design
		Capacity Management
	Order	Order Management and Processing
		Omnichannel Order Fulfillment Strategy
		Order Prioritization and Optimization
	Source	Supplier Selection and Management
		Supplier Collaboration
		Automated Purchasing and Order Management
		Risk Management
	Transform	Smart Manufacturing and Process Improvement
		Real-time production monitoring
		Workforce management and planning
		Maintenance and Repair Management
	Fulfill	Order Processing and Delivery Performance
		Logistics Optimization
		Emergency Planning in Delivery Process
		Dynamic route planning
	Return	Return and Product Reprocessing
		Return Management and Transparency
		Feedback and Quality Improvement
		Automating Return Processes

Fig. 3.1 Strategic decisions in digital supply chains based on the SCOR model

management, and enable strategic decisions to be made more effectively (Gardas & Narwane, 2024). Moreover, the SCOR Model offers a critical structure to increase the effectiveness of processes in digital supply chains and aims to better manage these processes with digital tools by systematically addressing digital supply chain processes (Guo & Mantravadi, 2024). This structure increases the standardization and transparency of processes, making decision-making processes more efficient (Kamble et al., 2023). In addition, in digital supply chains, the SCOR Model increases the accuracy of demand forecasts based on data analysis, improves supplier and customer relationships, and encourages continuous improvement by monitoring the overall performance of processes (Eyieyien et al., 2024).

Strategic decisions form the backbone of processes in digital supply chains due to their increasingly complex operations and vulnerable structure (Mishra et al., 2024). Making strategic decisions in digital supply chains is critical for businesses to maintain competitive advantage and succeed in the market. These decisions affect every stage of the supply chain, and strategic decisions are used to increase the accuracy of demand forecasts, optimize supplier selection, and streamline production processes (Gardas & Narwane, 2024). In addition, making strategic decisions most appropriately in digital supply chains ensures risk management in digital supply chain processes. It creates a protection shield against the vulnerable structure of the digital supply chain in order to predict risks with strategic decisions and develop proactive strategies against these risks (Rashid et al., 2024). For example, potential disruptions in the supply chain or supplier problems can be detected early through data analysis and monitoring systems. This increases the ability of businesses to respond to crises and minimizes potential losses. Strategic decisions play a critical role in determining the most appropriate strategies to minimize the effects of such risks (Guillot et al., 2024).

Moreover, the SCOR Model is essential in steering strategic choices in supply chain management by aligning operations with corporate objectives, digital transformation, and sustainability targets. The model assists organizations in defining, measuring, and enhancing their supply chain processes through data-driven insights and digital tools, guaranteeing that precise performance metrics supports strategic decisions.

From a strategic standpoint, the SCOR Model improves decision-making in crucial areas like technology investment, risk management, sustainability, and workforce development. For instance, companies can utilize the SCOR Model to identify the most effective digital technologies (AI, IoT, blockchain) to enhance operational efficiency, evaluate the balance between cost and sustainability, and cultivate a competent workforce for Industry 4.0. (Chanpuypetch et al., 2025). By incorporating digital tools and intelligent analytics, the SCOR Model aids leaders in making well-informed, proactive choices that enhance resilience, agility, and sustained success in the supply chain. Strategic decisions in digital supply chains are classified according to the SCOR model processes, and they are presented in Fig. 3.1.

Planning is crucial for strategic decisions in digital supply chains as it guarantees the alignment and synchronization of all supply chain phases. Utilizing digital tools for planning leads to quicker and more precise decision-making, enhancing operational effectiveness and cutting down on expenses (Cannas et al., 2024). The planning process in digital supply chains includes essential components such as demand forecasting and planning, resource management, capacity planning, and strategic network design. This process is supported by digital tools such as big data analytics and artificial intelligence, ensuring that all supply chain stages are managed harmoniously and effectively (Cannas et al., 2024). **Demand forecasting and planning** use digital analytics and modeling techniques to accurately predict customer demands and optimize stock levels. This results in rapid response to demand changes and improved inventory management (Dixit et al., 2024). **Resource planning** tools support resource planning to ensure the most efficient use of available resources (Brandenburg, 2024). This stage reduces costs and enables production and distribution processes to be managed more effectively. In addition, **strategic network design** involves optimizing the network structure of the supply chain through digital simulations and analysis (Hasan et al., 2024). Identifying warehouse and distribution center locations enhances the supply chain's overall performance and cost-effectiveness (Chien et al., 2024). Examining **capacity planning**, production, and logistics capacities with digital solutions ultimately guarantees that the supply chain is aligned with demand, boosting flexibility and efficiency (Naghshineh, 2024).

In the SCOR Model, the **Order process** is essential for strategic decision-making in digital supply chains, as it ensures the seamless processing, fulfillment, and prioritization of customer orders. Digital solutions significantly enhance these processes, improving efficiency, reducing errors, and optimizing overall supply chain performance. The **Order Management and Processing** process involves automating order handling, verifying transactions, and ensuring smooth execution through digital tools (Singu et al., 2024). AI-powered systems, and cloud-based platforms streamline **order entry, tracking, and invoicing**, minimizing manual intervention and increasing accuracy (Kagalwala et al., 2025). The integration of these technologies enhances real-time order visibility and enables businesses to respond rapidly to changes in demand. Another strategic decision is the **Omnichannel Order Fulfillment Strategy**, which ensures a cohesive and synchronized fulfillment process across multiple sales channels (Malik et al., 2024; Radanliev et al., 2024). Digital platforms enable real-time inventory synchronization, ensuring that orders are efficiently fulfilled from the nearest or most cost-effective location. IoT, AI, and blockchain facilitate accurate order routing, tracking, and returns processing, enhancing customer satisfaction and operational flexibility (Chien et al., 2024).

In the SCOR Model, the sourcing process is critical for strategic decisions in digital supply chains because this process optimizes all operations from supplier selection to risk management. Digital solutions enable more effective and efficient management of these processes, ultimately lowering costs and enhancing the overall performance of the supply chain. The **supplier selection and management process i**nvolve assessing, choosing, and overseeing supplier performance through digital tools (Sahoo et al., 2024). This process evaluates suppliers' compliance with the criteria and ensures the management of long-term relationships (Alora and Gupta, 2024). Another strategic decision is **supplier collaboration.** Digital systems enable more effective collaboration with suppliers, and these systems strengthen relationships with suppliers and harmonize business processes by easily sharing information, such as real-time data sharing, order updates, and performance reports through common platforms (Pettau & Montag, 2024).

Additionally, strategic decisions about **automated purchasing and order management** include digital automation of purchasing processes. These processes enable automatic processing, approval, and tracking of orders (Nasereddin, 2024). Finally, **risk management** is an essential issue as a strategic decision, and digital solutions are used to identify and manage risks in the supply chain. This process uses data analytics and monitoring systems to analyze potential supplier risks, predict disruptions in the supply chain, and minimize risks (Rashid et al., 2024). Risk management enables the detection of possible problems in advance and the creation of strategic intervention plans.

According to the SCOR Model, the "transform" process increases the efficiency of production processes by forming the basis of strategic decisions in digital supply chains. This process ensures that production decisions are taken in the light of digital data and analysis, ensuring the processes are more predictable and harmonious. In the strategic decision of **intelligent manufacturing and process improvement**, while brilliant production refers to the optimization of production processes with automation and digital technologies, process improvement refers to continuous analysis and improvement activities to identify and correct inefficiencies and errors in production processes (Wulandari et al., 2024). **Real-time production monitoring**, one of the most critical stages in the making process, enables instant equipment monitoring (Guo and Mantravadi, 2024). Through sensors and data collection systems, problems in production lines are detected immediately, and intervention is ensured, ensuring that processes continue without interruption (Ganesh et al., 2024). This method supports continuous monitoring and improvement of the performance in the production process.

Moreover, **workforce management and planning** refer to the effective management and planning of human resources required in production processes (Akbari et al., 2024). Digital tools enable the workforce to be optimized according to workload, skills, and production requirements, including planning to use the workforce efficiently and achieve production targets.

Within the SCOR Model framework, the "fulfil" process is critical in making strategic decisions in digital supply chains. This process supports the decisions necessary to effectively guide logistics and delivery management and optimize the performance of the supply chain (Abusalma et al., 2024). During deployment, strategic decisions include solutions integrated with digital technologies to increase customer satisfaction and operational efficiency (Gardas and Narwane, 2024). Therefore, it is essential to correctly determine and implement strategic decisions to ensure effective management and coordination at the final stage of the supply chain. One of the critical strategic decisions in the delivery process is **order processing and delivery performance.** Order processing and delivery performance define how to optimize customer order processing and delivery performance (Seiler, 2024). Using digital tools and techniques in these strategic decision processes in digital supply chains improves order processing and delivery performance, providing businesses with more flexibility and adaptability, thus increasing customer satisfaction and operational efficiency.

Moreover, logistics optimization in digital supply chain delivery processes can shorten delivery times, reduce costs, and increase customer satisfaction (Malatji, 2024). Furthermore, **emergency planning in the delivery process** aims to make strategic decisions in delivery processes, analyze risks in advance with digital tools and techniques, and effectively manage response plans (Dash & Dixit, 2024). Risk predictions can be made using digital techniques and tools, and rapid responses to emergency scenarios can be provided (Rashid et al., 2024). Lastly, for the delivery process, **dynamic route planning** includes strategic decisions in delivery processes, route optimization, and real-time adaptation strategies (Mubarik & Khan, 2024).

In digital supply chains, the returns process directly impacts customer satisfaction and operational efficiency in terms of strategic decision-making (Hasan et al., 2024). Effectively managing returns using digital tools increases overall supply chain performance by providing back logistics, product quality improvement, and cost reduction (Çıkmak et al., 2024). Therefore, the strategic decisions that must be taken to ensure a better returns process in digital supply chains are essential to increase efficiency and ensure customer satisfaction (Patil et al., 2024). Strategic decisions for **automating return processes** cover increasing efficiency and reducing workload by automating return and return processes with digital systems (Cannas et al., 2024). Moreover, **feedback and quality improvement** consist of increasing product quality and customer satisfaction by analyzing feedback from returned products with digital tools (Pettau & Montag, 2024). In addition, **return management and transparency** ensure transparency and make processes more manageable by monitoring return processes with digital systems (Garcia-Torres et al., 2024). Lastly, strategic decisions about **return and product reprocessing** mean using digital solutions to reprocess returned products and create value.

Along with the changing digital environment, strategic decisions are also changing to ensure the durability of digital supply chains' complex and critical structure. In this context, strategic decisions focused on the SCOR Model in digital supply chains must be carried out with digital tools and techniques for companies to maintain their competitiveness and sustainability. The following section explains the proposed framework for integrating strategic decisions based on the SCOR model in digital supply chains and digital tools and techniques.

3.3 Proposed Framework

Recently, digital supply chains have been of great precision and critical importance. Therefore, using digital tools and techniques is crucial in strategic decision-making. Digital solutions establish the necessary infrastructure to manage the complexity of the supply chain, optimize data flow, and execute processes faster and more accurately (Dixit et al., 2024). Thanks to these tools, the data provided in various areas, from demand forecasts to supplier performance, allows strategic decisions to be made more consciously and effectively.

Basing the decisions to be made in digital supply chains on the structured framework of the SCOR model provides the opportunity to analyze each supply chain process comprehensively and effectively apply digital tools and techniques at every stage (Asrol et al., 2024). Since the model describes each of the supply chain processes in detail, it makes the integration of digital technologies into these processes more accessible and practical. This integration enables decision-makers to make more informed and predictable strategic choices, thus increasing operational efficiency, reducing costs, and increasing customer satisfaction (Mubarik & Khan, 2024). In summary, the SCOR model is an essential guide in managing digital supply chains, enabling strategic decisions based on a solid foundation.

In digital supply chains, integrating SCOR model-based strategic decision-making processes with digital-based tools and techniques significantly improves the supply chain performance of organizations (Pettau & Montag, 2024). Digital technologies enable each step of the five primary processes of the SCOR model to be managed more quickly and precisely. This integration makes decision-making processes more predictable and effective using various digital tools, from data analytics to automation systems. Therefore, understanding which digital-based tools and techniques can be used to meet each strategic decision based on the dimensions of the SCOR model not only gives companies a competitive advantage and makes it easier to follow process changes in the changing technological environment.

Table 3.1 shows the strategic decisions that emerge from the changing decision-making processes in digital supply chains with digital transformation and the digital tools and techniques that can be used for each strategic decision.

Table 3.1 Integration between strategic decisions and digital-based tools and techniques in decision-making in digital supply chains based on the SCOR model

SCOR dimensions	Strategic decisions	AI	IoT	Business intelligence (BI)	Blockchain	Cloud computing	Related literature
Plan	Demand and Forecasting Planning	AI-driven Forecasting	IoT-enabled Demand Sensing	BI for Predictive Demand Analytics	Blockchain for Secure Demand Data	Cloud-based Demand Forecasting	Grover et al. (2024); Nweje and Taiwo (2025); Veena et al. (2024)
	Resource Planning	AI-based Resource Optimization	IoT-based Inventory Tracking	BI-driven Resource Performance Reports	Blockchain for Transparent Resource Allocation	Cloud-based Resource Planning	Kamali and Marzani (2025); Zrelli and Rejeb (2024)
	Strategic Network Design	AI-based Scenario Analysis	IoT for Real-time Network Monitoring	BI for Supply Chain Network Optimization	Blockchain for Secure Network Transactions	Cloud-enabled Network Optimization	Kamali and Marzani (2025); Veena et al. (2024)
	Capacity Planning	AI for Demand-Capacity Balancing	IoT-based Equipment Utilization	BI for Capacity Utilization Trends	Blockchain for Capacity Data Integrity	Cloud for Capacity Scalability	Choi and Powers (2025); Kamali and Marzani (2025)
Order	Order Management and Processing	AI-driven order processing automation	Real-time tracking of order status	Data analytics for order trends	Secure order validation	Cloud-based order management system	Kagalwala et al. (2025); Malik et al. (2024); Radanliev et al. (2024)
	Omnichannel Order Fulfillment Strategy	AI-driven dynamic inventory allocation	IoT-enabled real-time inventory visibility across multiple fulfillment centers	BI-powered customer demand analysis	Secure and transparent order tracking across multiple sales channels	Centralized inventory synchronization	Kamali and Marzani (2025); Veena et al. (2024)
	Order Prioritization and Optimization	AI-driven order prioritization algorithms based on delivery urgency	IoT-enabled real-time tracking of order	Order performance dashboards	Smart contracts for automated order approvals	Cloud-based order orchestration, real-time integration between processes	Grover et al. (2024); Nweje and Taiwo (2025); Veena et al. (2024)

Source	Supplier Selection and Management	AI for Supplier Scoring	IoT Sensors for Supplier Monitoring	BI for Supplier Market Analysis	Blockchain for Supplier Data Security	Cloud-based Supplier Portals	Grover et al. (2024); Nweje and Taiwo (2025); Veena et al. (2024)
	Supplier Collaboration	AI-based Supplier Relationship Management	IoT-based Collaborative Platforms	BI-driven Supplier Scorecards	Blockchain-enabled Smart Contracts	Cloud-based Collaboration Platforms	Allal-Chérif et al. (2021); Reed et al. (2025)
	Automated Purchasing and Order Management	AI-driven Procurement Automation	IoT-integrated Procurement Tracking	BI for Procurement Performance Analysis	Blockchain for Secure Purchase Orders	Cloud-based Procurement Solutions	Althabatah et al. (2023); Grover et al. (2024); Nweje and Taiwo (2025); Veena et al. (2024)
	Risk Management	AI for Risk Prediction	IoT-based Risk Sensors	BI for Supplier Risk Profiling	Blockchain for Risk Data Security	Cloud Risk Management Systems	Kagalwala et al. (2025); Malik et al. (2024); Radanliev et al. (2024)
Transform	Smart Manufacturing and Process Improvement	AI-powered Predictive Maintenance	IoT for Machine Health Monitoring	BI for Production Efficiency Analysis	Blockchain for Supply Chain Traceability	Cloud for Smart Manufacturing Data	Kagalwala et al. (2025); Kamali and Marzani (2025)
	Real-Time Production Monitoring	AI-based Production Analytics	IoT-based Real-time Production Insights	BI for Factory Performance Dashboards	Blockchain for Secure Production Records	Cloud-based Production Monitoring	Archana (2025); Soldatos et al. (2021)
	Workforce Management and Planning	AI-driven Workforce Scheduling	IoT for Employee Safety Monitoring	BI for Workforce Productivity Reports	Blockchain for Workforce Credential Verification	Cloud-enabled Workforce Management	Nayak et al. (2023); Thangaraja et al. (2024)

(continued)

Table 3.1 (continued)

SCOR dimensions	Strategic decisions	AI	IoT	Business intelligence (BI)	Blockchain	Cloud computing	Related literature
Fulfill	Order Processing and Delivery Performance	AI-based Order Prioritization	IoT-enabled Real-time Order Tracking	BI for Order Processing Trends	Blockchain for Order Verification	Cloud-based Order Processing	Grover et al. (2024); Nweje and Taiwo (2025); Veena et al. (2024)
	Logistics Optimization	AI-powered Demand Prediction	IoT for Fleet and Inventory Monitoring	BI for Logistics Cost Analysis	Blockchain for Transparent Logistics Data	Cloud-based Logistics Optimization	Owusu-Berko (2025); Singh (2025)
	Emergency Planning in the Delivery Process	AI for Emergency Risk Assessment	IoT for Disaster Response	BI for Emergency Supply Chain Planning	Blockchain for Emergency Response Traceability	Cloud-based Emergency Planning	Owusu-Berko (2025); Veena et al. (2024)
	Dynamic Route Planning	AI for Route Optimization	IoT for Live Route Tracking	BI for Route Efficiency Dashboards	Blockchain for Secure Routing Transactions	Cloud-based Dynamic Routing	Paramesha et al. (2024); Tong et al. (2025)
Return	Return Management and Transparency	AI-based Returns Analysis	IoT-based Return Condition Analysis	BI for Reverse Logistics Performance	Blockchain for Return Verification	Cloud-based Returns Management	Van Hoang (2024); Veena et al. (2024)
	Feedback and Quality Improvement	AI-powered Customer Sentiment Analysis	IoT-driven Quality Monitoring	BI-driven Customer Feedback Insights	Blockchain for Tamper-Proof Feedback	Cloud-based Feedback Analysis	Demszky et al. (2025); Nazli et al. (2025)
	Return and Product Reprocessing	AI-driven Remanufacturing Optimization	IoT for Smart Remanufacturing	BI for Return and Remanufacturing Trends	Blockchain for Product Authentication in Returns	Cloud for Remanufacturing Process Control	Owusu-Berko (2025);Van Hoang (2024); Veena et al. (2024)

As given in Table 3.1, the proposed framework introduces a novel approach to integrating strategic decision-making with digital-based tools and techniques within the SCOR model, addressing the growing need for data-driven, agile, and resilient supply chains. Unlike traditional SCOR implementations, which primarily focus on standardized process optimization, this framework incorporates digital transformation enablers such as AI, IoT, blockchain, predictive analytics, and automation to enhance decision-making across all SCOR dimensions. By mapping specific digital tools to each phase of the SCOR model (Plan, Order, Source, Transform, Fulfill, Return), the framework enables real-time visibility, predictive capabilities, and operational efficiency, ensuring supply chains are more adaptive to disruptions and dynamic market conditions.

The key contribution of this framework lies in its ability to bridge the gap between strategic decision-making and digital transformation by providing a structured, technology-driven approach to modern supply chain management. It enhances efficiency, risk mitigation, and sustainability by integrating real-time data analytics, automation, and AI-powered decision support systems into traditional SCOR processes. The validity and comprehensiveness of the framework are reinforced through its alignment with SCOR Version 14 (SCOR - SDS), industry best practices, and emerging digital trends. By offering a scalable and adaptable solution, this framework provides organizations with the tools to optimize supply chain performance, improve responsiveness, and achieve a competitive advantage in the digital era.

3.4 Discussions and Conclusions

In recent years, businesses have experienced significant strategic shifts due to digital transformation and supply chains. The rapid advancement of digital technologies, such as AI, IoT, Blockchain, BI, Cloud Computing etc. has increased the complexity of supply chain operations. However, these technological advancements have also created greater efficiency, agility, and competitiveness opportunities. In this dynamic landscape, strategic decision-making processes are pivotal in ensuring supply chains remain resilient, responsive, and data-driven.

This chapter explores how strategic decision-making processes have evolved in digital supply chains and the incorporation of digital tools and techniques using the SCOR Model. The SCOR provides a comprehensive framework for aligning Plan, Source, Make, Deliver, Return, and Enable processes with digital technologies, allowing businesses to make more data-driven and automated decisions. Harmoniously combining strategic decision-making processes with digital tools during the digital transformation process enhances supply chain efficiency and gives a competitive edge. The proposed framework in this chapter aims to provide an in-depth understanding of the evolution of digital supply chains in response to Industry 4.0, the transformation of decision-making processes, and their alignment with advanced digital tools and analytics-driven approaches.

Incorporating AI, IoT, Blockchain, Cloud Computing, and various digital advancements into the SCOR Model has resulted in considerable enhancements in supply chain management. Digital technologies and methods allow for real-time monitoring, predictive analysis, smart automation, and improved collaboration within supply chain networks. This model offers essential solutions for tackling significant issues like supply chain transparency, demand prediction, inventory management, logistics effectiveness, and adherence to sustainability requirements. Moreover, through digital transformation, companies can create improved pricing strategies, boost customer satisfaction, reduce supply chain issues, and secure a lasting competitive edge. Due to the highly dynamic and continuously changing characteristics of global supply chains, the suggested framework rooted in the SCOR Model acts as a strategic guide for companies to manage digital transformation, enhance efficiency, and promote ongoing improvement in contemporary supply chain management.

References

Abusalma, A., Oraini, B., Al-Daoud, K., & Alshurideh, M. (2024). The impact of supply chain performance on financial performance: Dimensions of the SCOR model. *Uncertain Supply Chain Management, 12*(3), 1409–1416.

Akbari, M., Kok, S. K., Hopkins, J., Frederico, G. F., Nguyen, H., & Alonso, A. D. (2024). The changing landscape of digital transformation in supply chains: Impacts of industry 4.0 in Vietnam. *The International Journal of Logistics Management, 35*(4), 1040–1072.

Allal-Chérif, O., Simón-Moya, V., & Ballester, A. C. C. (2021). Intelligent purchasing: How artificial intelligence can redefine the purchasing function. *Journal of Business Research, 124*, 69–76.

AlMulhim, A. F. (2021). Smart supply chain and firm performance: the role of digital technologies. *Business Process Management Journal, 27*(5), 1353–1372.

Alora, A., & Gupta, H. (2024). Supplier selection among manufacturing companies based on their MSME partners' supply chain finance adoption capability using BWM and VIKOR. *Journal of Business & Industrial Marketing, 39*(6), 1337–1351.

Althabatah, A., Yaqot, M., Menezes, B., & Kerbache, L. (2023). Transformative procurement trends: Integrating industry 4.0 technologies for enhanced procurement processes. *Logistics, 7*(3), 63.

APICS. (2022). SCOR Digital Standard. Retrieved from: https://www.ascm.org/globalassets/ascm_website_assets/docs/intro-and-front-matter-scor-digital-standard2.pdf

Archana, T. (2025). IoT and AI integration for real-time supply chain monitoring. In *Ecological and human dimensions of AI-based supply chain* (pp. 339–372). IGI Global Scientific.

Asrol, M., Marimin, M., Machfud, M., & Yani, M. (2024). An intelligent decision support system is modeling for improving agroindustry's supply chain performance: a case study. *International Journal of Information and Decision Sciences, 16*(2), 134–168.

Ayyildiz, E., & Taskin Gumus, A. (2021). Interval-valued Pythagorean fuzzy AHP method-based supply chain performance evaluation by a new extension of SCOR model: SCOR 4.0. *Complex & Intelligent Systems, 7*(1), 559–576.

Brandenburg, M. (2024). Operations and supply chain planning. In J. Sarkis (Ed.), *The Palgrave handbook of supply chain management* (pp. 293–312). Springer.

Brau, R. I., Sanders, N. R., Aloysius, J., & Williams, D. (2024). Utilizing people, analytics, and AI for decision making in the digitalized retail supply chain. *Journal of Business Logistics, 45*(1), e12355.

Brintrup, A., Kosasih, E., Schaffer, P., Zheng, G., Demirel, G., & MacCarthy, B. L. (2024). Digital supply chain surveillance using artificial intelligence: definitions, opportunities and risks. *International Journal of Production Research, 62*(13), 4674–4695.

Cannas, V. G., Ciano, M. P., Saltalamacchia, M., & Secchi, R. (2024). Artificial intelligence in supply chain and operations management: a multiple case study research. *International Journal of Production Research, 62*(9), 3333–3360.

Chanpuypetch, W., Kritchanchai, D., Niemsakul, S., & Niemsakul, J. (2025). Distribution logistics network optimization with a multi-distribution center considering carbon emission: A case study in the petrochemical industry in Thailand. *Procedia Computer Science, 253*, 237–246.

Chien, C. F., Kuo, H. A., & Lin, Y. S. (2024). Smart semiconductor manufacturing for pricing, demand planning, capacity portfolio and cost for sustainable supply chain management. *International Journal of Logistics Research and Applications, 27*(1), 193–216.

Choi, S., & Powers, T. (2025). Optimizing health system volume flexibility: key determinants and strategies during the COVID-19 crisis. *Journal of Health Organization and Management*. https://doi.org/10.1108/JHOM-08-2023-0249

Çıkmak, S., Kantoğlu, B., & Kırbaç, G. (2024). Evaluation of the effects of blockchain technology characteristics on SCOR model supply chain performance measurement attributes using an integrated fuzzy MCDM methodology. *International Journal of Logistics Research and Applications, 27*(6), 1015–1045.

Dash, B. P., & Dixit, V. (2024). Disaster supply chain with information and digital technology integrated in its institutional framework. *International Journal of Production Research, 62*(8), 3003–3022.

Demszky, D., Liu, J., Hill, H. C., Sanghi, S., & Chung, A. (2025). Automated feedback improves teachers' questioning quality in brick-and-mortar classrooms: Opportunities for further enhancement. *Computers & Education, 227*, 105183.

Dixit, V. K., Malviya, R. K., Kumar, V., & Shankar, R. (2024). An analysis of the strategies for overcoming digital supply chain implementation barriers. *Decision Analytics Journal, 10*, 100389.

Eyieyien, O. G., Adebayo, V. I., Ikevuje, A. H., & Anaba, D. C. (2024). Conceptual foundations of Tech-Driven logistics and supply chain management for economic competitiveness in the United Kingdom. *International Journal of Management & Entrepreneurship Research, 6*(7), 2292–2313.

Ganesh, M., Deo, S., & Devalkar, S. K. (2024). Leveraging digital technology to improve monitoring and planning in public sector supply chains: Evidence from India's food security program. *Production and Operations Management, 33*(4), 861–879.

Garcia-Torres, S., Rey-Garcia, M., & Sáenz, J. (2024). Enhancing sustainable supply chains through traceability, transparency and stakeholder collaboration: A quantitative analysis. *Business Strategy and the Environment, 33*(7), 7607–7629.

Gardas, R., & Narwane, S. (2024). An analysis of critical factors for adopting machine learning in manufacturing supply chains. *Decision Analytics Journal, 10*, 100377.

Grover, V., Balusamy, B. B., Milanova, M. G., & Felix, A. Y. (2024). *Blockchain, IoT, and AI technologies for supply chain management: Apply emerging technologies to address and improve supply chain management*. Springer Nature.

Guillot, R., Dubey, R., & Kumari, S. (2024). B2B supply chain risk measurement systems: a SCOR perspective. *Journal of Business & Industrial Marketing, 39*(3), 553–567.

Guo, D., & Mantravadi, S. (2024). The role of digital twins in lean supply chain management: review and research directions. *International Journal of Production Research, 63*(5), 1851–1872.

Hasan, R., Kamal, M. M., Daowd, A., Eldabi, T., Koliousis, I., & Papadopoulos, T. (2024). Critical analysis of the impact of big data analytics on supply chain operations. *Production Planning & Control, 35*(1), 46–70.

Kagalwala, H., Radhakrishnan, G. V., Mohammed, I. A., Kothinti, R. R., & Kulkarni, N. (2025). Predictive analytics in supply chain management: The role of AI and machine learning in demand forecasting. *Advances in Consumer Research, 2*, 142–149.

Kamali, S., & Marzani, A. (2025). A demand-capacity approach to define failure thresholds in anomaly detection monitoring systems. *Journal of Dynamic Disasters, 1*(1), 100004.

Kamble, S. S., Mor, R. S., & Belhadi, A. (2023). Big data analytics for supply chain transformation: A systematic literature review using SCOR framework. In S. S. Kamble, R. S. Mor, & A. Belhadi (Eds.), *Digital transformation and industry 4.0 for sustainable supply chain performance* (pp. 1–50). Springer.

Li, Q., Zhang, H., Liu, K., Zhang, Z. J., & Jasimuddin, S. M. (2024). Linkage between digital supply chain, supply chain innovation and supply chain dynamic capabilities: An empirical study. *The International Journal of Logistics Management, 35*(4), 1200–1223.

Malatji, M. (2024). Accelerating the African continental free trade area through optimization of digital supply chains. *Engineering Reports, 6*(2), e12711.

Malik, J. A., Hussain, A., Shah, H., Saleem, M., Alsanoosy, T., & Chaudhary, U. M. D. (2024). Optimizing agricultural risk management with hybrid block-chain and fog computing architectures for secure and efficient data handling. In M. G. Sumithra, M. Sathyamoorthy, M. Manikandan, R. K. Dhanaraj, & M. Ouaissa (Eds.), *Computational intelligence in internet of agricultural things* (pp. 309–337). Springer Nature.

Mishra, N. K., Pande Sharma, P., & Chaudhary, S. K. (2024). Redefining agile supply chain practices in the disruptive era: a case study identifying vital dimensions and factors. *Journal of Global Operations and Strategic Sourcing, 18*(1), 64–90.

Mubarik, M. S., & Khan, S. A. (2024). Digital supply chain and industry 4.0 technologies. In M. S. Mubarik & S. A. Khan (Eds.), *The theory, methods and application of managing digital supply chains* (pp. 121–132). Emerald.

Naghshineh, B. (2024). Additive manufacturing technology adoption for supply chain agility: a systematic search and review. *International Journal of Production Research*, 1–33.

Nasereddin, A. (2024). A comprehensive survey of contemporary supply chain management practices in charting the digital age revolution. *Uncertain Supply Chain Management, 12*(2), 1331–1352.

Nayak, A., Satpathy, I., Khang, A., & Patnaik, B. C. M. (2023). Artificial intelligence-driven workforce planning for the gig economy era: Setting a new direction. In A. Khang, B. Jadhav, V. A. Hajimahmud, & I. Satpathy (Eds.), *Synergy of AI and fintech in the digital gig economy* (pp. 81–95). CRC Press.

Nazli, K., Jumani, S., & Masum, R. (2025). The effectiveness of AI-powered corrective feedback on students' writing performance. *Pakistan Social Sciences Review, 9*(1), 391–403.

Nweje, U., & Taiwo, M. (2025). Leveraging Artificial Intelligence for predictive supply chain management, focus on how AI-driven tools revolutionize demand forecasting and inventory optimization. *International Journal of Science and Research Archive, 14*(1), 230–250.

Owusu-Berko, L. (2025). Advanced supply chain analytics: Leveraging digital twins, IoT and blockchain for resilient, data-driven business operations. *World Journal of Advanced Research and Reviews, 25*(2), 1777–1799.

Ozbiltekin-Pala, M., Koçak, A., & Kazancoglu, Y. (2023). A proposed circular-SCOR model for supply chain performance measurement in manufacturing industry during COVID-19. *International Journal of Quality & Reliability Management, 40*(5), 1203–1232.

Pal, K. (2024). Data-driven decision making for achieving sustainability in apparel business value chain. In *Data-driven decision making for long-term business success* (pp. 395–414). IGI Global.

Paramesha, M., Rane, N. L., & Rane, J. (2024). Big data analytics, artificial intelligence, machine learning, internet of things, and blockchain for enhanced business intelligence. *Partners Universal Multidisciplinary Research Journal, 1*(2), 110–133.

Patil, K., Garg, V., Gabaldon, J., Patil, H., Niranjan, S., & Hawkins, T. (2024). Firm performance in digitally integrated supply chains: a combined perspective of transaction cost economics and relational exchange theory. *Journal of Enterprise Information Management, 37*(2), 381–413.

Pettau, T., & Montag, L. (2024). The role of digital platforms as circularity broker: an updated SCOR perspective on circular supply chain performance. In S. Baumann (Ed.), *Handbook on digital platforms and business ecosystems in manufacturing* (pp. 257–277). Edward Elgar.

Radanliev, P., De Roure, D., Maple, C., Nurse, J. R., Nicolescu, R., & Ani, U. (2024). AI security and cyber risk in IoT systems. *Frontiers in Big Data, 7*, 1402745.

Raja, V., Muralidhar, D., Mythrayan, B., Prathiksha, K. G., Venkateshwaran, S. P., & Sivakumar, V. (2024). Implementation of digital twin in supply chain and logistics. In *Recent advances in material, manufacturing, and machine learning* (pp. 340–345). CRC Press.

Ramesh, K., Renjith, P. N., Balasubramani, S., Bennet, M. A., Saritha, S., Pandey, D., & Kanike, U. K. (2024). Leveraging cutting-edge technologies and innovative strategies to optimize the IoT and AI integration in supply chain management. In D. Pandey, B. K. Pandey, & U. K. Kanike (Eds.), *Utilization of AI technology in supply chain management* (pp. 159–173). IGI Global.

Rashid, A., Rasheed, R., Ngah, A. H., Pradeepa Jayaratne, M. D. R., Rahi, S., & Tunio, M. N. (2024). Role of information processing and digital supply chain in supply chain resilience through supply chain risk management. *Journal of Global Operations and Strategic Sourcing, 17*(2), 429–447.

Reed, C., Wynn, M., & Bown, R. (2025). Artificial intelligence in digital marketing: Towards an analytical framework for revealing and mitigating bias. *Big Data and Cognitive Computing, 9*(2), 40.

Ricárdez-Estrada, J., Garay-Rondero, C. L., Romero, D., Wuest, T., & Pinto, R. (2024). Identifying digital supply chain capabilities. *Procedia Computer Science, 232*, 1182–1191.

Sadeghi, K., Ojha, D., Kaur, P., Mahto, R. V., & Dhir, A. (2024). Explainable artificial intelligence and agile decision-making in supply chain cyber resilience. *Decision Support Systems, 180*, 114194.

Sahoo, S. K., Goswami, S. S., & Halder, R. (2024). Supplier selection in the age of industry 4.0: a review on MCDM applications and trends. *Decision Making Advances, 2*(1), 32–47.

Seiler, A. (2024). *The impact of order processing automation on customer care and sales logistics* (Sage business cases originals). Sage.

Singh, B. (2025). Revolutionizing supply chains for optimized demand planning, inventory management, and logistics: An in-depth analysis of AI and ML solutions in the modern era. In E. Sabri (Ed.), *Supply chain transformation through generative AI and machine learning* (pp. 103–128). IGI Global Scientific.

Singu, S. K., Jibril, I., Meyer, F., Ngobi, M., & Kelly, A. (2024). *Data management and system optimization*. Cari Journals.

Soldatos, J., Despotopoulou, A., Kefalakis, N., & Ipektsidis, B. (2021). Blockchain based data provenance for trusted artificial intelligence. In J. Soldatos & D. Kyriazis (Eds.), *Trusted artificial intelligence in manufacturing: A review of the emerging wave of ethical and human centric AI technologies for smart production* (pp. 1–29). now.

Thangaraja, T., Maharudrappa, M., Bakkiyaraj, M., Johari, L., & Muthuvel, S. (2024). AI-powered HR technology implementation for business growth in industrial 5.0. In T. Kajla, P. Kansra, & N. Singh (Eds.), *Multidisciplinary applications of extended reality for human experience* (pp. 171–200). IGI Global.

Tong, X., Hamzei, M., & Jafari, N. (2025). Towards secure and efficient data aggregation in blockchain-driven IoT environments: A comprehensive and systematic study. *Transactions on Emerging Telecommunications Technologies, 36*(2), e70061.

Van Hoang, T. (2024). Impact of integrated artificial intelligence and internet of things technologies on smart city transformation. *Journal of Technical Education Science, 19*(Special Issue 01), 64–73.

Veena, G., Balamurugan, B. B., Mariofanna, M., & Yovan, F. A. (2024). Blockchain, IoT, and AI technologies for supply chain management: Apply emerging technologies to address and improve supply chain management.

Wulandari, R. A., Haristi, M., & Sormin, Y. S. A. (2024). Improving service supply chain performance with the Supply Chain Operations Reference (SCOR) approach; Case study: Indonesia government testing laboratory. *World Journal of Advanced Research and Reviews, 22*(3), 742–756.

Zhao, D., & Wang, X. (2024). Evolutionary game of digital decision-making in supply chains based on system dynamics. *RAIRO-Operations Research, 58*(1), 475–510.

Zrelli, I., & Rejeb, A. (2024). A bibliometric analysis of IoT applications in logistics and supply chain management. *Heliyon, 10*(16).

Chapter 4
Transportation 4.0: Leveraging Industry 4.0 Technologies to Enhance Economic, Environmental, and Social Performance

Markus Biehl

Abstract The transportation sector is undergoing a transformative phase driven by the integration of Industry 4.0 technologies, collectively termed as Transportation 4.0 (T4.0). This chapter explores the relevance and potential of these advanced technologies—such as the Internet of Things, cyber-physical systems, big data, machine learning and artificial intelligence—in enhancing the sector's economic, environmental, and social performance. The rapid evolution of global trade and shifts in consumer demand, accelerated by the COVID-19 pandemic, have underscored the necessity for efficient, reliable, and sustainable transportation solutions. This chapter provides an overview of T4.0 technologies, with extensions into Logistics 4.0, and examines how these technologies can address these challenges by improving supply chain visibility, fleet management, data processing and automation. Key applications as well as their impacts on triple bottom line performance are discussed. Despite the significant benefits, the deployment of some of the new technologies present challenges, such as data security, high implementation costs, and the need for skilled labor. The chapter concludes by identifying areas for future research and emphasizing the importance of addressing these challenges to fully realize the potential of T4.0.

Keywords Industry 4.0 · Logistics 4.0 · Transportation 4.0 · Internet of things (IoT) · Warehouse automation

---

M. Biehl (✉)
Schulich School of Business, York University, Toronto, ON, Canada
e-mail: mbiehl@schulich.yorku.ca

Y. Khojasteh, S. M. M. Jahan Abadi (eds.), *Emerging Technologies in Supply Chains*, International Series in Operations Research & Management Science 373, https://doi.org/10.1007/978-3-032-01218-0_4

4.1 The Changing Landscape of Transportation

Until the end of the eighteenth century, transportation relied on humans and animals as well as non-mechanized conveyances like carts and sail ships, limiting trade to local and regional levels. The industrial revolution introduced canals, railways, steam ships, and motorized coaches, significantly increasing transportation speed and volume. This also boosted manufacturing output and reduced costs, allowing for a wider market reach to take advantage of economies of scale. Further advancements included the unitization of loads with pallets in the 1930s and containers in the 1950s, which greatly increased throughput in the supply chain and, as a result, global trade demand. Recent decades have seen an explosion in global trade due to reduced regulations and trade barriers, and boosted by efficient, large-scale transportation (Rodrigue, 2024). The World Trade Organization (2020) notes that the value of exported merchandise rose from US $6000 billion to US $19,000 billion between 2000 and 2018, driving the need for increased transportation capacity.

Demand patterns have also changed. During the COVID-19 pandemic, many consumers shifted spending from experiences to products, thus further increasing volumes. At the same time, they now demand near-instantaneous delivery to their homes, rather than going to a physical store, making warehousing and first-mile/last-mile delivery important parts of the supply chain. This prompted shippers and transporters to move from a very efficient use of resources to express delivery to individuals via drop-shipping or milk runs (Wong et al., 2024).

Another factor is the increasing unpredictability of demand patterns. During COVID-19 factories and transportation capacities were temporarily shut down while companies experienced demand peaks. This combination resulted in severe delays and dramatic spikes in shipping costs (The Economist, 2023a), with manufacturers and shippers unsure how to react (Sundarakani et al., 2024). With Just-in-Time (JIT) as the dominant method of transportation, the supply chain was ill-equipped to deal with this scenario. While the transportation industry has largely recovered by now, it has not yet returned to a steady state. This is partly due to countries recalibrating where to source certain technologies to address production and transportation risks ("near-sourcing"), political risks ("friend-sourcing" and the re-imposition of tariffs) (The Economist, 2023b, 2023c, 2025), and partly because flows have not balanced out yet physically or due to changes in production/demand (The Economist, 2024a).

A final factor impacting transportation is the pressing need to address its environmental externalities. The transportation sector accounts for over 20% of global greenhouse gas emissions, with commercial transportation responsible for more than half of this total (Tiseo, 2023, 2024). Global warming affects shipping through infrastructure damage from rising temperatures, extreme weather events, disappearing Arctic permafrost (Bielecki et al., 2024; Crownhart, 2023), flooding of ports from rising sea levels, reduced aircraft performance from lower air density, and lower water levels in inland waterways (Englert et al., 2023). Transportation also has negative social impacts through pollution, noise, and the use of space and time

(Rodrigue, 2024). Consequently, significant improvements are necessary, and logistics customers are already starting to seek out logistics service providers that are capable of meeting these challenges (Qureshi et al., 2024).

The fluctuating volume, shifting demand patterns, increasing unpredictability and need to make transportation more sustainable prompt the question of how to improve transportation across multiple dimensions. These dimensions include typical performance indicators—speed (transit time), reliability, accessibility, capability, and security (Coyle et al., 2011)—summarized as economic performance, but also social and environmental performance.

Exciting new technologies are being developed to help the transportation sector address its challenges. These technologies, summarized as Industry 4.0, are part of the fourth industrial revolution (Jagtap et al., 2021). While the first two revolutions focused on mechanization and mass production, the third and fourth are rooted in information technology—automation and robotization, respectively (Rodrigue, 2024). Industry 4.0 integrates advanced digital technologies into fully automated manufacturing processes, creating smart factories. It is characterized by the use of Cyber-physical Systems (CPS), the Internet of Things (IoT), cloud computing, and Artificial Intelligence (AI), fundamentally changing how products are designed, manufactured, supplied, and serviced (Schwab, 1997). Üster (2024) identifies speed, the breadth and depth of its application, and its systematic impact on organizations, industries and society as defining elements of this collection of new technologies. Depending on the degree of its implementation, processes may become able to self-regulate, with a combination of central and local decision-making (Malagon-Suarez & Orjuela-Castro, 2023).

In this chapter, transportation refers to the movement of goods from one point to another. Figure 4.1 distinguishes between the information and physical activities

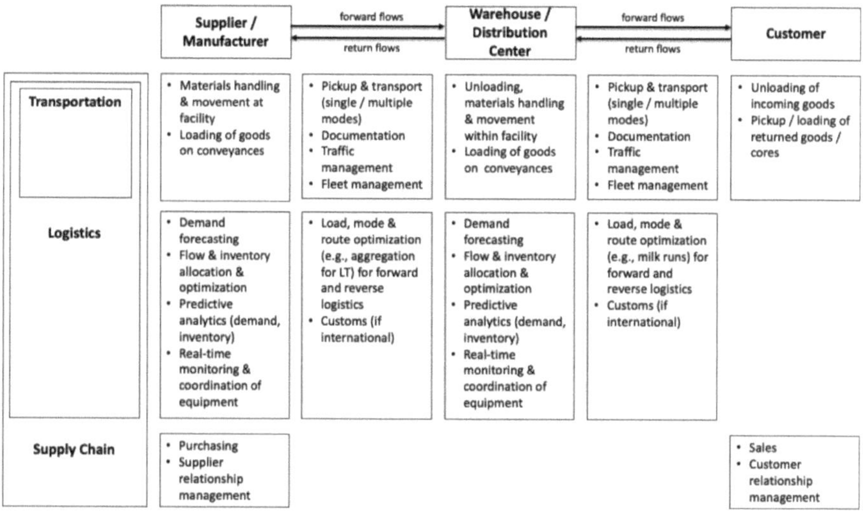

Fig. 4.1 Transportation, logistics and supply chain management

involved in transportation, logistics, and supply chain management. Transportation encompasses the handling and movement of goods within and between facilities, along with supporting activities that facilitate these operations. While transportation is a core component of logistics, logistics extends beyond transportation by coordinating the various elements of the logistics chain (Malagon-Suarez & Orjuela-Castro, 2023). Therefore, non-transportation activities within logistics are information-driven, focusing on optimizing inventory levels and the flow of goods through real-time monitoring, demand forecasting, and route optimization. Supply chain management, in turn, takes a broader perspective by integrating customer and supplier interactions to provide a comprehensive, end-to-end view of the supply chain.

While Fig. 4.1 depicts discrete activities at each stage of the supply chain, these stages are inherently interconnected. A single truck, for example, can manage both forward and reverse flows across multiple logistics stages while relying on the same maintenance program and traffic management software. Shipping optimization occurs at both the macro level—where modes and lanes are selected across the entire logistics chain—and the local level, where delivery routes are adjusted in real time based on traffic conditions. Although transportation and logistics functions overlap, logistics serves as the integrative framework, coordinating transportation activities such as warehouse storage allocation, retrieval and staging optimization, efficient loading sequences, and multimodal transport planning. At a broader level, supply chain management extends beyond logistics by driving systemic improvements, including product design for better transport efficiency and collaboration with suppliers and customers to optimize quality control and flow management. As a result, while this chapter focuses on transportation, it naturally intersects with logistics applications that directly shape transportation processes.

The term Transportation 4.0 (T4.0) derives from Industry 4.0. T4.0 can be thought of as a sub-system of Logistics 4.0, which itself is a subsystem of Supply Chain 4.0 (see Fig. 4.1). Industry 4.0, extending its technologies into logistics to apply them to areas such as manufacturing (factory logistics), warehousing and pickup and delivery (Wong et al., 2024).

This chapter attempts to address the following question:

Which Industry 4.0 technologies are relevant to transportation, how are they used, and are they capable of improving the sector's triple bottom line (economic, environmental and social) performance?

It does so by first providing an overview of the technologies—both hardware and software—used in T4.0. In the third section we investigate how these technologies are used for operations in transportation. The fourth section discusses their potential impact on triple bottom line performance, followed by a section that addresses issues and potential drawbacks. The final section offers conclusions and avenues for further research.

4.2 Transportation 4.0 Technologies

Goods are loaded, unloaded, put away, stored, retrieved, packaged and transported. Information about the goods (size, weight, value, origin, destination, routing, customs, and financial details) is transmitted with the goods (bill of lading) and separately (e.g., customs documents, invoices). The information that is available to plan, predict and control the movement of goods, including handling, transshipment, routing, and exception management, is crucial to logistics performance. IoT devices, CPS, big data (BD), machine learning (ML) and AI collect and analyze this information (Wong et al., 2024). Robots, automated guided vehicles (AGVs) and autonomous vehicles (AVs) also hold significant potential (Bielecki et al., 2024). Table 4.1 shows which activities associated with transportation are impacted by these technologies.

In this section these technologies are introduced—first the hardware, then the software that frequently uses inputs from the hardware and may even issue instructions back to it.

4.2.1 Hardware

At the core of the technological innovations driving Industry 4.0—and by extension, T4.0—are advancements in hardware. These include sensors, controllers, vehicles, and other conveyance systems, all of which are integrated with information systems that process their inputs to optimize operations. This subsection first examines connected sensors and controllers, key components of the IoT, followed by an exploration of transportation equipment that leverages these technologies. While the primary focus is on sensors, controllers function similarly by actuating equipment based on sensor inputs.

4.2.1.1 Sensors: The Internet of Things

IoT devices are fundamental to T4.0, serving as small physical units capable of collecting and transmitting data, often in real time (Ferraro et al., 2023). Examples include cameras or inductive loops that are embedded in roads and count vehicles or measure the speed and density of traffic. IoT devices also encompass transponders

Table 4.1 Industry 4.0 technologies used in Transportation 4.0

	IoT, CBS	Robots, AGVs, AVs	BD, AI and ML
Materials handling and inventory management	x	x	
Transport (long-haul)	x	x	x
Pickup and delivery (PUD)	x	x	
Transport, fleet and exception management	x		x

that prioritize public transportation, sensors used in manufacturing to measure variables such as position, temperature and color (Wikipedia, 2024). Specialized acoustic sensors can monitor rail tracks (Efthymiou & Ponis, 2021) and sonar systems assist maritime navigation (Facchini et al., 2019). Force and proximity sensors detect pressure and distance, which is important for robots that handle goods.

In transportation, RFID tags are a particularly important type of IoT as they play a crucial role by providing real-time data updates for items such as packaging, containers, forklifts, and vehicles. These tags function either continuously (active tags) or when scanned (passive tags). Compared to traditional 2D or 3D barcodes, RFID tags offer several advantages (Bogataj et al., 2017):

- Wireless readability without line-of-sight requirements.
- The ability to store additional information, such as timestamps and transport details.
- Integration with sensors to capture environmental data, including temperature, humidity, and acceleration.

Beyond RFID, other IoT-enabled devices include environmental sensors for monitoring warehouse conditions, fuel consumption, equipment vibrations and GPS tracking. Controllers may also provide data directly to warehouse management systems (Jagtap et al., 2021).

4.2.1.2 Automated and Autonomous Vehicles

Advancements in sensor technology have paved the way for widespread automation in transportation. Automated vehicles, including trucks, drones and locomotives, are equipped with LiDAR, radar, cameras, and various sensors that collect real-time data about the vehicle (including ships and aircraft) and its surroundings. These systems facilitate communication with infrastructure and other vehicles, enabling varying levels of automation and autonomy (Sadaf et al., 2023).

Robots ensure safe and efficient material handling, for example, in lieu of conveyors, or to take goods off conveyors. Collaborative robots, or "cobots," work alongside human operators to accomplish a variety of tasks. While not an automated device, automotive manufacturers such as ford use exoskeletons to help humans with heavy loads or repetitive tasks (Ford, 2018).

AGVs are widely used in logistics to transport parts, SKUs, and pallets. Companies like Amazon employ AGVs capable of lifting entire shelving units and transporting them to human workers for order fulfillment before returning them to storage (Knight, 2015). Similarly, conveyor systems, though not autonomous, utilize sensors to monitor item movement, weight, and flow (Ferraro et al., 2023), with advanced facilities like FedEx's Toronto airport hub handling up to 12,000 packages per hour across a 3.6 km network of conveyors (Mott Macdonald, n.d.).

Aerial transportation is also undergoing rapid innovation, with companies like FedEx and Amazon piloting drones for last-mile deliveries. The recent approval by the FAA for Amazon to operate delivery drones beyond the visual line of sight marks a significant milestone, allowing for expanded pilot programs (Shakir, 2024).

4.2.2 Software

Sensors and controllers alone cannot revolutionize transportation and manufacturing; software systems that process all this and other data are equally crucial. Software coordinates devices and supports broader planning and decision-making. This section provides an overview of these software systems and their role in enabling T4.0, starting with cyber-physical systems and then exploring AI and related technologies.

4.2.2.1 Cyber-Physical Systems (CPS)

A defining feature of CPS is the seamless integration of physical devices, ranging from micro-level components such as sensors and actuators to macro-level systems like robots, conveyor belts and AGVs. These devices rely on both embedded and remote computing. Embedded systems process sensor inputs and control motors and actuators, creating a continuous feedback loop where physical processes influence computational decisions and vice versa (Modica et al., 2021). Remote computing—whether cloud-based or on-premises—can coordinate multiple devices, such as picking robots, AGVs or conveyor systems.

While primarily applied in warehousing, CPS also plays a critical role in transportation systems (Lyu et al., 2023; Winkelhaus & Grosse, 2020). Here, CPS analyze real-time traffic information, leveraging infrastructure-embedded sensors (Efthymiou & Ponis, 2021). CPS also play a vital role in predictive maintenance by analyzing data from sensors that monitor fuel consumption, weight, position, and vibration (Lyu et al., 2023). Furthermore, CPS are essential for autonomous vehicles, such as self-driving cars and drones, which rely on numerous sensors that provide data for real-time decision-making (Markov & Vitliemov, 2020).

4.2.2.2 Big Data, Machine Learning and Artificial Intelligence

BD, ML, and AI are closely interconnected concepts that are transforming T4.0. AI encompasses advanced computational methods that simulate human cognitive functions by using algorithms to analyze large data sets, recognize patterns, and make data-driven decisions (Chen et al., 2021; Draksler et al., 2023). ML, a subset of AI, trains algorithms on historical data to identify patterns and make predictions, employing techniques such as neural networks to improve performance over time (Lyu et al., 2023). AI systems continuously learn and can be deployed across embedded systems, local networks, and large-scale organizational infrastructures (Boorová et al., 2024).

BD involves the collection, storage, and analysis of massive data volumes, which are often gathered automatically from diverse sources such as credit card transactions, internet searches, social media activity, loyalty programs, mapping applications, vehicle sensors, RFID tags, and traffic monitoring systems (Chen et al., 2021; Draksler et al., 2023; von Stietencron et al., 2022). This vast data pool is used to train and test ML algorithms, which, in turn, power AI systems to monitor and optimize transportation processes. The convergence of IoT, BD, CPS, ML, and AI forms a cohesive ecosystem that enables seamless operations across the supply chain that allows to improve predictions and make real-time adjustments over current systems.

The successful implementation of these technologies often requires cloud-based systems (CBS), which provide the scalability and processing power needed for BD applications and logistics optimizations at scale. CBS also facilitate seamless interconnectivity between different applications across stationary and mobile locations (Andres et al., 2024; Mihai, 2022). As such, CBS are a fundamental component of T4.0, supporting real-time decision-making and enhancing operational efficiency.

Understanding this integration allows us to explore the use of these technologies as part of T4.0.

4.3 Transportation 4.0 Applications

This section discusses the applications of Industry 4.0 technologies to T4.0. The section consists of two subsections: their application to transportation (movement) and warehousing (storage and movement). Throughout this section, the focus is on CPS and AI (including BD and ML) while understanding that physical processes need input—often in real-time—from technologies described in the IoT section.

4.3.1 Transportation

Understanding the scope of these technologies reveals their numerous applications in transportation. Key areas of implementation include the following (see Table 4.2):

- Route optimization & real-time traffic management
- Fleet management
- Supply chain visibility
- Control of autonomous vehicles
- Demand forecasting

Real-time traffic management rely on traffic sensors, including in-pavement and overhead inductive or visual sensors. Traffic management uses traffic density data to dynamically optimize flow by adjusting speeds, switching traffic lights, lane

Table 4.2 Applications of CPS, Big Data and AI in transportation

<table>
<tr><th>Article</th><th>Route optimization</th><th>Real-time traffic mgmt.</th><th>Fleet mgmt.</th><th>SC visibility</th><th>Autonomous vehicles</th><th>Demand forecasting</th></tr>
<tr><td>von Stietencron et al. (2022)</td><td>Based on traffic patterns and weather conditions (AI)</td><td></td><td rowspan="3">Predictive maintenance using data from IoT sensors (CPS, BD, AI)</td><td></td><td></td><td></td></tr>
<tr><td>Facchini et al. (2019)</td><td></td><td></td><td></td><td></td><td></td></tr>
<tr><td>Chen et al. (2021)</td><td></td><td></td><td></td><td></td><td></td></tr>
<tr><td>Khan et al. (2022)</td><td></td><td></td><td rowspan="2">Optimization (AI, ML, BD)</td><td></td><td></td><td></td></tr>
<tr><td>Lyu et al. (2023)</td><td></td><td rowspan="2">Manage real-time traffic data (AI)</td><td></td><td></td><td>(AI, BD)</td></tr>
<tr><td>Modica et al. (2021)</td><td></td><td></td><td></td><td></td><td></td></tr>
<tr><td>Markov and Vitliemov (2020)</td><td></td><td>Monitor traffic conditions (AI)</td><td>Improve fleet efficiency and manage vehicles (AI)</td><td></td><td>Navigation and decision-making for autonomous vehicles (AI)</td><td></td></tr>
<tr><td>Tang et al. (2024)</td><td></td><td></td><td>Improve vehicle performance (BD, AI)</td><td></td><td></td><td></td></tr>
<tr><td>Kucukaltan et al. (2022)</td><td></td><td></td><td></td><td rowspan="4">Through real-time tracking (CPS, BD)</td><td></td><td></td></tr>
<tr><td>Draksler et al. (2023)</td><td></td><td></td><td></td><td></td><td></td></tr>
<tr><td>Helo and Thai (2024)</td><td></td><td></td><td></td><td></td><td>(CPS)</td></tr>
<tr><td>Üster (2024)</td><td></td><td></td><td></td><td></td><td></td></tr>
<tr><td>Tubis and Grzybowska (2022)</td><td></td><td></td><td></td><td></td><td></td><td rowspan="2">By analyzing trends and patterns (BD, AI)</td></tr>
<tr><td>Boorová et al. (2024)</td><td>By analyzing historical data, traffic, weather (BD, ML, AI)</td><td>By analyzing traffic, weather (BD, ML, AI)</td><td></td><td></td><td></td></tr>
</table>

directions and prioritizing public transportation (Efthymiou & Ponis, 2021; Lyu et al., 2023; Markov & Vitliemov, 2020; Modica et al., 2021). Route optimization requires knowledge of the vehicle's position, intended path, and destination. AI-based navigation systems use real-time traffic and weather data to optimize routes while considering restrictions like weight, height, and cargo type (von Stietencron et al., 2022). A good example of a T4.0 application is a European trucking firm, which reduced fuel costs by 15% using an AI system that monitors vehicle performance and driver behavior, advising drivers on better techniques in real-time (Chui et al., 2018).

Route optimization further requires real-time knowledge of a vehicle's position, intended path, and destination. AI-based navigation systems use such data plus weather data to calculate optimal routes while considering restrictions such as vehicle weight, height, and cargo type (von Stietencron et al., 2022). One example is an ML-based routing system developed by MIT. This system identifies patterns in BD and offers real-time routing adjustments in cities where constantly changing traffic conditions often render pre-optimized routes ineffective (Hinkel, 2024).

Fleet management has several facets. Three authors (Khan et al., 2022; Lyu et al., 2023; Markov & Vitliemov, 2020) discuss the optimization of fleet management, including overall fleet efficiency improvement. Others (Chen et al., 2021; Facchini et al., 2019; Markov & Vitliemov, 2020; Tang et al., 2024; Von Hippel & Tyre, 1995) describe how IoT, BD, CPS, ML and AI support predictive maintenance and vehicle performance. Real-time data analysis, enabled by CPS, plays a critical role in reducing downtime and extending asset lifespans (Lyu et al., 2023). A practical example of such technology in action is Carisbrooke Shipping, which uses fleet optimization software to improve the operational efficiency of its 31 vessels (Wärtsilä Corp., 2022). These advancements demonstrate how technology-driven approaches are transforming traditional fleet management practices.

An increasingly important field within transportation, particularly within factories and in spread-out countries such as Australia, Russia, China, the USA, or Canada, is the use of *autonomous vehicles* (Markov & Vitliemov, 2020). Currently, autonomous vehicles used outside warehouses are mostly restricted to driving outside of urban areas, whereby drivers take vehicles to launching points, vehicles (often road trains with 2–3 trailers in Australia) then drive to their destinations close to another urban center, and drivers picking them up from there for retrieval to their final destinations. Drones are also increasingly being deployed for medical deliveries, humanitarian aid missions, and military logistics, particularly in areas that are difficult to traverse (Levy, 2022; Smith, 2024).

Despite the progress in autonomous freight transportation, the industry remains in an experimental phase, with companies such as Waymo One, Nuro, and Amazon Scout testing self-driving vehicles for last-mile delivery (Ackerman, 2024; Patrao, 2022). However, challenges related to technical feasibility and cost have led some companies to scale back their efforts (Ackerman, 2024; Korosec, 2023). In contrast, Canadian startup Waabi is progressing cautiously, with plans to gradually phase out safety drivers from its autonomous trucking operations next year (Castaldo, 2024).

Supply chain visibility represents another significant application of AI in transportation, achieved through real-time tracking powered by IoT and CPS (Draksler et al., 2023; Helo & Thai, 2024; Kucukaltan et al., 2022). These technologies enable real-time monitoring of vehicles and loads (containers, pallets or even individual cases) and managing exceptions such as delays or potential spoilage due to variations in temperature, pressure, or chemical conditions. Radio frequency Identification (RFID) is particularly valuable in cold chain logistics and hazardous goods transportation, ensuring compliance with handling requirements while providing critical data on product conditions. AI can also provide autonomous decision-making (Jagtap et al., 2021; Lyu et al., 2023 ; Winkelhaus & Grosse, 2020), and CBS facilitate seamless coordination across stationary and mobile locations, supporting collaboration among supply chain partners and enabling data sharing across organizations (Andres et al., 2024; Mihai, 2022). As remote computing also brings together data from one or more warehouses, it can facilitate autonomous local decision-making (Lyu et al., 2023; Winkelhaus & Grosse, 2020). Helo and Thai (2024) highlight cases where transport containers are tracked directly throughout entire distribution systems, allowing for visibility by shippers and customers alike.

Finally, two authors mention *demand forecasting* as an important application of AI for transportation (Lyu et al., 2023; Tubis & Grzybowska, 2022). Demand forecasting using BD and AI models is more accurate than traditional regression-based forecasting because it considers many more factors. For example, a study by Mamede et al. (2023) found that long short-term memory networks (LSTM) outperformed autoregressive integrated moving average models in 94% of scenarios across multiple distribution centers in Brazil, as AI is effective in capturing complex patterns in data that traditional models may miss (Boorová et al., 2024; Mamede et al., 2023). McKinsey reports that AI-based forecasting that considers underlying drivers, rather than just past patterns, further improves performance. Helo and Thai (2024) describe a case in which milk-run reverse logistics trucks are equipped with edge devices that provide the factory with advanced information of incoming inventory—a type of system that predicts demand for production capacity.

The most mentioned categories in the literature are fleet management (seven papers) and route optimization/traffic forecasting (four papers). This may indicate that autonomous deliveries are not yet widespread or commercially viable and that demand forecasting may not be seen as a core element of the transportation process. The latter is debatable as the economics of pickup and delivery (i.e., less than truckload) routes, as well as the provision of long-haul capacities, depend on demand predictions.

Table 4.3 Applications of CPS, Big Data and AI in warehousing

<table>
<tr><th>Article</th><th>Inventory management</th><th>Real-time monitoring</th><th>Warehouse automation</th><th>Predictive analytics</th></tr>
<tr><td>Tubis and Grzybowska (2022)</td><td>Using AI</td><td></td><td></td><td>Using historical data to predict demand (BD)</td></tr>
<tr><td>Lyu et al. (2023)</td><td>Using BD, AI</td><td></td><td rowspan="2">Uses CPS, ML and AI</td><td>To predict demand and optimize operations (BD, AI)</td></tr>
<tr><td>Khan et al. (2022)</td><td rowspan="2">Using ML and AI</td><td></td><td></td></tr>
<tr><td>Markov and Vitliemov (2020)</td><td></td><td>Uses AI for automate sorting and packing</td><td></td></tr>
<tr><td>Boorová et al. (2024)</td><td></td><td></td><td>Uses robotics, ML and AI</td><td></td></tr>
<tr><td>Facchini et al. (2019)</td><td></td><td></td><td></td><td>For predictive maintenance and analytics (IOT, CPS, AI)</td></tr>
<tr><td>Chen et al. (2021)</td><td></td><td></td><td></td><td rowspan="2">Using real-time data collection from IoT devices (IOT, BD, AI)</td></tr>
<tr><td>von Stietencron et al. (2022)</td><td></td><td></td><td></td></tr>
<tr><td>Tang et al. (2024)</td><td></td><td rowspan="5">… of warehouse operations (IOT, BD)</td><td></td><td>Using IoT data (IOT, BD)</td></tr>
<tr><td>Draksler et al. (2023)</td><td></td><td></td><td></td></tr>
<tr><td>Helo and Thai (2024)</td><td></td><td></td><td></td></tr>
<tr><td>Daftari (2024)</td><td></td><td>Using AGVs</td><td></td></tr>
<tr><td>Üster (2024)</td><td></td><td></td><td></td></tr>
<tr><td>Modica et al. (2021)</td><td></td><td>… of freight operations (AI)</td><td></td><td></td></tr>
</table>

4.3.2 *Warehousing*

In warehousing, the following processes are impacted by AI (see Table 4.3):

- Inventory management and optimization
- Real-time monitoring of operational aspects & warehouse automation
- Predictive analytics

One primary application of Industry 4.0 technologies in warehousing is the *optimization of inventory* using predictive analytics (forecasting) (Lyu et al., 2023; Markov & Vitliemov, 2020) as well as inventory handling, including decisions on storage, retrieval, demand prediction, reorder points and safety stocks (Khan et al., 2022; Markov & Vitliemov, 2020). The integration of IoT sensors, historical and real-time data and AI algorithms in ERP and warehouse management systems (WMS) also allows for improved automated reordering (Khan et al., 2022; Markov & Vitliemov, 2020).

Real-time monitoring is crucial for effective warehouse management, and Industry 4.0 technologies provide the tools needed for continuous tracking of inventory, equipment, and operational processes. A variety of data is processed in real-time, enabling centralized control and decision-making, thus enabling warehouse automation (Helo & Thai, 2024; Tang et al., 2024). Markov and Vitliemov (2020) report the ability to automatically assess inventory through both counting (flow-based) and validation (e.g., weight based) (Markov & Vitliemov, 2020). The addition of augmented reality (AR) glasses helps pickers speed up their work and avoid mistakes in some of DHL's warehouses (Boorová et al., 2024). Helo and Thai's (2024) case of trucks that are equipped with edge devices is also an example of real-time monitoring.

Automation in warehouses is significantly enhanced by data from sensors. Large-scale systems process information from thousands of sensors to identify patterns, predict trends and optimize operations (Modica et al., 2021). Not only stock keeping units (SKUs) are monitored and controlled, but also machinery such as conveyor belts, AGVs, and autonomous mobile robots (Facchini et al., 2019). Automated storage and retrieval using these technologies, including robots, simplify the organization of a warehouse and correct errors on the fly (Boorová et al., 2024; Jagtap et al., 2021). This control allows for optimized storage location assignments, SKU retrieval and staging in reverse order for delivery runs.

Predictive analytics leverages BD and AI to forecast trends and optimize warehouse operations, including demand forecasting, inventory management and maintenance scheduling. By analyzing historical data and real-time inputs from IoT sensors, predictive models can identify patterns and anticipate future needs, allowing warehouses to adjust strategies accordingly. For instance, demand forecasting models can predict inventory requirements based on seasonal trends and market conditions, improving inventory planning and reducing stockouts or excess inventory. Predictive maintenance models can anticipate equipment failures and schedule maintenance before breakdowns occur, ensuring continuous and efficient operations (Chen et al., 2021; Facchini et al., 2019; Tubis & Grzybowska, 2022).

The application of Industry 4.0 technologies in warehousing within the framework of T4.0 has led to substantial improvements. At the level of the supply chain, their use may also result in new business opportunities, including revenue streams, and result in changed transportation strategies as the Industry 4.0 technologies uncover previously unconsidered links, logic, and knowledge (Üster, 2024)

The next section will discuss these performance improvements along the three dimensions of the triple bottom line.

4.4 Transportation 4.0 and Triple Bottom Line Performance

One major issue facing the world is the significant contribution of the transportation sector to greenhouse gas emissions and other environmental and social impacts. Therefore, it is crucial to consider T4.0 technologies in terms of economic, environmental, and social performance. This section will explore T4.0's potential for improving these three dimensions on the sector's triple bottom line performance. The findings are summarized in Table 4.4 and described in more detail in the balance of this section.

4.4.1 *Economic Performance*

Enhanced Operational Efficiency and Cost Reduction As shown above, the use of Industry 4.0 technologies has significantly improved logistics operations by optimizing product flow across warehouses, transit points, and transportation networks (Ferraro et al., 2023), including pickup and delivery decisions, thus enhancing ser-

Table 4.4 Benefits of Transportation 4.0

	Key contributions	Technologies involved	Benefits/outcomes
Economic performance	Optimized routing	IoT, AI, digital twins	Cost reduction, faster delivery, customized service
	Real-time risk prediction	Cloud-based systems	Improved efficiency and space utilization
	Automation in warehousing	AGVs, smart robots	Faster storage and fulfillment, better supply chain visibility
	Predictive maintenance	BD systems	Enhanced customer satisfaction
Environmental performance	Resource optimization	Routing optimization	Lower emissions
	Fuel and energy reduction	Traffic and weather analysis	Reduced pollution and waste
	Sharing resources	IoT-enabled monitoring	Efficient multimodal transportation
	Drones for last-mile delivery	Drone systems	Sustainable resource use
Social performance	Improved safety for workers	Collaborative robots	Safer working conditions
	Accessible work for disabled individuals	Exoskeletons, co-bots	Enhanced training and decision-making support
	Stress reduction	Augmented reality	Benefits to public health through reduced pollution
	Reduced exposure to hazards	IoT-enabled risk detection	Public health benefits

vice levels while reducing operational costs (Sun et al., 2021). McKinsey reports that such optimization can improve accuracy by 10–20%, reduce inventory costs by 5%, and increase revenue by 2–3% (Chui et al., 2018). For example, Amazon's AI-powered robots autonomously retrieve and transport items, improving speed, utilization, and reliability while reducing labor costs in its fulfilment centers (Alım & Kesen, 2020; Kanerika, 2024). Similarly, DHL's implementation of LocusBots has led to a threefold productivity increase across multiple sites by improving the efficiency of order picking and reducing human error (Boorová et al., 2024). When connected to a warehouse management system, AI algorithms can also trigger automated reordering based on improved predictions (Khan et al., 2022; Markov & Vitliemov, 2020).

Improved Supply Chain Visibility and Decision-Making Quick access to information enables organizations to swiftly detect and address deviations in supply chain operations. Real-time forecasting and control of events and risks allows for proactive responses to minimize delays and disruptions. Advanced technologies, such as digital twins, simulate physical flows in real-time and recommend adjustments during adverse events, enabling autonomous decision-making or supporting human operators (Ferraro et al., 2023; Sundarakani et al., 2024; Wong et al., 2024). Visibility facilitates improved automated reordering, reduced stockouts and transaction costs and improved space utilization and error rates (Efthymiou & Ponis, 2021; Ünal et al., 2023). It helps optimize operations and mitigate risks, resulting in improved cash flow (Mihai, 2022; Üster, 2024) and allows companies like UPS ensure on-time deliveries and high service levels (Toxigon, 2024).

Scalability and Flexibility Automation technologies allow companies to scale operations efficiently to meet varying demand while maintaining flexibility in warehouse operations (Ferraro et al., 2023; Wong et al., 2024). Automation reduces labor dependency and increases throughput, as demonstrated by UK retailer John Lewis, which expanded storage capacity and improved order fulfillment efficiency with autonomous robots (Jones, 2024; Reporter, 2024). This added effective capacity also improves volume flexibility. Cloud computing not only reduces the need for expensive physical infrastructure, allowing smaller firms to compete in larger markets (Üster, 2024). Its capacity is also scalable, thus ensuring that companies remain easily operational during peak demand periods.

Asset Utilization and Predictive Maintenance Predictive maintenance powered by AI and BD analytics extends the lifespan of high-value assets such as trucks, railways and AGVs by detecting potential failures before they occur. This proactive approach minimizes downtime, lowers maintenance costs, and ensures consistent operational output (Mihai, 2022; Wong et al., 2024). For example, with the help of NVIDIA and Accenture, KION Group AG's has achieved a significant reduction in downtime (Spies & Derksen, 2025). Predictive maintenance also occurs in warehouses, similarly leading to reduced costs and disruptions (Modica et al., 2021).

Customer Satisfaction and Service Level Improvements Automation and AI-powered decision-making not only enhance internal operations but also improve customer service by ensuring faster, more accurate order fulfillment and delivery. Enhanced forecasting accuracy enables firms to respond dynamically to market changes and better meet customer demand, ultimately leading to higher satisfaction levels (Andres et al., 2024; Markov & Vitliemov, 2020). Similarly, cloud-based logistics platforms, such as those used by DHL, result in enhanced traceability of SKUs and fleets, which in turn enhances supply chain transparency and product security, ensuring safer and more reliable deliveries (Patil, 2024).

The integration of IoT, AI, BD, and digital twins has transformed economic performance in transportation and logistics. These technologies enhance real-time monitoring, predictive analytics, and optimization, improving reliability, fleet routing, and efficiency. Automation reduces costs by minimizing inventory, spoilage, and resource waste, while CBS enhance communication and coordination. AGVs and smart robots boost accuracy, speed, and scalability in warehouses, and predictive maintenance reduces downtime and extends asset life. Together, these advancements drive profitability, strengthen competitiveness, and enhance customer satisfaction as firms increasingly adopt these capabilities (Qureshi et al., 2024).

4.4.2 Environmental Performance

Reduced Fuel Consumption and Emissions T4.0 systems play a crucial role in enhancing environmental sustainability, particularly in transportation, where the greatest benefits are realized. Optimizing routes, analyzing traffic and weather conditions, monitoring vehicles and machinery, and managing load assignments and timing can significantly improve resource efficiency. These improvements help reduce fuel consumption, greenhouse gas emissions, pollution and waste, including spoilage and shrinkage (Ferraro et al., 2023). Such benefits apply to single-leg, multi-leg, and multimodal transportation, as well as reverse logistics operations such as waste collection and recycling (Bielecki et al., 2024; Sun et al., 2021; Wong et al., 2024). Unmanned aerial vehicles (UAVs), used either independently or alongside delivery trucks, offer additional fuel savings by reducing travel distances otherwise covered by trucks (Sun et al., 2021). Collaborative distribution, where companies share resources for pickup and delivery operations, further increases capacity utilization and minimizes unnecessary trips, contributing to sustainability (AWI, 2024). For instance, DHL's advanced route optimization software, which incorporates traffic and weather data while using shared transportation resources and load assignments, has resulted in lower fuel consumption and emissions (N.N., 2024).

Reduced Energy Use Within warehouses, better predictions and operational optimizations contribute to reducing energy consumption and resource usage. Automated systems powered by AI and IoT devices perform tasks faster and more accurately than humans, thus reducing delays, damage, errors and service times, all of which contribute to lower energy use and less waste (Sun et al., 2021; Wong et al., 2024). Ocado, an online supermarket, has reported significant reductions in energy consumption and waste through the use of highly automated warehouse systems with robotic assistance (Ali & Phan, 2022). Cloud-based logistics platforms further enhance operational efficiency by improving coordination and reducing the need for excessive physical infrastructure, leading to better energy management both in computing and across supply chains (Üster, 2024).

Alignment with Corporate Sustainability Goals Companies like DHL use BD analytics to continuously refine routing algorithms, leading to lower fuel consumption and emissions while maintaining high service levels (Boorová et al., 2024). These efforts contribute to broader corporate environmental goals, ensuring compliance with sustainability regulations and enhancing brand reputation through greener operations.

Challenges and Environmental Trade-Offs Despite the advantages of T4.0 technologies, their adoption also presents environmental challenges. The increased reliance on automation and AI systems can lead to higher energy consumption, which may offset some of the environmental gains achieved through efficiency improvements. AI systems, in particular, can be very energy intensive. Furthermore, the widespread use of lithium batteries in robotic devices, UAVs, and electric vehicles has negative environmental implications, particularly in terms of production and disposal (Ferraro et al., 2023; IER, 2020). Companies must carefully balance the benefits of automation with its potential downsides, seeking sustainable solutions such as recycling programs and energy-efficient technologies.

In summary, the integration of advanced logistics technologies, such as optimized routing, resource sharing, and automation, significantly enhances environmental performance by reducing emissions, fuel consumption, and waste. Collaborative efforts, like shared delivery networks, promise substantial environmental benefits. While challenges such as increased energy use and material impacts remain, these technologies represent a move toward sustainable logistics systems.

4.4.3 Social Performance

T4.0 technologies significantly improve *workplace safety* by minimizing exposure to hazardous conditions and reducing the risk of injuries. Sensors continuously monitor risks, triggering preventive actions, while automated and autonomous systems handle high-risk tasks, including hazardous material management (Ferraro et al., 2023; Wong et al., 2024). For instance, Nokia's industrial devices enhance situational awareness in high-risk environments such as mines, ports, and chemical facilities, improving worker safety (Nokia, n.d.). Similarly, Ford's deployment of EksoVest, a wearable exoskeleton, helps reduce worker fatigue and injury risks in assembly plants (Ford, 2018). IoT-enabled cargo tracking further enhances security by preventing fraud and counterfeiting, ensuring safer operations across the supply chain (Ferraro et al., 2023).

T4.0 technologies, such as exoskeletons and collaborative robots (cobots), enhance *worker capabilities and inclusivity* by making physically demanding tasks more accessible. While exoskeletons are not yet widely adopted, they can support strenuous tasks and enable disabled individuals to perform jobs that would otherwise be inaccessible (Wong et al., 2024). Cobots, designed for safe human interaction, are extensively used for heavy and repetitive tasks, thus avoiding injuries. Beyond industrial settings, cobots also play a vital role in caregiving, particularly in assisting seniors, thereby enhancing quality of life and promoting inclusivity (Ross, 2024).

The adoption of T4.0 technologies requires a more skilled workforce, nudging average wages upwards while likely reducing the overall headcount. Technical rather than manual workers tend to also experience better working conditions. Workers may further benefit from better *training and onboarding* processes. Virtual reality (VR) and AR systems are increasingly used for immersive training experiences, helping workers acquire necessary skills safely and efficiently. Wearable IoT devices provide real-time safety information, while AR solutions offer hands-free procedural guidance and troubleshooting support from remote experts (Kaminsky, 2019). Companies such as LightGuide employ AR systems to assist workers in complex assembly processes, such as assembling car headliners and performing quality control, with potential applications in transportation tasks such as fleet maintenance and package packing (LightGuide, 2022).

T4.0 technologies *improve health* by reducing public exposure to pollution, noise, and transportation hazards through optimized routing and efficient fleet management (Ferraro et al., 2023; Sun et al., 2021). They also contribute to cleaner and quieter urban environments. Drones have demonstrated significant social benefits by delivering supplies to remote areas, such as during the response to the 2015 Nepal Earthquake (Füellenbach, 2024). Furthermore, improved scheduling and routing contribute to a better work-life balance for employees by ensuring adherence to planned working hours, reducing stress, and enhancing overall well-being (Sun et al., 2021).

In summary, T4.0 technologies significantly enhance social sustainability by improving safety, inclusivity, and operational efficiency while fostering healthier and more inclusive workplaces. Although many of these benefits are already realized, the full potential of T4.0 technologies, particularly in transportation, will continue to develop as they become more widely adopted and refined.

Having discussed the triple bottom line benefits of T4.0 technologies, we will now investigate potential issues and drawbacks stemming from their deployment.

4.5 Potential Issues with T4.0 Technologies

The introduction of Industry 4.0 technologies to transportation and warehousing comes with a set of challenges that need to be addressed. The main challenges discussed in the literature encompass data security and privacy concerns, human readiness, the high cost and infrastructure needs and technical issues.

The widespread use of IoT devices in transportation and logistics raises significant *data security and privacy* concerns. These devices collect vast amounts of operational data, providing critical insights but also increasing vulnerability to unauthorized access (The Economist, 2024b). Many IoT manufacturers lack expertise in cybersecurity, as their devices were originally not designed for connected applications, making them easier targets for hackers (Greengard, 2019). Unauthorized access can compromise data integrity and create backdoors into larger systems, posing financial, environmental, and safety risks. IoT devices used in transportation are particularly susceptible to physical tampering, unauthorized access, and data breaches. For example, RFID tags can be cloned or manipulated, threatening data security and automated vehicle safety. Ensuring both informational and physical security is essential to prevent such vulnerabilities (Kugler, 2023), especially in large-scale commercial applications where sensors are integrated into extensive networks.

Despite these risks, there is no universally accepted security standard for IoT devices. While some agencies enforce partial regulations, these standards are often outdated, inconsistent, or ignored by manufacturers due to a lack of market demand for security (Greengard, 2019; Kugler, 2023). Governments, such as those in the European Union and the United States, are considering regulatory measures to mandate "security by design" in IoT development (Greengard, 2019).

To mitigate risks, organizations should enforce minimum security standards for IoT devices, implement a zero-trust access model to authenticate users and devices before each session, and use network micro-segmentation to limit unauthorized access. Additionally, encrypting connections and data traffic can further enhance security (Kugler, 2023).

As T4.0 technology becomes more pervasive, the need for a *skilled workforce* capable of interacting with these advanced systems becomes critical. This includes not only technical skills for operating and maintaining these systems but also safety training to ensure safe interactions with robots and automated vehicles. Ongoing training and skill development are essential to address the skills gap (Wong et al., 2024).

T4.0 technologies require significant *financial investments and infrastructure* (Khan et al., 2022; Saniuk, 2022), including robust information systems and fast and reliable data connections for cloud processing. It is also difficult to integrate multi-source data for this purpose (Sundarakani et al., 2024) as reengineering of processes and systems that are frequently necessary to accomplish this—particularly for SMEs (Qureshi et al., 2024).

The high pace of innovation and associated costs favor not only larger firms but also wealthier countries, leading to *disparities* in operational efficiencies and technological capabilities (Khan et al., 2022; Malagon-Suarez & Orjuela-Castro, 2023; Saniuk, 2022; Wong et al., 2024). Regional differences in infrastructure, connectivity, economic conditions, regulatory environments and digital literacy further contribute to these disparities (Wong et al., 2024). For example, an empirical study found that almost 70% of Columbian companies knew at least some advanced logistics technologies, but almost two thirds do not use any (Malagon-Suarez & Orjuela-Castro, 2023). Addressing such disparities requires improving infrastructure, regulatory frameworks, and workforce skills across regions.

Several *technical challenges* need to be addressed for successful T4.0 implementation, including connectivity issues with RFID and IoT devices, limited wireless range and interoperability among devices from different manufacturers (Jagtap et al., 2021). Addressing these challenges requires advancements in technology standards and practices to ensure seamless integration and operation of diverse systems.

In conclusion, addressing the challenges of data security, workforce readiness, financial investment, and technical issues is critical for the successful implementation of T4.0. Overcoming these obstacles will require advancements in technology standards, regulatory frameworks, and infrastructure improvements to ensure seamless integration and operation.

4.6 Outlook

The advent of T4.0, driven by Industry 4.0 technologies, is driving a significant transformation in the transportation sector. This chapter has outlined the need for T4.0, examined various trends and challenges impacting the sector, and explored the potential of T4.0—such as IoT, CPS, BD, ML and AI—to transform transportation and warehousing.

The current JIT transportation model struggles with uncertainty and disruptions. A more efficient, flexible, and resilient transportation system is needed, especially given shifting consumer behaviors and global warming. Technologies like IoT, CPS, and AI enhance efficiency, reliability, and responsiveness. T4.0 can improve effective capacity, increasing the speed and efficiency of processing SKUs or containers and providing greater flexibility. T4.0 enables real-time tracking, predictive maintenance, route optimization, fleet management, warehouse automation and supply chain visibility, leading to significant operational improvements. Predictive analytics assist with on-time performance and adapting to shifting trade and customer demand patterns, addressing trends in demand volumes and patterns while quickly meeting customer requirements.

The World Bank estimates that international shipping alone needs to reduce its emissions intensity by 61% by 2030 and 91% by 2040 (Englert et al., 2023). While T4.0 can help decrease emissions, the question is whether the improvements based on T4.0 alone are large enough to meet these constraints. Along the social dimension, T4.0 provides some significant benefits, including better work conditions and safety. These benefits may be traded off with a reduced workforce as automatic or autonomous vehicles and devices very slowly displace human work.

The adoption of T4.0 is not without challenges. Issues such as data security and privacy, human readiness and skills gap, high costs and infrastructure needs, and technical interoperability pose significant hurdles. Addressing these challenges is crucial for successful implementation and widespread adoption.

Several questions and areas for further research and action remain as these technologies are developed, refined and deployed. Security and privacy frameworks need to be standardized and regulated, especially for IoT devices and CPS. Cost-effective implementation is necessary for small and medium-sized enterprises (SMEs) and in developing regions. Many SMEs are crucial for road transportation developing regions frequently serve as major manufacturing hubs. Exploring funding models, scalable solutions, and collaborative approaches could bridge gaps between regions and company sizes. Additionally, more research is needed to quantify T4.0's environmental and social impacts to determine whether it provides a significant net benefit to humanity.

On the non-technical side, a further investigation of the human impacts of adopting T4.0 is essential. This includes exploring training programs, skill development initiatives, and strategies to address the skills gap and enhance human readiness for interacting with advanced systems. Finally, assessing the environmental impact of T4.0 technologies and identifying ways to enhance their sustainability is critical. This includes evaluating the net effect on greenhouse gas emissions, exploring green technologies, and developing practices that align with global sustainability goals.

T4.0 holds great promise for transforming the transportation sector. The potential benefits are substantial, but careful consideration and action are required to address the associated challenges. By doing so, the transportation industry can achieve greater efficiency, resilience and sustainability, benefiting commerce, humanity and the natural environment. The successful deployment of T4.0 will depend on our ability to navigate these challenges and leverage these technologies for the greater good.

References

Ackerman, E. (2024). This year, autonomous trucks will take to the road with no one on board. Retrieved June 12, 2024, from https://spectrum.ieee.org/this-year-autonomous-trucks-will-take-to-the-road-with-no-one-on-board.

Ali, I., & Phan, H. M. (2022). Industry 4.0 technologies and sustainable warehousing: a systematic literature review and future research agenda. *The International Journal of Logistics Management*, 644–662. https://doi.org/10.1108/IJLM-05-2021-0277

Alım, M., & Kesen, S. E. (2020). Smart warehouses in Logistics 4.0. In T. Paksoy, C. G. Kochan, & S. S. Ali (Eds.), *Logistics 4.0—Digital transformation of supply chain management* (pp. 186–201). CRC Press.

Andres, B., Diaz-Madroñero, M., Soares, A. L., & Poler, R. (2024). Enabling technologies to support supply chain Logistics 5.0. *IEEE Access*, 43889–43906. https://doi.org/10.1109/ACCESS.2024.3374194

AWI. (2024). How can the sharing model bring changes to the logistics industry? AWI. Retrieved January 15, 2025, from https://www.awi-usa.com/post/how-can-the-sharing-model-bring-changes-to-the-logistics-industry.

Bielecki, M., Wisniewski, Z., Dukic, G., & Dujak, D. (2024). Supply Chain 4.0: What the supply chains of the future might look like. *Engineering Management in Production and Services, 16*, 77–92. https://doi.org/10.2478/emj-2024-0006

Bogataj, D., Bogataj, M., & Hudoklin, D. (2017). Mitigating risks of perishable products in the cyber-physical systems based on the extended MRP model. *International Journal of Production Economics, 194*, 113–125. https://doi.org/10.1016/j.ijpe.2017.11.004

Boorová, B., Mijušković, V. M., Aćimović, S., & Durđić, D. (2024). Application of artificial intelligence in logistics 4.0: DHL case study analysis. *Ekonomika preduzeća, 72*, 292–304. https://doi.org/10.5937/EKOPRE2406292B

Castaldo, J. (2024). Waabi raises $200-million to launch driverless commercial trucks in Texas. *The Globe and Mail B3*.

Chen, Y.-T., Sun, E. W., Chang, M.-F., & Lin, Y.-B. (2021). Pragmatic real-time logistics management with traffic IoT infrastructure: Big data predictive analytics of freight travel time for Logistics 4.0. *International Journal of Production Economics, 238*, 1–27. https://doi.org/10.1016/j.ijpe.2021.108157

Chui, M., Manyika, J., Miremadi, M., Henke, N., Chung, R., Nel, P., & Malhotra, S. (2018). Notes from the AI frontier: Applications and value of deep learning. Retrieved June 13, 2024, from https://www.mckinsey.com/featured-insights/artificial-intelligence/notes-from-the-ai-frontier-applications-and-value-of-deep-learning.

Coyle, J. J., Novack, R. A., Gibson, B. J., & Bardi, E. J. (2011). *Transportation: A supply chain perspective* (7th ed.). Sout-Western Cengage Learning.

Crownhart, C. (2023). Shipping brief. The UN just set a net-zero goal for shipping. Here's how that could work. Retrieved May 30, 2024, from https://www.technologyreview.com/2023/07/11/1076080/un-climate-goals-global-shipping/.

Daftari, A. (2024). Logistics 4.0: a systematic investigation leading to an innovative logistics system. *International Journal of Management and Commerce Innovations, 12*, 151–157. https://doi.org/10.5281/zenodo.13291881

Draksler, T. Z., Cimperman, M., & Obrecht, M. (2023). Data-driven supply chain operations—The pilot case of postal logistics and the cross-border optimization potential. *Sensors, 23*(3), 1624. https://doi.org/10.3390/s23031624

Efthymiou, O. K., & Ponis, S. T. (2021). Industry 4.0 technologies and their impact in contemporary logistics: A systematic literature review. *Sustainability, 13*(21), 11643. https://doi.org/10.3390/su132111643

Englert, D., Rorjon, I., Salgmann, R., & Sulikova, S. (2023). A new climate deal for shipping: Three decades to zero. *World Bank Blogs*. Retrieved May 28, 2024, from https://www.ft.com/content/365781a7-af36-429f-8122-37d93e1f125d.

Facchini, F., Olesków-Szłapka, J., Ranieri, L., & Urbinati, A. (2019). A maturity model for Logistics 4.0: An empirical analysis and a roadmap for future research. *Sustainability, 12*(1), 86. https://doi.org/10.3390/su12010086

Ferraro, S., Cantini, A., Leoni, L., & De Carlo, F. (2023). Sustainable Logistics 4.0: A study on selecting the best technology for internal material handling. *Sustainability, 15*(9), 7067. https://doi.org/10.3390/su15097067

Ford. (2018). Ford rolls out exoskeleton wearable technology globally to help lessen worker fatigue, injury. Retrieved July 27, 2024, from https://media.ford.com/content/fordmedia/fna/us/en/news/2018/08/07/ford-rolls-out-exoskeleton-wearable-technology-globally-to-help-.html.

Füellenbach, G. (2024). Drones for medical delivery: A lifeline for remote communities | 2024-04-25T18:11:49.439Z. *Drone Pros*. Retrieved January 16, 2025, from https://www.dronepros.co/post/drones-medical-delivery.

Greengard, S. (2019). Deep insecurities: The internet of things shifts technology risk. *Communications of the ACM, 62*(5), 20–22. https://doi.org/10.1145/3317675. Association for Computing Machinery.

Helo, P., & Thai, V. V. (2024). Logistics 4.0—Digital transformation with smart connected tracking and tracing devices. *International Journal of Production Economics, 275*, 109336–109349. https://doi.org/10.1016/j.ijpe.2024.109336

Hinkel, L. (2024). 3 questions: Enhancing last-mile logistics with machine learning. *MIT News*. Retrieved July 27, 2024, from https://news.mit.edu/2024/matthias-winkenbach-enhancing-last-mile-logistics-machine-learning-0416.

IER. (2020). The environmental impact of lithium batteries. IER. Retrieved January 15, 2025, from https://www.instituteforenergyresearch.org/renewable/the-environmental-impact-of-lithium-batteries/.

Jagtap, S., Bader, F., Garcia-Garcia, G., Trollman, H., Fadiji, T., & Salonitis, K. (2021). Food Logistics 4.0: Opportunities and challenges. *Logistics, 5*(1), 2. https://doi.org/10.3390/logistics5010002

Jones, S. (2024). John Lewis 'optimistic' £400m investment in automation will pay off this Christmas. *The Grocer*. Retrieved January 15, 2025, from https://www.thegrocer.co.uk/news/john-lewis-optimistic-400m-investment-in-automation-will-pay-off-this-christmas/698462.article.

Kaminsky, G. (2019). 4 ways industry 4.0 technology can improve workforce safety. Retrieved January 16, 2025, from https://www.ptc.com/en/blogs/ar/4-ways-technology-can-improve-worker-safety.

Kanerika. (2024). Why AI in warehouse management is essential for competitive advantage. *Kanerika*. Retrieved January 14, 2025, from https://kanerika.com/blogs/ai-in-warehouse-management/.

Khan, S., Singh, R., Haleem, A., Dsilva, J., & Ali, S. S. (2022). Exploration of critical success factors of Logistics 4.0: A DEMATEL approach. *Logistics, 6*(1), 13. https://doi.org/10.3390/logistics6010013

Knight, W. (2015). Inside Amazon's warehouse, human-robot symbiosis. *MIT Technology Review*. Retrieved July 27, 2024, from https://www.technologyreview.com/2015/07/07/248370/inside-amazons-warehouse-human-robot-symbiosis/.

Korosec, K. (2023). Another blow for self-driving trucks as former industry leader abandons the US. Retrieved June 12, 2024, from https://techcrunch.com/2023/12/05/another-blow-for-self-driving-trucks-as-former-industry-leader-abandons-the-u-s/.

Kucukaltan, B., Saatcioglu, O. Y., Irani, Z., & Runa, O. (2022). Gaining strategic insights into Logistics 4.0: expectations and impacts. *Production Planning & Control, 33*(2–3), 211–227. https://doi.org/10.1080/09537287.2020.1810760

Kugler, L. (2023). Standards to secure the sensors that power IoT. *Communications of the ACM*, 14–16. https://doi.org/10.1145/3591215. Association for Computing Machinery.

Levy, M. G. (2022). Drones have transformed blood delivery in Rwanda. Retrieved June 12, 2024, from https://www.wired.com/story/drones-have-transformed-blood-delivery-in-rwanda/.

LightGuide, C. (2022). 6 uses of augmented reality for manufacturing in every industry. *LightGuide*. Retrieved January 16, 2025, from https://www.lightguidesys.com/resource-center/blog/6-uses-of-augmented-reality-for-manufacturing-in-every-industry/.

Lyu, J., Zhou, F., & He, Y. (2023). Digital technique-enabled container logistics supply chain sustainability achievement. *Sustainability, 15*(22), 16014. https://doi.org/10.3390/su152216014

Malagon-Suarez, C. P., & Orjuela-Castro, J. A. (2023). Challenges and trends in logistics 4.0. *Ingeniería*. https://doi.org/10.14483/23448393.18492

Mamede, F. P., da Silva, R. F., de Brito Junior, I., Yoshizaki, H. T., Hino, C. M., & Cugnasca, C. E. (2023). Deep learning and statistical models for forecasting transportation demand: A case study of multiple distribution centers. *Logistics, 7*(4), 86. https://doi.org/10.3390/logistics7040086

Markov, K., & Vitliemov, P. (2020). Logistics 4.0 and Supply Chain 4.0 in the automotive industry. *IOP Conference Series: Materials Science and Engineering, 878*(1), 012047. https://doi.org/10.1088/1757-899X/878/1/012047

Mihai, N. (2022). Digital logistics: historical background and current key development trends. In Ž. Simanavičienė (Ed.), *Digital technologies in the contemporary economy*. Mykolas Romeris University.

Modica, T., Colicchia, C., Tappia, E., & Melacini, M. (2021). Empowering freight transportation through Logistics 4.0: a maturity model for value creation. *Production Planning & Control, 34*(12), 1149–1164. https://doi.org/10.1080/09537287.2021.1988176

Mott Macdonald. (n.d.). Helping FedEx serve Canada's busiest airport. Retrieved July 27, 2024, from https://www.mottmac.com/en-US/article/12109/projects/toronto-pearson-fedex-sorting-facility.

N.N. (2024). *The role of 4.0 technologies in improving sustainability at logistics facilities*. Politecnico di Milano School of Management. Retrieved January 15, 2025, from https://www.som.polimi.it/en/the-role-of-4-0-technologies-in-improving-sustainability-at-logistics-facilities/.

Nokia. (n.d.). Nokia strengthens situational awareness and worker safety in industrial environments with expanded device portfolio. Retrieved January 15, 2025, from https://www.nokia.com/about-us/news/releases/2024/10/09/nokia-strengthens-situational-awareness-and-worker-safety-in-industrial-environments-with-expanded-device-portfolio/.

Patil, G. (2024). IoT in warehouse: Benefits and use case. *IdeaUsher*. Retrieved January 15, 2025, from https://ideausher.com/blog/iot-warehouse-management-solutions-and-use-cases/.

Patrao, J. (2022). Amazon is working with communities to build the future of Scout. Retrieved June 12, 2024, from https://www.aboutamazon.com/news/innovation-at-amazon/amazon-is-working-with-communities-to-build-the-future-of-scout.

Qureshi, K. M., Mewada, B. G., Kaur, S., Khan, A., Al-Qahtani, M. M., & Qureshi, M. R. N. M. (2024). Investigating industry 4.0 technologies in logistics 4.0 usage towards sustainable manufacturing supply chain. *Heliyon*. https://doi.org/10.1016/j.heliyon.2024.e30661

Reporter, ET Business. (2024). Robots help John Lewis stack its Christmas gifts high. Retrieved January 15, 2025, from https://www.thetimes.com/business-money/technology/article/robots-help-john-lewis-stack-its-christmas-gifts-high-nl8x7rtlr.

Rodrigue, J.-P. (2024). *The geography of transport systems* (6th ed.). Routledge.

Ross, L. (2024). 8 examples of cobots in everyday life. Retrieved January 15, 2025, from https://www.thomasnet.com/insights/the-year-of-the-cobot/.

Sadaf, M., Iqbal, Z., Javed, A. R., Saba, I., Krichen, M., Majeed, S., & Raza, A. (2023). Connected and automated vehicles: Infrastructure, applications, security, critical challenges, and future aspects. *Technologies, 11*(5), 117. https://doi.org/10.3390/technologies11050117

Saniuk, A. (2022). The Logistics 4.0 implementation supported by the balanced scorecard method. *European Research Studies Journal, XXV*(1), 198–207.

Schwab, K. (1997). The fourth industrial revolution. Currency.

Shakir, U. (2024). Amazon's Prime Air delivery drones can fly even farther after FAA approval. Retrieved June 12, 2024, from https://www.theverge.com/2024/5/30/24167776/amazon-prime-air-delivery-drones-faa-bvlos-approval.

Smith, N. (2024). US Army's Gabriel Drone revolutionizes tactical resupply amid global tensions. Retrieved June 13, 2024, from https://www.msn.com/en-us/news/technology/us-armys-gabriel-drone-revolutionizes-tactical-resupply-amid-global-tensions/ar-BB1n41Tp.

Spies, C., & Derksen, J. (2025). KION teams with NVIDIA and Accenture to optimize supply chains with AI-powered robots and digital twins. Retrieved January 15, 2025, from https://

www.businesswire.com/news/home/20250106399453/en/KION-Teams-with-NVIDIA-and-Accenture-to-Optimize-Supply-Chains-with-AI-Powered-Robots-and-Digital-Twins.

Sun, X., Solvang, W. D., Wang, Y., & Wang, K. (2021). The application of Industry 4.0 technologies in sustainable logistics: a systematic literature review (2012–2020) to explore future research opportunities. *Environmental Science and Pollution Research, 29*, 9560–9591. https://doi.org/10.1007/s11356-021-17693-y

Sundarakani, B., Manikas, I., & Gunasekaran, A. (2024). The role of digital transformation in achieving sustainable supply chain management in Industry 4.0: an editorial review perspective. *International Journal of Logistics Research and Applications, 27*(6), 843–851. https://doi.org/10.1080/13675567.2024.2332123

Tang, Y. M., Chau, K. Y., Kuo, W. T., & Liu, X. X. (2024). IoT-based information system on cold-chain logistics service quality (ICCLSQ) management in Logistics 4.0. *Information Systems Frontiers, 26*(2), 689–708. https://doi.org/10.1007/s10796-023-10393-7

The Economist. (2023a). Global shipping costs are returning to pre-pandemic levels. *The Economist*. Retrieved May 28, 2024, from https://www.economist.com/graphic-detail/2023/01/09/global-shipping-costs-are-returning-to-pre-pandemic-levels.

The Economist. (2023b). Europe can't decide how to unplug from China. *The Economist*. Retrieved May 28, 2024, from https://www.economist.com/international/2023/05/15/europe-cant-decide-how-to-unplug-from-china.

The Economist. (2023c). Attempts to make supply chains "resilient" are likely to fail. *The Economist*. Retrieved May 28, 2024, from https://www.economist.com/special-report/2023/10/02/attempts-to-make-supply-chains-resilient-are-likely-to-fail.

The Economist. (2024a). Boom times are back for container shipping. *The Economist*. Retrieved July 22, 2024, from https://www.economist.com/business/2024/06/27/boom-times-are-back-for-container-shipping.

The Economist. (2024b). Ubiquitous technical surveillance has made spying more difficult. *The Economist*. Retrieved July 2, 2024, from https://www.economist.com/technology-quarterly/2024/07/01/ubiquitous-technical-surveillance-has-made-spying-more-difficult.

The Economist. (2025). The unpredictability of Trump's tariffs will increase the pain. *The Economist*. Retrieved March 27, 2025, from https://www.economist.com/leaders/2025/03/27/the-unpredictability-of-trumps-tariffs-will-increase-the-pain.

Tiseo, I. (2023). Distribution of carbon dioxide emissions produced by the transportation sector worldwide in 2022, by sub sector. Retrieved May 28, 2024, from https://www.statista.com/statistics/1185535/transport-carbon-dioxide-emissions-breakdown/.

Tiseo, I. (2024). Distribution of carbon dioxide emissions worldwide in 2022, by sector. Retrieved May 28, 2024, from https://www.statista.com/statistics/1129656/global-share-of-co2-emissions-from-fossil-fuel-and-cement/.

Toxigon. (2024). Optimizing warehouse operations with AI: Practical tips and real-world examples. *Toxigon*. Retrieved January 14, 2025, from https://toxigon.com/optimizing-warehouse-operations-with-ai.

Tubis, A., & Grzybowska, K. (2022). In search of Industry 4.0 and Logistics 4.0 in small-medium enterprises—A state of the art review. *Energies, 15*(22), 8595. https://doi.org/10.3390/en15228595

Ünal, Ö. A., Erkayman, B., & Usanmaz, B. (2023). Applications of artificial intelligence in inventory management: A systematic review of the literature. *Archives of Computational Methods in Engineering, 30*(4), 2605–2625. https://doi.org/10.1007/s11831-022-09879-5

Üster, Z. (2024). Supply chain management and logistics in the Industry 4.0 environment: A theoretical evaluation. *Quantrade Journal of Complex Systems in Social Science, 6*(1), 94–107. https://doi.org/10.5281/zenodo.11408049

Von Hippel, E., & Tyre, M. (1995). How learning by doing is done: Problem identification in novel process equipment. *Research Policy, 24*(1), 1–12.

von Stietencron, M., Hribernik, K., Lepenioti, K., Bousdekis, A., Lewandowski, M., Apostolou, D., & Mentzas, G. (2022). Towards Logistics 4.0: an edge-cloud software framework for big data analytics in logistics processes. *International Journal of Production Research, 60*(19), 5994–6012. https://doi.org/10.1080/00207543.2021.1977408

Wärtsilä Corp. (2022). Carisbrooke Shipping significantly reduces CO2 emissions with Wärtsilä's fleet optimisation solution. Retrieved July 27, 2024, from https://www.wartsila.com/media/news/15-11-2022-carisbrooke-shipping-significantly-reduces-co2-emissions-with-wartsila-s-fleet-optimisation-solution-3270465.

Wikipedia. (2024). Internet of things. Retrieved June 4, 2024, from https://en.wikipedia.org/wiki/Internet_of_things.

Winkelhaus, S., & Grosse, E. H. (2020). Logistics 4.0: a systematic review towards a new logistics system. *International Journal of Production Research, 58*(1), 18–43. https://doi.org/10.1080/00207543.2019.1612964

Wong, W. P., Anwar, M. F., & Soh, K. L. (2024). Transportation 4.0 in supply chain management: State-of-the-art and future directions towards 5.0 in the transportation sector. *Operations Management Research*. https://doi.org/10.1007/s12063-024-00471-7

World Trade Organization. (2020). World trade statistical review 2020.

Chapter 5
Semiconductor Supply Chain Change Management: A Semantic Web Based Knowledge Graph

Patrick Moder, Nour Ramzy, and Hans Ehm

Abstract In order to operate complex supply chains, it is necessary to react to various changes. Product change management affects operations along the entire product lifecycle: from early development to manufacturing, sales and marketing operations until discontinuation of products. Various stakeholders need to make decisions when such changes occur. The main contribution is that we introduce a semantic web based knowledge graph in the realm of product changes, which is interpretable to humans and machines and allows a common understanding of the domain. The proposed Supply Chain Change Management knowledge graph is instantiated with real-world data from the semiconductor domain. In addition, we find that the definition of rule sets enhances the expressivity of the knowledge graph. We verify the quality of the knowledge graph in terms of visualization, deduction of implicit knowledge, and detection of inconsistent data through reasoning mechanisms. Standardized interfaces enable integration into related supply chain ontologies. The main implication for practice is that our proposed solution can help automating supply chain change processes that are currently time-consuming and prone to errors.

Keywords Supply chain change management · Semiconductor supply chain · Inconsistency identification · Knowledge graph · Semantic web

---

P. Moder (✉)
Supply Chain Engineering Innovation, Infineon Technologies AG, Neubiberg, Germany

merconics GmbH & Co. KG, Bergkirchen, Germany
e-mail: patrickmoder@merconics.com

N. Ramzy · H. Ehm
Supply Chain Engineering Innovation, Infineon Technologies AG, Neubiberg, Germany

Y. Khojasteh, S. M. M. Jahan Abadi (eds.), *Emerging Technologies in Supply Chains*, International Series in Operations Research & Management Science 373, https://doi.org/10.1007/978-3-032-01218-0_5

5.1 Motivation

Digital transformation of business and society offers enormous growth potential. Digitalization is observed as an enabler of fundamental innovations. With regard to supply chain environments, smart networks and other Industry 4.0 approaches are emerging to connect entities from end-to-end and along the entire product lifecycle. Involving data is considered playing a central role in future supply chains, if not already emerging today. Relying on data-informed decisions is all the more important when highly automated and complex systems are considered. Consequently, an increasing amount of data points needs to be automatically gathered, processed and made accessible to relevant stakeholders. Furthermore, a growing number of adjacent domains tries to be part of such a data-driven interconnected system. Particular emphasis is put on customer interfaces, in order to maintain and expand a satisfied customer base that requests fast and facile interaction possibilities. These boundary conditions give rise to solutions that are flexible, scalable as well as versatile and hence overcome emerging complexity issues (Mujica et al., 2017).

Semiconductor manufacturers manage and operate highly automated and complex supply chains. It is necessary to react to various dynamic conditions, as recently observed after heavy demand fluctuations driven by the pandemic outbreak. With multiple disciplines involved in decision making throughout the semiconductor supply chain, facile collaboration is a prerequisite. Hence, the exchanged information needs to be understandable and consistent. These prerequisites include communication between humans as well as machines and require a language that is understandable and interpretable for both. These requirements are even more significant in changing scenarios, for instance when product changes are managed: to react quickly and reliably, decision makers take into account the whole product lifecycle from early development stages towards manufacturing, sales and operations until the discontinuation of products. Various domain experts and external stakeholders are in charge of handling the change operation and need to act in order to proceed in the supply chain change process. This complex decision process frequently leads to longer process durations. To shorten the time it takes to derive a reliable decision in practice, it is necessary to guarantee data consistency, clear visualizations of the relevant knowledge and smart decision support for all experts involved.

The slow adoption of digital technologies—such as knowledge graphs—is among other reasons caused by persistent uncertainty about specific use cases. Additionally, issues with complex relations and large data sets are not yet overcome comprehensively. We address this shortcoming by introducing a semantic web based knowledge graph of the entire value chain in the semiconductor supply chain change management. It serves as decision support solution for several stakeholders and provides an efficient solution for defining and maintaining the vocabulary. We map objects, processes, roles and relations that are relevant to supply chain change management. For augmenting the existing representation, we select specific use cases that give rise to more detailed information, serve as demonstration samples for data integration and enable exploitation. We consider this knowledge graph as advancement towards several of the design principles for digitalization in supply chain. In order to guarantee certainty that the processed data is of desirable high quality, recent research investigates approaches to detect inconsistencies. This research stream includes inconsistency classification and mechanisms for automated identification. Furthermore, applying semantic web based concepts for interoperable decision support is of interest in recent studies. We augment these research streams by designing a specific knowledge graph for supply chain change management. It has well defined interfaces to existing enterprise models and allows enhanced decision support with (1) dynamic visualization of relevant entities, relations and interfaces, (2) identification of data inconsistencies, and (3) seamless access for multiple stakeholders along the entire product value chain.

We structure this article as follows. First, we introduce preliminary principles of knowledge graphs in supply chain applications and the semiconductor-specific supply chain change management process. Furthermore, we provide an overview of relevant literature that comprises the state of science and technology, identify the research gap and formulate our research questions. Moreover, we introduce the developed knowledge graph and evaluate it with regards to the formulated research questions. We close this article with our conclusion and present arising future research interest.

5.2 Research Environment

The concept of change arouses increasing interest and is respected as a serious fact that enterprises and markets need to face in order to stay competitive. The dominance and radical nature of change are facilitated by modern economic trends, such as dynamic markets, global customers and reduced innovation cycle times. Changes in this sense equal to both strategic changes that affect business processes and operative changes that affect the fabricated assets. The latter include product design changes and manufacturing process changes, consequentially influencing the product itself. In any of these cases, as products increase in complexity and quantity, changes give rise to a reliable information management for tracing affected dependencies. (Eigner & Stelzer, 2009; Scheer et al., 2005)

For suppliers of microelectronics, the J-STD-046 standard is defining rules and principles for notifying customers in case of product changes, formulated by a joint consortium of JEDEC Solid State Technology Association, Electronic Components Association (ECIA) and the Association Connecting Electronics Industries (IPC), according to JEDEC (2016). Within our case company—a global semiconductor manufacturer—, the change management process is aligned with JEDEC J-STD-046 and furthermore structured on different organizational layers.

By definition, organizations with product purchases within the bygone 2 years are required to be notified in case those products or its manufacturing processes are subject to major changes. The customer is able to provide requirements for qualifying the change, such as reports, product samples for test or supplementary data. Customers are manifold and contain backend subcontractors, silicon foundries, manufacturing companies, and commercial end customers. Only in case the change gets accepted, the product is released to be shipped.

The product change process occupies a core position with the product lifecycle. Since this process involves many internal or external sub-processes, stakeholders and iterations, it frequently gets time-consuming due to slow customer feedback or the mere quantity of changes handled in parallel. It is difficult to guarantee the accessibility of real-time data that is emerging at different stages in the process. This difficulty is evoked by the presence of multiple data sources and various necessary access points. Inconsistencies may appear when changes at a certain entity do not lead to an immediate change at related entities, thus causing alignment errors. Uncertainty exists when the impact of a product change on the entire supply chain and lifecycle system is unsure. Hence, inconsistency and uncertainty have negative impact on core performance measures such as margin or efficiency.

Modelling and knowledge management are core principles of supply chain management. Although there exists a wide variety of rich models that capture supply chain concepts, oftentimes they rather cover specific subsystems and lack interoperability. In order to make intelligent decisions that are beneficial for multiple supply chain members, involved parties (disciplines, process owners, stakeholders, e.g.) need to align on their—sometimes contradicting—objectives. Supply chain interoperability requires data integration. Knowledge graphs are a promising solution in this regard. Moreover, their semantic foundation enables scalability towards adjacent models. As such, knowledge graphs support interoperability between processes, components or stakeholders that are involved in decision making for supply chain management (Ramzy et al., 2022).

In our study, we develop and visualize the change process as part of the product lifecycle with all entities and their interdependencies. The resulting knowledge graph can therefore address the challenges outlined above. It allows an interpretable representation of the product change process that includes data from heterogeneous sources (e.g., enterprise resource planning, master data, and product lifecycle information) or reflecting perspectives of various supply chain stakeholders. The knowledge graph supports continuous data integration and ensures that automatic updates accurately reflect the current state of the supply chain. By applying reasoning, inconsistencies in the developed knowledge graph are identified on data level. Furthermore, inconsistencies are classified according to their respective nature. Decision support to supply chain experts is examined by examining queries that detect the impact of certain elements relevant to the supply chain change management process.

5.3 Related Literature

We present a selection of related work that is relevant to our research topic. Based on the literature analysis, we specify the research gap we aim to close, which leads to the formulation of our research questions.

Chabot et al. (2014) present a method to support existing product lifecycle management solutions. It enables population of the applied knowledge base by means of semantic web. Furthermore, it provides certainty of completeness and coherency regarding the knowledge base. The approach is not yet fully utilizable since it lacks both an architecture that takes into account the entire domain and a more user-friendly interface. Moreover, the method is not dedicated to solve issues in supply chain change management tasks.

Gausemeier (2011) describes an approach for decision support in change management by introducing a respective user interface. It concentrates on product change notifications and related processes and aims for overcoming uncertainties regarding technical requirements as well as organizational relations. The developed platform generates product data and models, utilizing an open-source data management repository. Decisions to be supported include whether a product is discontinued or changed and whether the changing product needs to be produced on stock. These decisions take into account contractual details, predicted order scenarios and other variables. Applying the proposed prototype, decision processes are secured, illustrated transparently and may be reused for future projects. The conceptual approach lacks integrated visualization of relevant data and is limited to a certain set of variables and implemented processes. Moreover, it is not equipped with features to guarantee consistency among the relevant data sets.

Ye et al. (2008) present a semantic web based ontology for supply chain management in order to overcome inconsistent knowledge exchange between supply chain stakeholders. The model is constructed as generic representation of the supply chain management domain and provides visualization for enhanced usability. The approach is demonstrated by integrating it in a printer supply chain. The theoretical model distinguishes fragments of the supply chain, such as its structure, activity, resources, items and management. These are regarded as the core concepts of the ontology. The approach concentrates on knowledge exchange and illustrates representations to facilitate communication between domain experts. It however leaves room for mapping coexisting information models, is limited to a minor complex scenario and is not dedicated to provide decision support.

Jarratt et al. (2004) provide an overview of visualization techniques for product changes in design stages of the product lifecycle. The presented approaches include matrices that model component dependencies to identify risks of changes. Moreover, connection plots visualize a product with its components as nodes and arcs. The tools are further analyzed according to their capability of illustrating complexity. Although it is not considered most current research in this field, the conclusion drawn by the authors still holds as it is not entirely given that visualizations feature a seamless interactivity for the end-user.

Our analysis of recent scientific studies leads to the current open research questions that the presented knowledge graph addresses. We develop a semantic web based supply chain change management ontology that is connected to a holistic product lifecycle and supply chain representation. Although some prior research is already conducted in the field of inconsistency identification by means of semantic web, many challenges remain, such as scalability and integration of multiple relevant perspectives. To illustrate these challenges, consider a global semiconductor manufacturer with hundreds of suppliers and customers across different regions. Each tier in the supply chain generates large volumes of transactional data including orders, shipments, and returns. Moreover, the case company integrates data from various sources such as enterprise resource planning systems, supplier databases, sensors, and industry reports. Each source uses different data formats and terminologies. The large amount of data that needs to be integrated, processed, and stored in this dynamic environment changes frequently. To scale decision support for internal product change processes that includes these different sources and perspectives, the knowledge graph must handle this large amount of data without performance degradation. Hence, it requires efficient data storage solutions and processing algorithms to ensure real-time or near-real-time updates and queries. The system must support continuous data ingestion and updating mechanisms to ensure that the knowledge graph accurately reflects the current state of the supply chain. In addition, integrating heterogeneous data sources into a cohesive knowledge graph is complex. It involves aligning different data models, resolving schema mismatches, and ensuring semantic consistency across the integrated data. The knowledge graph must integrate and reconcile these diverse perspectives into a unified representation. Ideally, integrating data across different domains is supported by a comprehensive ontology that can represent the entire lifecycle and various aspects of the supply chain. For example, it handles domain-specific vocabularies, ensures interoperability, and integrates cross-functional data in a meaningful way.

In particular, our proposed approach aims for supporting the complex process of product changes in the semiconductor industry and enabling proper automation interfaces to meet the requirements of a digitally transformed supply chain. Instantiated with real-world data, the developed ontology lays the foundation for our proposed Supply Chain Change Management knowledge graph. Inconsistency identification will ensure high data and relationship quality within the developed representation. Both organizational and technical data sets are subject to integration. Furthermore, we present an easy-to-use visualization. The presented decision support solution is planned to be scalable and will support experts that are in charge of handling and implementing semiconductor product changes.

For tackling the challenges described above, our study aims to evaluate the following research questions. (1) How can we taxonomically classify and represent the supply chain change domain, while ensuring interactive utilization? We suggest a dynamic visualization containing relevant entities, relations and interfaces based on semantic web technologies. (2) Which inconsistencies exist in a complex supply chain environment among individuals and how may these be detected? We suggest to generate and integrate necessary information, identify possible inconsistencies on data level, classify them and provide methods on how to detect them. (3) Does the described approach enable decision support in terms of tracing impacts on affected elements? We suggest to ensure decentralized, seamless accessibility of correlating elements and specify which solution path is endorsed in case of an open decision. We propose a Supply Chain Change Management ontology, i.e., SCCM ontology, that establishes the base for an SSCM knowledge graph as it is instantiated with real-world data. Moreover, we equip the SCCM knowledge graph with SPARQL queries to access dynamic visualization, inconsistency identification and decision support for supply chain change management at our case company, a global semiconductor manufacturer.

5.4 Model Development

5.4.1 Method

To answer the research questions, we base our methodology on the ontology engineering procedures described by Noy and McGuinness (2001) and Hitzler et al. (2009). The initial stage for developing our SCCM ontology contains the specification of scope, application and users of the ontology as well as which questions to be answered. We use knowledge bases that serve as a foundation and contain expert treasure of experience, primarily for a better understanding of the final application. Additionally, semi-structured expert interviews, process documentations, data bases and change management tools at the case company serve as reliable information sources. Afterwards, valid and mature ontologies with similar scope are analyzed and possibly fragments of these concepts are reutilized. Within our scope, we reuse PROV-O (Provenance Ontology) and FOAF (Friend Of A Friend). PROV-O serves as part of the core classes, yet its properties are not integrated due to formal reasons with regard to the naming convention (Lebo et al., 2013). FOAF provides an interface to a social graph and for implementing personal data (Brickley & Miller, 2015). In a next step, we specify key terms that are important for problem solving or understanding the scope-specific concept. We include basic supply chain elements, change management concepts, stakeholders and the process landscape. Being an iterative process, some terms disappear or newly emerge and their importance perception incrementally changes as the development proceeds. In order to define taxonomy, properties and restrictions, we follow a mixed approach, iteratively moving top-down with very basic core vocabulary, continuing bottom-up with increasingly detailed concepts and again observing it from a top level. In the last step it is required to create instances that are regarded as minimum unit within the ontology for a given application. Individuals need to be assigned to a class and may be given specific values. For the SCCM ontology, we distinguish two flavors of individuals. First, we manually generate rather static individuals that are needed for a frictionless operation, such as milestones or process steps. Second, we transform data sets to individuals with a variety of data properties and values, which is supported by tools.

5.4.2 Modelling Environment

As development environment, we initially apply the open-source platform *Protégé* in order to define taxonomy and properties. One challenge that arises in ontology development is that data integration and processing performance can be limited. Therefore, we then use the Java based *Apache Jena* as framework, utilizing application programming interfaces (API) such as RDF API and SPARQL API as well as the SPARQL server *Fuseki*. For visualization, we use the *Protégé* plug-in *OntoGraf* during the verification phase with subject matter experts. However, this visualization approach allows only limited functionalities by users. We aim to provide experts with a more interactive decision support that can be tailored to their needs and enables more sophisticated features. Hence, for enhanced usability, we visualize it via the graph platform *Neo4j* that furthermore enables interfaces for data modeling, data import and queries. For reasoning purposes, we chose *Pellet* with its capabilities when facing advanced OWL versions, or when Description Logic or more exotic XML datatypes are involved. Since conjunctive queries may be of interest in the future and many of the datatypes may be analytically reasoned according to their pattern (nomenclature, identification), this requirement is important in our case. Otherwise, the decision support tool would only support limited scalability toward other datatypes that are potentially present in the many heterogeneous data sources. For this project, *Pellet* is implemented in both *Jena* and *Protégé*. Although possible to be used in *Neo4j* directly, it is quite cumbersome, and thus the OWL file is first reasoned in a development environment and then exported to *Neo4j*. Additionally, we use SHACL (Shapes Constraint Language) as rule language and validation tool for RDF graphs. It includes rules as well as shapes for nodes and properties. One major challenge in many industry-related ontology development projects arises at the stage of instantiation. From various data sources across the company, sometimes only unclean data is available for integration. Hence, before deploying the relevant data sets, we clean up their possibly unclean and inconsistent nature by using the *OpenRefine* web application. Afterwards, we integrate the data sets by using the *Protégé* plug-in *Cellfie* to match instances, data properties and values, respectively.

5.4.3 Model Description

We introduce the core taxonomy of the ontology in Fig. 5.1, showing the top-level concepts of the SCCM ontology. Parts of it are reused from PROV-O as indicated by a prefix, whereas other parts are developed based on the engineering procedure described above. Each of the headed boxes contains itself a sub-ontology with its respective sub-classes as listed in the box. Directed edges represent the object properties that connect classes, red boxes represent datatypes. Similarly, we develop the taxonomies for the four core classes and their relationships. Furthermore, we

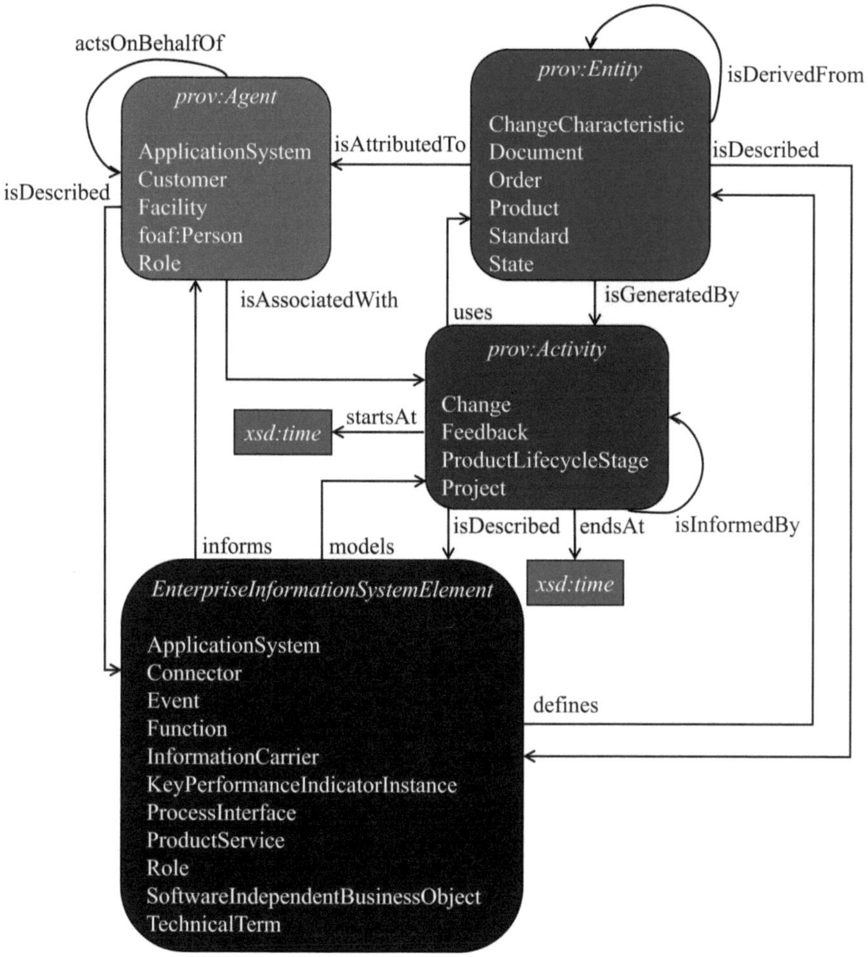

Fig. 5.1 Core concepts of the developed SCCM ontology

structure the affiliation of data points and the relationship between data sets, which equals to the object properties defined earlier. Some of the data points equal each other and are thus subject to be part of a rule set that is to be defined. Rule sets are inevitable to provide a profound linkage between the different data property resources. The linked data types are selected since each enables a unique assertion on individual relations. Here, we define taxonomical rules to cross-classify the defined entities and data based rules that link data property resources to each other. Figure 5.2 depicts the dynamic holistic visualization of the SCCM ontology, utilizing the *Neo4j* platform.

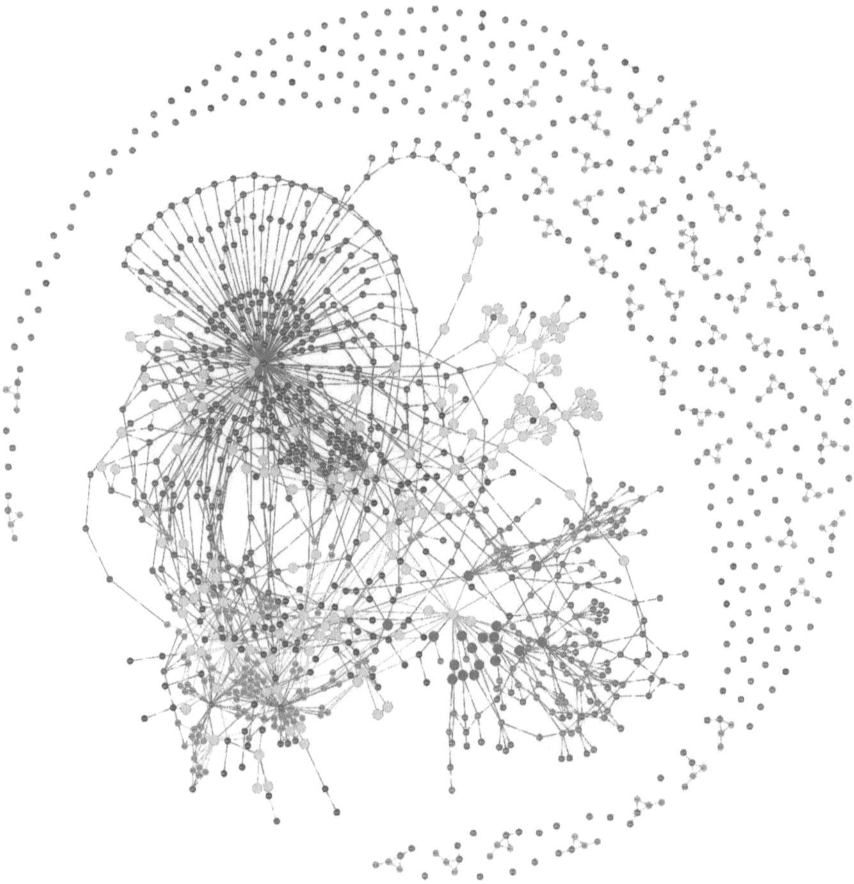

Fig. 5.2 Holistic visualization of SCCM ontology

5.5 Results

We define test scenarios to show the performance of the SCCM ontology with regard to the three aspects of (1) visualization, (2) inconsistency detection, and (3) decision support that match with the three research questions. For evaluating these aspects, we select a random data set that in our case contains supply chain change data as of April 10, 2019. For the modeled SCCM ontology solution, we integrate a manually randomized selection of data points for each of the data property domains. We extract current working routines and respective drawbacks from domain expert interviews. The scenarios S we define simulate frequent tasks of stakeholders involved in the supply chain change process. S_1 is from a supply chain planner perspective, who is interested in which products correspond to product change notification projects that are not yet archived and hence still need attention. S_2 is from a production planner perspective, who is interested in which customers and respectively which products are affected, having a certain last delivery date. S_3 is from a product owner perspective, who is interested in customer logistics manager's locations to take care of customers that have not yet agreed on a product change notification. We then provide queries to provide information in the scenarios and compare it to the current solutions. To demonstrate our solution, we present queries that address the three scenarios and answer our research questions, i.e., we showcase visualization capabilities for S_1, inconsistency identification for S_2, and decision support S_3. Nevertheless, more scenarios can be defined and likewise all three capabilities of the SCCM knowledge graph can be shown for multiple scenarios.

5.5.1 *Visualization*

Visualization is structured in a prior- and post-development part and is validated from a user perspective. Prior to the development, we interrogate the perception of current available tools, namely the usability and visualization quality of 2D tables and process frameworks. For example, the query for S_1 is provided in Listing 5.1.

Listing 5.1: Visualization for S_1

```
MATCH ( pcn: ns0_PPCN) - [: n s 0 _ d e s c r i b e s] -> ( c:
ns0_Change) - [: n s 0 _

 a f f e c t s] -> ( p: ns0_Product)
WHERE ( pcn. ns0_PCNIsArchived) = false
RETURN p. ns0 _hasSPNumber AS SPNumber, pcn. ns0_hasDocNo AS
PCN_DocNo,
  c. ns0_hasChangeID AS ChangeID
```

Both the existing tool and our proposed solution are capable of a 2D table representation containing the desired information by setting filters or sorting for comparison preferences. We present such query results in Fig. 5.3. While the current solution is more intuitive, for more complex queries, we observe benefits of the SCCM ontology due to increased performance and flexibility. Displayed as graph-like triple structure, the suggested visualization allows for dynamic user-interaction with requested data points. The dynamic triple visualization is depicted in Fig. 5.4. Enhanced information on the queried nodes can also be included in a textual representation that allows for machine processing and interpretation. Another textual visualization is depicted in Fig. 5.5. Summarizing the visualization validation, both approaches (the existing tool at the case company and our proposed solution) are

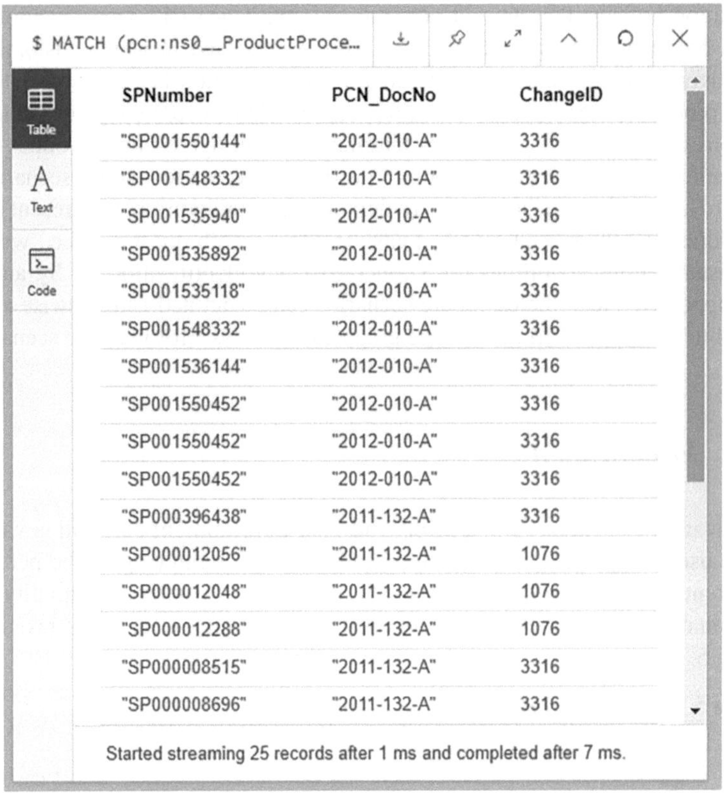

$ MATCH (pcn:ns0__ProductProce...

SPNumber	PCN_DocNo	ChangeID
"SP001550144"	"2012-010-A"	3316
"SP001548332"	"2012-010-A"	3316
"SP001535940"	"2012-010-A"	3316
"SP001535892"	"2012-010-A"	3316
"SP001535118"	"2012-010-A"	3316
"SP001548332"	"2012-010-A"	3316
"SP001536144"	"2012-010-A"	3316
"SP001550452"	"2012-010-A"	3316
"SP001550452"	"2012-010-A"	3316
"SP001550452"	"2012-010-A"	3316
"SP000396438"	"2011-132-A"	3316
"SP000012056"	"2011-132-A"	1076
"SP000012048"	"2011-132-A"	1076
"SP000012288"	"2011-132-A"	1076
"SP000008515"	"2011-132-A"	3316
"SP000008696"	"2011-132-A"	3316

Started streaming 25 records after 1 ms and completed after 7 ms.

Fig. 5.3 Extract of SCCM knowledge graph based query results for S_1

capable of displaying the desired information. Filtering, sorting and comparison of values is given for both approaches by following different paths. This evaluation is mainly dependent on personal user preferences. However, the SCCM approach extends beyond information retrieval and enables users to interact dynamically with the search results. Furthermore, it provides machine processable and interpretable data as triples.

5.5.2 *Inconsistency Identification*

Inconsistencies are identified on a data level. This presented validation part is considered as a use case scenario to an inconsistency classification framework that is presented by Basirati et al. (2018). We base our inconsistency classification on their suggestion to distinguish groups of notational, convention-based, system-specific and domain-specific inconsistencies as well as their respective relation as additional

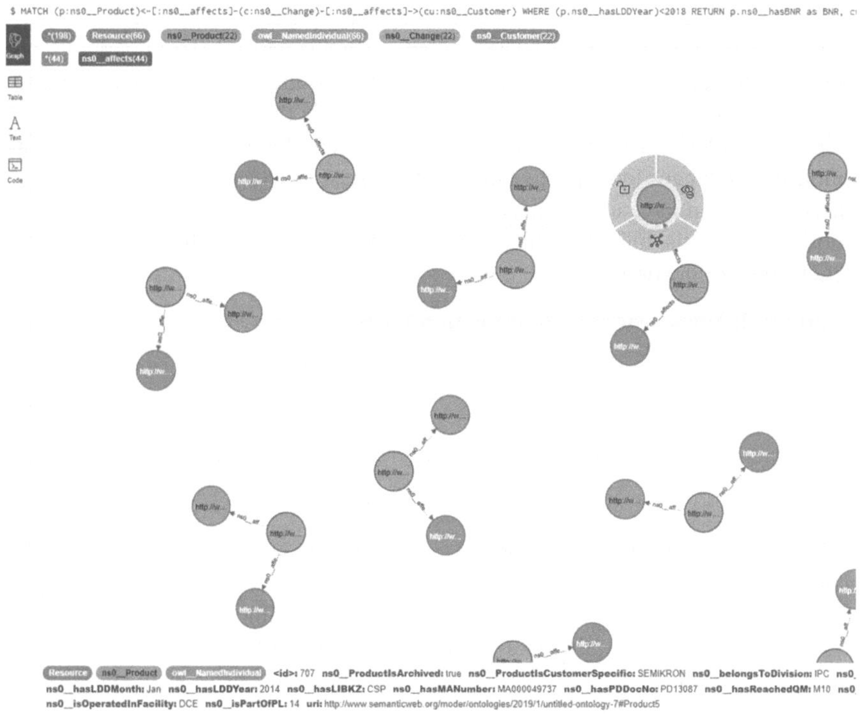

Fig. 5.4 Extract of dynamic SCCM knowledge graph based visualization with query results for S_2

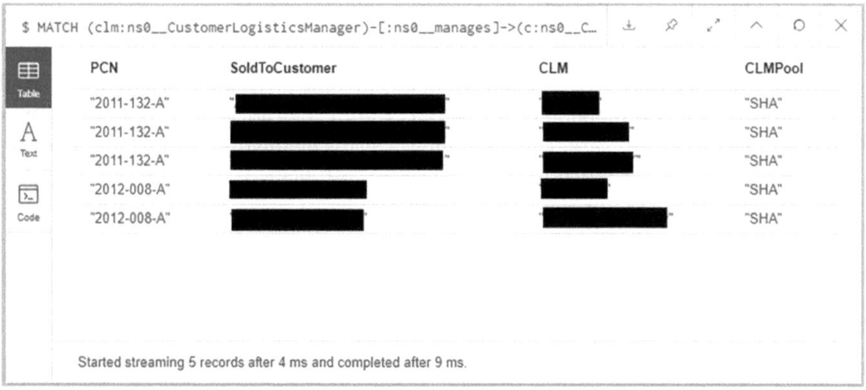
$ MATCH (clm:ns0__CustomerLogisticsManager)-[:ns0__manages]->(c:ns0__C...

PCN	SoldToCustomer	CLM	CLMPool
"2011-132-A"	[redacted]	[redacted]	"SHA"
"2011-132-A"	[redacted]	[redacted]	"SHA"
"2011-132-A"	[redacted]	[redacted]	"SHA"
"2012-008-A"	[redacted]	[redacted]	"SHA"
"2012-008-A"	[redacted]	[redacted]	"SHA"

Started streaming 5 records after 4 ms and completed after 9 ms.

Fig. 5.5 Extract of SCCM knowledge graph based query results for S_3. Customers and CLM pool members are anonymized to respect data privacy

dimension: existence, equivalence, correspondence, refinement, satisfaction (Basirati et al., 2018). We define case-specific samples for each group and provide a query for detecting a commonly critical inconsistency within the presented model. It returns all individuals that are not meeting a certain system-specific equivalence condition. As mentioned above, every available product is signified by its sales name. However, sometimes various designations appear in different storage locations. In the dataset on hand, the product change notification containing a product link and the product having a sales name shall contain the same value. This requirement is considered crucial since deviations may cause fatal results, e.g., when communicating changes to respective customers. The query shown in Listing 5.2 finds all products that have a certain sales name and are contained in a product change notification, yet the linked product does not correspond to the indication in the product change notification.

Listing 5.2: Inconsistency Identification for S_2

```
MATCH ( pcn: ns0_PPCN) – [] – (: ns0_Change) – []
– ( p: ns0_Product)
WHERE pcn. ns0_containsProductLink <> p. ns0_hasSalesName
RETURN pcn. ns0_containsProductLink, p. ns0_hasSalesName
```

The query results in a table showing the product links and sales names that do not match and is therefore an exemplary approach to detecting inconsistencies, applicable to the classification described above. One possible result is presented in Fig. 5.4. In tabular form, the query identifies three records that are inconsistent, as presented in Table 5.1. With the results on hand, the recognized data errors can be investigated further and consequently be cleaned.

5.5.3 *Decision Support*

For examining the third research question, an exemplary decision problem is to find the path that is recommendable in case a certain event appears or optimization is required. Again, *Neo4j* serves as user interface for evaluation, proposing decision supporting queries for stakeholders. The proposed decision support affects above mentioned S_3 scenario, where a customer has not yet agreed on a product change notification. The result is a recommendation for which products a product change notification agreement is meaningful to trigger based on the motivation for a

Table 5.1 Query results for inconsistency identification in S_2

ProductLink	SalesName
“FP100R12KT4_B11”	“AUIRS4427STR”
“FP75R12KT4_B11”	“AUIRS2012STR”
“FP100R12KT4_B11”	“AUIRS4427STR”

product change. Furthermore, it proposes a set of customer logistic managers that may support the agreement request, based on locations of both the respective customer and customer logistics manager. The query is presented in Listing 5.3 and the respective results are shown below in Table 5.1.

Listing 5.3: Decision Support for S_3

```
MATCH ( clm: ns0_CustomerLogisticsManager) - [: ns0_manages] ->
 ( c: ns0_Customer) - [: ns0_isNotifiedBy] -> ( pcn: ns0_PPCN)
MATCH ( support: ns0_CustomerLogisticsManager) - [: ns0_
manages] ->
 (cu: ns0_Customer)
WHERE pcn. ns0_isAgreed <> " false"
AND pcn. ns0_containsMotivation contains " cost"
AND c. ns0_hasSoldToName contains " Example Enterprise DUS"
AND support. ns0_arisesFromCLMPool = 'WAR'
RETURN "YES !" AS HelpRequired ?, pcn. ns0_hasDocNo AS PCN, c.
 ns0_hasSoldToName AS Customer, clm. ns0_arisesFromCLMPool AS
 CustomerLogisticsManager, collect ( support. ns0_hasCLMName) AS
 Possible Support
```

Above query supports the decision on how to proceed with the outstanding agreement of a product change notification. It tells that a trigger is economically beneficial if a cost reduction is part of the change motivation. Moreover, it proposes a set of customer logistics managers that might support the triggering process according to their location, since Example Enterprise DUS implies the company's site Dusseldorf and WAR implies Warstein, which are both German towns and relatively close. We present the query results for S_3 in Fig. 5.5. Similar to the notion in our first research question, we are able to add more filters to the query, such as product properties or types of relationships. Moreover, it could also be enhanced with more contextual information that may seem relevant for the respective decision problem. The query results in the output shown in Table 5.2. The results may be used to identify CLM pool members at the case company who can efficiently support the customer's outstanding product change notification acceptance. In order to reflect the expressiveness of the validation procedure we provide answers to above defined research questions. The sample size of the conducted experiment is rather

Table 5.2 Query results for decision support in S_3. CLM pool members are presented as "#n" to respect data privacy

Trigger	PCN	SoldToCustomer	CLM		Support	
"YES!"	"2012-010-A"	"Example Enterprise DUS"	"WAR"	"#1",	"#7", "#23", "#25",	"#27"

small, hence it is not considered representative. With regard to (1) visualization, the solution we present shows a representation of the supply chain change domain in the form of a semantic web based knowledge graph. Our taxonomy takes into account both technical and organizational aspects and is built up as an ontology. By means of using a semantic web (OWL) based description of entities and their relations, we deduct a net-like illustration that is conceived as knowledge graph. In order to add interaction features for human users, we illustrate the model in tools such as *Neo4j* and *OntoGraf* as interface. While the latter ensures basic understanding of the taxonomy that is important for domain experts, *Neo4j* provides a virtual environment at high expressivity. With regard to (2) inconsistency identification, the nature of detected inconsistencies matches with the classification scheme proposed by Basirati et al. (2018). In order to identify inconsistencies on data level, queries are inevitable. However, it is necessary to integrate data into the knowledge representation in order to receive meaningful results of the queries. For this purpose, existing data sets with various data points are transformed and integrated. One exemplary query is presented, including its results. With regard to (3) decision support capabilities, we find that decision support requires predefined rules and constraints that provide more meaning to the sheer representation and reasoning approaches. We show exemplary that for one explicitly modeled decision a decision support may be implemented. The decision support is based on the development of a taxonomy and implemented data access. However, it is regarded as a prototypical demonstration rather than an industrially applicable tool. For this purpose, more data that go beyond the described use case and more interrelations to other enterprise models are necessary.

5.6 Conclusions

5.6.1 Discussion

This work provides a detailed documentation of a semantic web based knowledge graph that represents the supply chain change management process. Objects and processes thus become understandable and matchable on intra- and inter-organizational levels. It serves as use case scenario and accomplishes certainty that this approach is beneficial for users. The model has standardized interfaces that enable integration in adjacent supply chain representations. Thus it is scalable towards similar use cases towards other enterprise models. The proposed knowledge graph furthermore supports supply chain change experts by means of information extraction through complex queries as well as deduction of implicit knowledge by reasoning mechanisms. As such, it constitutes a step towards digitalization as the SCCM knowledge graph provides a tool for automating internal processes that are currently time-consuming.

Additionally, the knowledge graph's ability to facilitate real-time data integration and analysis empowers organizations to respond to changes in the supply chain, improving overall agility and resilience. Using semantic web technologies, the model fosters enhanced collaboration and transparency among different stakeholders, ensuring that decisions are informed by a holistic view of the supply chain and its change management. In addition, the inherent ability of the knowledge graph to evolve and adapt to new data sources and emerging dynamics makes it a sustainable solution for long-term supply chain management. This adaptability is crucial in an era where supply chains are increasingly dynamic, necessitating flexible and robust digital infrastructures. Ultimately, this work contributes to the broader goal of achieving a digitalized and interconnected supply chain ecosystem, paving the way for more efficient and transparent supply chain practices.

5.6.2 *Summary of Findings*

We introduce the development of a scalable knowledge graph that is considered with the supply chain change management process. Detected requirements within the use case domain and investigations of capabilities in related research lead to the need of our proposed solution. We model the knowledge graph according to the circumstances discovered at our case company, a global semiconductor manufacturer, so that it can serve as a *lingua franca* within its supply chain domain. The procedure of development gets introduced in detail. Moreover, we present utilized tools and technological concepts for practical implications. The SCCM ontology is based on semantic web technologies, which ensure readability and interpretability among various users and machines. We thus consider having a common understanding of the domain as one key benefit of our study. By integrating and normalizing real data in the model, the proposed SCCM knowledge graph depicts relevant entities for this use case, including basic individuals and dependencies throughout different layers. The definition of rule sets enhances the expressivity of the knowledge graph. Its quality is verified and further evaluated with regard to three research questions that target (1) visualization, (2) inconsistency identification and (3) decision support. We show that an interactive visualization is possible and that it may have advantages over the current solution. We classify potential inconsistencies based on Basirati et al. (2018). Moreover, we provide a query option to detect such inconsistencies. Applying the model as decision support solution additionally contributes to the validation procedure.

5.6.3 Managerial Implications

With our study we address current issues in industry. Long production ramp up times for changed products can be shortened by providing a transparent process model. Difficulties in collaboration and knowledge sharing between interdisciplinary teams is addressed by an extensive knowledge base. The SCCM knowledge graph thus bridges the digital divide that leads to information asymmetries across stakeholders. The aim for flexible data integration increases complexity and leads to highly distributed and poorly clustered data sets. These obstacles are addressed by integrating necessary knowledge, enabling its classification and automating information discovery. The proposed decision support facilitates actions in complex environments and will accelerate tardy reactions on disruptive events. The weak human-machine interaction is improved by a *lingua franca* that both parties are able to understand and interpret. Since digitalization leads to increasing volume, variety, and dynamics of data, this model enables knowledge extraction that is independent of data size. Furthermore, by enhancing visibility and traceability across the supply chain, managers can make more informed decisions. This transparency is crucial to identify potential bottlenecks in supply chain change management and to proactively address them before they escalate to significant issues. Our proposed SCCM knowledge graph facilitates real-time data access and analysis, which improves responsiveness to market changes, supplier issues, or internal value chain disruptions, leading to more agile and competitive supply chain change management. Using the capabilities of the SCCM knowledge graph, managers can foster a culture of continuous improvement and transparency, ultimately driving long-term success and efficiency in their operations.

5.6.4 Limitations and Future Research

The model we describe is subject to further research, as it requires improvements to establish a real industry application. In order to ensure broad-scale trust to the concept, it is necessary to obtain validation by experts from different enterprises or subject domains. The implementation of novel technologies that enhance trust in the concept is of further interest. For instance, recently emerging blockchain technologies may add elements that improve trust of the user by finding a consensus about certain naming conventions or recommended decisions. Moreover, establishing relations to different parts of the company may lead to a more powerful exploitation. An in-depth connection to the manufacturing domain representation may guarantee automated control of production strategies taking into account current customer release decisions. For this need, data sets require automatic ontology connection and orchestration, guaranteeing dynamic real-time processing. More research is proposed on automatic detection of inconsistencies, which leads to increased information quality. Furthermore, investigating a larger amount of scenarios may provide more certainty to decision support solutions. This extension may be further supported by more complex rule based reasoning and semantic search approaches. Future application areas consist of product tracking and tracing after sales takes place, among others. Such capabilities open up opportunities to tailor marketing efforts with regard to expected application cases. Moreover, in situ prediction of maintenance operations may be enabled through a comparison of ambient conditions to the specified product working range.

References

Basirati, M. R., Zou, M., Bauer, H., Kattner, N., Reinhart, G., Lindemann, U., Böhm, M., Krcmar, H., & Vogel-Heuser, B. (2018). Towards systematic inconsistency identification for product service systems. In *DS 92: Proceedings of the DESIGN 2018 15th international design conference* (pp. 2811–2820).

Brickley, D., & Miller, L. (2015). FOAF vocabulary specification 0.99. 2014. Namespace Document. Retrieved November 23, 2018, from http://xmlns.com/foaf/spec/.

Chabot, B., Gautreau, P., & Sommacal, B. (2014). How to improve PLM approach efficiency based on knowledge engineering, knowledge management and semantic web technologies domains? In *IFIP international conference on product lifecycle management* (pp. 365–376). Springer.

Eigner, M., & Stelzer, R. (2009). *Product lifecycle management: Ein Leitfaden für product development und life cycle management*. Springer.

Gausemeier, J. (2011). Prozessorientierte entscheidungsunterstützung im Änderungswesen. *Zeitschrift für wirtschaftlichen Fabrikbetrieb, 106*(6), 389–390.

Hitzler, P., Krotzsch, M., & Rudolph, S. (2009). *Foundations of semantic web technologies*. Chapman and Hall/CRC.

Jarratt, T., Keller, R., Nair, S., Eckert, C., & Clarkson, P. J. (2004) Visualization techniques for product change and product modelling in complex design. In *International conference on theory and application of diagrams*. Springer (pp. 388–391).

JEDEC. (2016). *J-STD-046: Customer notification standard for product/process changes by electronic product suppliers*. Standard, JEDEC Solid State Technology Association.

Lebo, T., Sahoo, S., McGuinness, D., Belhajjame, K., Cheney, J., Corsar, D., Garijo, D., Soiland-Reyes, S., Zednik, S., & Zhao, J. (2013). *Prov-o: The prov ontology*. World Wide Web Consortium.

Mujica, G., Portilla, J., & Riesgo, T. (2017). Deployment strategies of wireless sensor networks for IoT: Challenges, trends, and solutions based on novel tools and HW/SW platforms. In G. Keramidas, N. Voros, & M. Hübner (Eds.), *Components and services for IoT platforms* (pp. 133–154). Springer.

Noy, N. F., & McGuinness, D. L. (2001). *Ontology development 101: A guide to creating your first ontology*. Stanford University.

Ramzy, N., Auer, S., Ehm, H., & Chamanara, J. (2022). Sens: Semantic synthetic benchmarking model for integrated supply chain simulation and analysis. *arXiv Preprint*. https://doi.org/10.48550/arXiv.2205.06484

Scheer, A.-W., Boczanski, M., Muth, M., Schmitz, W.-G., & Segelbacher, U. (2005). *Prozessorientiertes product lifecycle management*. Springer.

Ye, Y., Yang, D., Jiang, Z., & Tong, L. (2008). Ontology-based semantic models for supply chain management. *The International Journal of Advanced Manufacturing Technology, 37*(11), 1250–1260.

Part II
Blockchain Applications for Enhancing Observability and Traceability

Chapter 6
Blockchain-Based Smart Contracting Services for Digital Supply Chain Operations in Crowdsourced Manufacturing

Xuejian Gong, Mulang Song, Roger J. Jiao, and Linda Zhang

Abstract Enabled by digital supply chain operations, crowdsourced manufacturing leverages cloud platforms to coordinate a distributed network of manufacturers for efficient product fulfillment. Smart contracting and crowdsourcing services are pivotal in managing manufacturing and logistics tasks within digital supply chains. However, the decentralized nature of these processes presents significant challenges in handling complex, real-time data streams. This chapter introduces an innovative smart contracting service system designed to support stakeholders across the crowdsourcing process in making informed, real-time decisions. By utilizing blockchain-based smart contracts, the system ensures secure, transparent, and efficient management of distributed data, real-time monitoring, and supply chain information sharing. A three-layer digital supply chain system architecture is proposed to facilitate this environment, supported by a distributed data streaming model built on IoT infrastructure. A case study in the tank trailer manufacturing industry demonstrates the system's effectiveness and highlights its potential to enhance operational performance in crowdsourced manufacturing environments.

Keywords Blockchain · Smart contracting · X-as-a-service · Crowdsourcing · Manufacturing logistics · Digital supply chain

X. Gong · M. Song · R. J. Jiao (✉)
School of Mechanical Engineering, Georgia Institute of Technology, Atlanta, GA, USA
e-mail: rjiao@gatech.edu

L. Zhang
IÉSEG School of Management, Paris-Lille, France

Y. Khojasteh, S. M. M. Jahan Abadi (eds.), *Emerging Technologies in Supply Chains*, International Series in Operations Research & Management Science 373, https://doi.org/10.1007/978-3-032-01218-0_6

6.1 Introduction

Aligned with the X-as-a-Service vision in the era of digital transformation (Jiao et al., 2022), crowdsourced manufacturing services offer great market potential across manufacturing supply chains and across various sectors, including sustainable manufacturing and consumer goods (Grabowska, 2020). Crowdsourced manufacturing operates through digital supply chain operations, providing comprehensive information services for efficient product fulfillment via cloud-based platforms that coordinate multiple manufacturer crowds (Gong, 2021). Smart contracting and crowdsourcing services are essential for digital manufacturing supply chains to manage various manufacturing and logistics tasks, including resource allocation, process tracking, data management, and supply contracting (Gong et al., 2021). However, digital supply chains face increasing complexity, particularly in managing intricate manufacturing and logistics processes and data streams, which are further complicated by manufacturers' dispersed nature within the supply chain (Leukel et al., 2011). Several critical challenges have emerged in the implementation of crowdsourced manufacturing.

Despite its potential, crowdsourced manufacturing faces significant challenges that hinder broader adoption and operational efficiency. These challenges demand robust mechanisms to ensure data integrity, trust verification, and effective coordination across distributed operations. Similar to untrusted business processes studied by Weber et al. (2016), crowdsourced manufacturing requires reliable verification of transactions and process execution when coordinating multiple stakeholders. Distributed networks face data management challenges (Li, Barenji, et al., 2018), crowdsourced manufacturing needs secured accessible systems for managing complex manufacturing and logistics data streams across its network of participants. The decentralized nature of operations introduces challenges in supply chain sustainability and transparency identified by Kouhizadeh and Sarkis (2018), making it essential to develop protocols that can verify process execution while protecting stakeholder interests and data security. Monitoring and maintaining quality standards across distributed operations pose additional challenges. Qu et al. (2016) emphasize the importance of IoT-based systems for real-time production monitoring. Similarly, crowdsourced manufacturing requires comprehensive monitoring capabilities to uphold operational efficiency and quality standards across geographically dispersed locations.

To tackle these challenges and harness the full potential of crowdsourcing, it's imperative to develop strategic solutions for smart contracting services. This involves implementing advanced data management systems capable of handling the complexities of process data streams effectively. Leveraging cutting-edge technologies like artificial intelligence and blockchain can significantly contribute to ensuring data accuracy, security, and real-time analysis. Furthermore, fostering collaboration and communication among dispersed manufacturers and logistics providers is crucial. Establishing a networked environment that facilitates seamless information exchange and supports collaborative decision-making will undoubtedly boost the efficiency of crowdsourcing and streamline manufacturing operations. In the age of digital transformation, digital supply chains hold immense promise, fundamentally changing how manufacturing tasks are executed. By addressing data management challenges and promoting collaboration among manufacturers, smart contracting services can pave the way for a more agile, efficient, and interconnected manufacturing ecosystem.

Blockchain technology has emerged as a potent solution for establishing highly distributed security databases that contain immutable and tamper-proof records, public ledgers, and digital events shared among multiple entities. This technology relies on smart contracting to forge cooperative relationships between stakeholders (Crosby et al., 2016). The process involves encrypting information into blocks using hash functions and disseminating these blocks to all network participants, ensuring secure key protection for process data while enabling accessibility for every decision-making agent within the supply chain.

The application of blockchain extends to connecting engineering software and facilitating data sharing among a network of companies (Xu et al., 2016). Decision agents can effectively monitor and verify untrusted processes in collaborative scenarios based on manufacturing activity models (Weber et al., 2016). Through the integration of cloud database technology, blockchain enables query and retrieval services for various manufacturing data sources, including machine-level data, manufacturing resources data, production data, and logistics data (Li, Barenji, et al., 2018). Consequently, implementing blockchain in the product fulfillment process acts as a catalyst, propelling manufacturing into a smart and service-oriented direction (Chen et al., 2022).

This chapter proposes the development of blockchain-based smart contracts and services to manage distributed data, thereby enabling manufacturing crowdsourcing. The subsequent sections delve into critical technical aspects of the blockchain-based information service system for manufacturing crowdsourcing. A three-layered digital supply chain system architecture is presented in Sect. 6.3 to facilitate a crowdsourced manufacturing cloud environment. Section 6.4 discusses the implementation of smart contracting technologies, laying the foundation for cooperative relationships among decision agents in crowdsourcing. Section 6.5 elaborates on the development of a distributed data streaming architecture that organizes crowdsourcing product fulfillment data based on IoT hardware. To validate the functionality of the proposed information service system and assess the potential for operational excellence in a crowdsourcing environment, a case study in the tank trailer manufacturing industry is presented in Sect. 6.6. Finally, the chapter concludes by reflecting on the proposed methods and information service system, summarizing contributions, and suggesting future development opportunities.

Overall, the chapter focuses on leveraging blockchain-based smart contracting services to revolutionize data management in manufacturing and supply chain information sharing, thereby facilitating the shift towards a service-oriented and intelligent manufacturing landscape. By exploring various technical aspects and conducting a practical case study, the chapter aims to advance the understanding and implementation of blockchain-based information service systems for manufacturing crowdsourcing.

6.2 Related Work

6.2.1 Crowdsourcing of Manufacturing for Cyber-Physical Product Fulfillment

As a visionary emerging technology, smart manufacturing is reshaping the manufacturing industry landscape through the integration of sensors, computing platforms, communication technologies, and data-intensive modeling, as well as predictive engineering (Kusiak, 2018). This transformation is driven by the evolution of the knowledge exchange marketplace, the emergence of sharing economies on the manufacturing shop floor, and the growing trend towards democratization and openness. Smart manufacturing is distinguished by its decentralized, service-oriented, and platform-based nature (Kusiak, 2019).

Originating from network manufacturing, which relied on centralized crowdsourcing, open manufacturing embraces a philosophy of decentralization. It leverages blockchain and edge computing technologies to establish a cross-enterprise framework for sharing knowledge and services (Agostinho et al., 2016; Li, Wang, et al., 2018). From a supply chain perspective, open manufacturing untangles the design, logistics, and service layers from physical assets (Kusiak, 2020). To achieve this capability, open manufacturing enterprises are poised to adopt the X-as-a-service model, where X represents various facets such as manufacturing, supply chain, and logistics. Several technical challenges have emerged as significant barriers to achieving a fully open manufacturing environment, such as crowdsourcing contract design, reconfiguration of diversified supply chains, and the development of distribution coordination mechanisms.

Additionally, the influence of social media and the Internet is reshaping the relationship between consumption and manufacturing in industries across four critical dimensions: production socialization, the shifting roles of consumers, the driving force behind production innovation, and the rapid dissemination of information through social media channels (Hamalainen & Karjalainen, 2017; Jiang et al., 2016). Within the realm of this novel manufacturing paradigm, a substantial body of research has emerged, advancing the discourse on social manufacturing. This research encompasses a range of topics, including the development of blockchain-based tracking systems for self-organizing processes (Leng et al., 2019), RFID-based execution systems designed for inter-enterprise monitoring and dispatching (Ding et al., 2016), the formulation of socialized production network generation frameworks (Jiang & Ding, 2018), and the establishment of coordination mechanisms between outsourcers and suppliers (Guo & Jiang, 2019; Leng et al., 2017), among others. These research endeavors collectively advocate the idea that small and medium-sized service-oriented enterprises can aggregate themselves into various types of horizontal manufacturing communities, thereby enhancing their bargaining power and mutual profits through initial clustering and self-organization (Ding et al., 2016). However, it is worth noting that current research in this domain has yet to fully explore the intricacies of coordinating manufacturing networks, establishing collaborative-negotiation contracts among stakeholders, and conducting evaluations for solution validation and quality control.

Cloud manufacturing builds upon the successful implementation of cloud computing, where diverse resources and capabilities are intelligently integrated into the broader Internet and managed automatically using IoT technologies (Tao et al., 2014). In this manufacturing landscape, users search for and access qualified manufacturing cloud services from relevant clouds based on their specific requirements. They then organize these services into a virtual manufacturing environment or solution to complete their manufacturing tasks (Tao et al., 2011). While researchers in cloud manufacturing have focused on cloud-based technologies, limited attention has been given to the development of manufacturing networks tailored to specific markets and cooperative architectures for different business entities involved in the product fulfillment workflow. The reconfigurability of digital supply chain system plays a pivotal role in enabling the swift and responsive construction of a product fulfillment network (Chida et al., 2019). This dynamic production network not only enables the creation of highly personalized product families (Tan et al., 2017) but also facilitates efficient sharing of manufacturing resources among federated production networks (Kadar et al., 2018).

The evolution from network manufacturing to crowdsourced manufacturing represents a shift from centralized to decentralized coordination approaches in smart manufacturing environments (Gong et al., 2021). Crowdsourced manufacturing leverages cyber platforms to coordinate design, production, and delivery tasks through open calls, building on the X-as-a-service model established in open manufacturing paradigms (Jiao et al., 2024). This platform-driven model enables collaborative manufacturing and service delivery while connecting distributed manufacturing resources through open calls (Song et al., 2024).

6.2.2 Digital Supply Chain Transformation and Logistics-as-a-Service

The trend of digitalization has revolutionized supply chain management, with technologies like RFID, GPS, and IoT being integrated into all aspects of operations, leading to significant improvements in cost and time efficiencies (Büyüközkan & Göçer, 2018; Huddar et al., 2017). Companies such as Amazon, Alibaba, and BMW are heavily investing in digitalizing their supply chains, leveraging technologies like digital mobility labs, e-freight, robotics, drones, and asset-light delivery services (Iddris, 2018; Whu, 2015). Digital supply chains are described as inter-organizational systems that digitize transaction and collaboration processes with supply chain partners (Xue et al., 2013). Supply chain viability plays a partial mediating role in digital supply chain practices, with digital technology orientation and digital supply chain platforms moderating supply chain viability (Yu et al., 2023).

The concept of a digital supply chain twin provides a computerized model representing real-time network states to manage disruption risks effectively (Ivanov & Dolgui, 2021). Unlike traditional digital twins, the supply chain digital twin is a spatiotemporal dynamic system, posing technological challenges in accurately capturing the complexity and dynamics of the entire supply chain (Ivanov et al., 2022; Zhang et al., 2021). Integration of data from various sources is crucial for creating a comprehensive model reflecting real-world supply chain behavior.

Cloud service-oriented computing revolutionizes supply chains by offering Logistics-as-a-Service (LaaS), allowing companies to outsource segments of their supply chain to third-party providers (Ivanov et al., 2022). LaaS provides instant access to expertise and infrastructure while mitigating the costs and risks associated with vertical supply chain development. Additionally, LaaS enables companies to optimize profits by leveraging excess capacity and offering it on the open market (Giannakis, 2011). Despite implementation challenges, LaaS has proven effective in driving company growth, reducing costs, and harnessing advanced technology across industries.

LaaS providers leverage advanced technologies like artificial intelligence, machine learning, and blockchain to enhance supply chain visibility, traceability, and collaboration (Leukel et al., 2011). Similar to third-party logistics providers, LaaS enables businesses to access third-party technology platforms. Cloud-based software is integral to the LaaS model, seamlessly fitting into operations and offering enhanced visibility. LaaS providers often utilize open cloud-based plug-and-play platforms to meet customer requirements for supply chain visibility, execution, and planning. Additionally, nearly all supply chain functions, including procurement, warehousing, distribution, fulfillment, and transportation, can be outsourced as on-demand services. The LaaS model allows companies with excess capacity to monetize it on the open market, maximizing profits (Maas et al., 2014).

6.2.3 Blockchain-Based Information Service Systems

The recent advancements in smart sensing and industrial IoT have ushered in a multitude of data sources that allow for comprehensive tracking of a product's lifecycle and facilitate remote quality checks (Tian, 2016). Leveraging low-latency networks and ubiquitous connectivity, real-time data on material flow within the factory can be seamlessly streamed to cloud databases for logistical coordination. This not only enhances equipment utilization rates but also smooths delivery peaks (Qu et al., 2016). Additionally, IoT-enabled manufacturing systems face interoperability challenges due to isolated data silos. Blockchain offers a viable solution by securely integrating and managing data across decentralized systems.

The adoption of crowdsourced manufacturing can be greatly facilitated through a platform-based blockchain structure for IoT. This technology enables interactive communication among manufacturers via smart contracts within a decentralized and peer-to-peer network, eliminating the need for intermediary trust (Bahga & Madisetti, 2016). By combining industrial IoT and blockchain, running smart information services has the capability to capture the real-time processes of manufacturers and assess their contributions to the entire supply chain. This integration allows for a more comprehensive and accurate evaluation of their role in the manufacturing ecosystem. The establishment of a product fulfillment block, which safeguards product fulfillment data and crowdsourcing contracts, enables transactions through an exchange system designed to serve the manufacturing industry within an open environment (Li, Wang, et al., 2018).

Furthermore, manufacturing information services should include the functionality of facilitating the exchange of the product fulfillment block, as well as fulfillment data and crowdsourcing contracts. Operating as a point-to-point system, a blockchain network operates without a centralized server system, thereby reducing information management and trust establishment costs. This decentralized approach relies on a consensus mechanism to validate transactions across nodes within the network, with options such as Proof of Work (PoW), Proof of Stake (PoS), and Proof of Authority (PoA) being commonly employed (Du et al., 2017). However, the scalability of these consensus mechanisms in handling high-frequency transactions for manufacturing remains a critical challenge.

Another crucial property inherent to blockchain technology is traceability, which meticulously records each cryptocurrency transaction among users, as well as information storage and message transmission within the blockchain network. Each block and its associated verified transactions are uniquely identified by hash values, serving both as a means to locate specific blocks or transactions and as a cryptographic method for accessing any block with a corresponding hash address (Zheng, Xie, et al., 2018).

Unchangeability is another cornerstone feature of blockchain technology. Once a block is mined, the transactions it contains can only be read, rendering them immune to editing or deletion (Nakamoto, 2008). This guarantees that information stored in the blockchain cannot be altered without consensus from the majority of nodes in the blockchain network. Alternatively, changes or updates to existing information can be transmitted via new transactions and stored in separate blocks.

Transparency ensures that every node or user within the blockchain network has equal access to the same blockchain ledger. The synergy of traceability, unchangeability, and transparency guarantees that any modifications made to the blockchain are detectable and traceable by all participants in the supply chain (Kouhizadeh & Sarkis, 2018). Similarly, for a manufacturing service involving multiple manufacturers, a blockchain-based information system can provide a secure and trust-worthy platform for all parties involved.

While blockchain-based information systems have shown great promise in addressing challenges in crowdsourced manufacturing and supply chain management, several gaps remain evident in the literature. Current blockchain implementations face limitations in scalability, particularly in handling the high transaction volumes required for real-time manufacturing environments. Additionally, the integration of blockchain with IoT systems is hindered by a lack of standardized protocols, which restricts seamless data sharing and interoperability. Despite theoretical advancements, real-world validation through large-scale case studies is limited, leaving questions about the practicality and scalability of these systems unanswered. Moreover, as blockchain networks become increasingly intertwined with IoT devices, ensuring robust data privacy and security against cyber threats remains a pressing challenge.

6.2.4 Technical Challenges of Crowdsourced Manufacturing Supply Chain Operations

The technical challenges of crowdsourced manufacturing supply chain operations build upon the foundational concepts discussed in Sects. 6.2.1, 6.2.2, and 6.2.3. Crowdsourced manufacturing introduces complex workflows, decentralized coordination, and dynamic resource allocation, requiring advanced systems to address these multifaceted issues.

1. Decentralized network management, specifically involving coordinating complex workflows, dynamic resource allocation, and trust and reputation management: LaaS for crowdsourced manufacturing involves numerous participants (manufacturers, logistics providers, etc.) with unique systems and processes. Orchestrating and managing these decentralized workflows across different platforms and organizations is key. Efficiently allocating resources (labor, equipment, etc.) based on real-time demand and capacity across a distributed network presents a significant challenge. Establishing trust and ensuring reliable performance within a decentralized network without central control requires robust reputation management mechanisms.
2. Secure data sharing and collaboration, in terms of granular access control, data privacy and compliance, and standardized data exchange formats: Implementing fine-grained access control to ensure appropriate data sharing and collaboration based on roles and permissions across the network is crucial. Meeting stringent data privacy regulations and ensuring compliance across different organizational boundaries poses a major challenge. Lack of standardized data formats and protocols for secure data exchange can create integration and communication hurdles.
3. Real-time visibility and analytics, in relation to end-to-end supply chain visibility, data aggregation and analysis, and cybersecurity threats: Obtaining a real-time, holistic view of all activities, inventory levels, and locations across the distributed network is critical for efficient decision-making. Integrating and analyzing data from diverse sources and formats in real-time to predict disruptions, optimize routes, and make informed decisions is a complex task. Securing

real-time data streams and ensuring data integrity throughout the network is critical to prevent cyberattacks and potential disruptions.

Addressing these critical technical challenges is crucial for businesses to successfully implement and leverage the full potential of LaaS for efficient, resilient, and secure crowdsourced manufacturing supply chain management. Cloud integration can be expected to offer a cost-effective business model for interoperable digital supply chains. Interviews with industry practitioners reveal that supply chain integration through blockchain technology can achieve disruptive transformation in digital supply chains and networks, in which smart contracting is one important blockchain functionality contributing to digital supply chain operations.

6.3 Crowdsourcing Information Services of a Digital Manufacturing Supply Chain

Figure 6.1 shows a conceptual system architecture based on industrial IoT and blockchain technology to provide information exchange services for cloud platform-driven crowdsourced manufacturing. The architecture consists of three layers: the resource layer, the cyber layer, and the contracting layer, as illustrated in.

The contracting layer serves as the core of the information service system and facilitates cooperation relationships among decision agents, such as project managers and manufacturing service providers. Blockchain technology is employed to construct contract and fulfillment databases. The contract database records crowdsourcing contracts, including task assignment results and essential profiles of manufacturing service providers. Project managers and service providers can access and read these contracts to collaborate effectively. The blockchain-based fulfillment database collects product fulfillment data from the planning and streaming databases in the lower cyber layer. It traces the fulfillment process, including decision-making results and activity logs, allowing managers and service providers to supervise the fulfillment process and exchange relevant information.

The cyber layer acts as an intermediary between the contract layer, which represents collaboration relationships among decision agents, and the resource layer, which executes the manufacturing service at the physical level. It manages product specifications and supports the derivation of crowdsourcing contracts, providing guidance for task execution and testing. A digital twin of the physical crowdsourcing network is established in the cyber layer in the form of a cyber production system model, enabling real-time monitoring of crowdsourcing task execution through smart sensing IoTs in the resource layer. The streaming data collected in the streaming database allows for the derivation of decision-making results, which are then stored in the planning database.

The resource layer provides the hardware framework for the platform-driven crowdsourced manufacturing community. Manufacturing resources can be reconfigured based on the crowdsourcing task derivation, following the decomposition

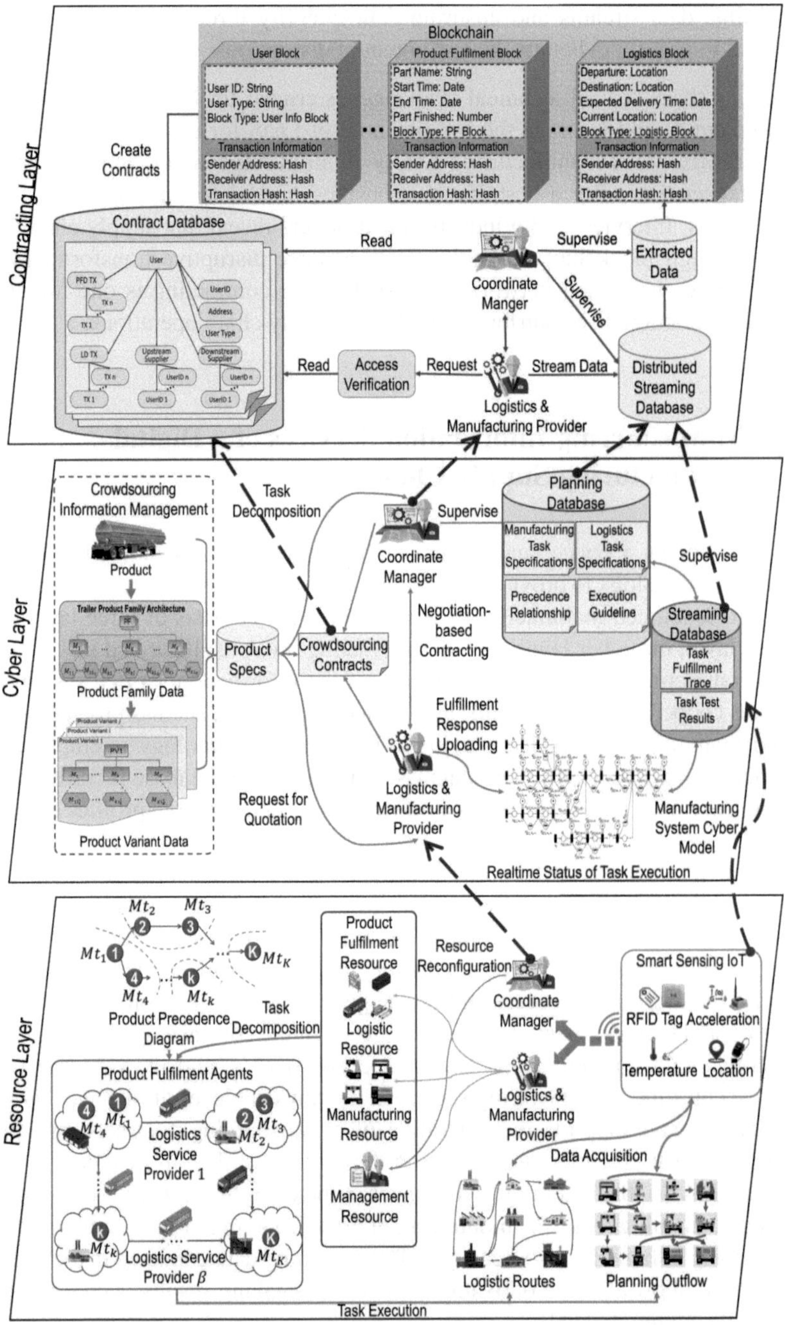

Fig. 6.1 System architecture of smart contracting services for manufacturing digital supply chain operations

result of the product precedence diagram. Smart sensing IoTs are applied in the resource layer to monitor material flow between and within manufacturers, visualizing logistics and process routes. The fulfillment data stream allows project managers and service providers to have real-time insights into the status of crowdsourcing task execution.

The research tasks in the information service system for crowdsourced manufacturing can be divided into two fields based on this architecture. The first category involves using smart contracting technologies to organize the crowdsourcing network, ensuring smooth collaboration among decision agents. The second category focuses on collecting and managing product fulfillment data in a distributed manner, ensuring the availability and accessibility of critical information for all stakeholders. By adopting this architecture and addressing the research tasks in these two categories, the information service system for platform-driven crowdsourced manufacturing can be effectively developed and optimized.

6.3.1 Contracting Layer

The contracting layer serves as the top layer of the system, providing services for managing user information, product fulfillment data, and logistics data on a blockchain network. In this network, each block contains unstructured data from the cyber layer, along with block information that records the transaction process. Within the blockchain, there are three types of blocks: user blocks, product fulfillment blocks, and logistics blocks. All three block types include the sender's account address, the address of the receiver (blockchain), and the transaction hash.

The user block's unstructured data stores information about supply chain participants, such as their names and user types (e.g., project manager, logistics provider, or manufacturing provider). Each user block represents one participant in the supply chain. In the initial blocks of the blockchain, user blocks are always present since collecting user information precedes the fulfillment data collection in the supply chain.

The product fulfillment block contains manufacturing data extracted from the manufacturing provider, representing the manufacturing process over a specific time interval. Similar to the user block, it includes both block information and unstructured data. The unstructured data within the product fulfillment block records details such as part names, start time, end time, and the number of finished parts to monitor the product fulfillment progress.

The logistics block stores data provided by the logistics provider. Its unstructured data includes information on departure location, destination location, current location, and expected arrival time, helping track and manage the logistics of the supply chain.

To upload data to the blockchain network, the streaming product fulfillment data from the logistics and manufacturing providers are first uploaded to the distributed streaming database for information extraction. This step is essential for two reasons: (1) to reduce data size and (2) to facilitate reading and managing product fulfillment progress. The coordinate manager oversees the entire uploading process. By utilizing this contracting layer and the blockchain network, the system efficiently manages user information, product fulfillment data, and logistics data for seamless coordination and tracking within the platform-driven crowdsourced manufacturing environment.

The blockchain mechanism ensures that stored information is immutable and traceable. However, accessing and managing unstructured data on the chain can be challenging. To address this, the designed system employs a process of structuralizing and storing raw data. The database creation is periodic as users continuously update product fulfillment progress in the blockchain network. The database contains crucial information, including participants' details, adjacent manufacturing providers, logistics providers, manufacturing task fulfillment progress, and logistics task fulfillment progress, as illustrated in Fig. 6.1a.

Full read access to the database is granted to the project manager, allowing them to retrieve and analyze relevant data. On the other hand, the manufacturing provider and logistics provider have limited data access. They can only view the product fulfillment progress uploaded by the upstream and downstream participants, ensuring the confidentiality and security of sensitive information. By implementing this structuralized data storage approach, the system enhances data accessibility and facilitates effective management of product fulfillment progress within the platform-driven crowdsourced manufacturing environment.

6.3.2 *Cyber Layer*

The Cyber layer serves as the intermediate layer of the entire system, As shown in Fig. 6.1b. Product specifications are derived from the product, product family data, and product variant data. The product family defines alternative feature choices for similar products, while the product variant data represents different combinations of the product. Together, these components form the product specification, which outlines the requirements for each part, the necessary manufacturing and assembly processes, and the details of each product variant.

To establish the supply chain for a specific product variant, the product specification is decomposed into tasks and shared with the project manager. Logistics and manufacturing providers are then engaged by broadcasting the specifications. A crowdsourcing contract is formed through a negotiation-based contracting mechanism, involving the project manager, logistics provider, and manufacturing provider. The negotiation process encompasses factors such as profit sharing, expected product delivery date, order quantity, and product quality.

Based on the results provided by the logistics and manufacturing providers, a planning database is created, with supervision from the project manager to ensure alignment. The planning database serves as a comprehensive handbook for task execution, containing manufacturing and logistics task specifications, precedence relationships, and execution guidelines.

During task execution, the streaming database captures the fulfillment data stream from smart sensing IoTs in the resource layer. This real-time data collection enables effective monitoring and tracking of the ongoing processes. By establishing this well-structured and interconnected Cyber layer, the system facilitates efficient task execution, collaboration among decision agents, and successful product fulfillment in the platform-driven crowdsourced manufacturing environment.

6.3.3 Resource Layer

The Resource layer constitutes the foundational level of the system. In this layer, crowdsourcing tasks are generated and distributed to each participant based on their available resources, while real-time stream data of each physical entity involved in product fulfillment is collected and uploaded. The resources available to the project manager, logistics provider, and manufacturing provider are reconfigured as a primary step. These resources can be broadly categorized as logistics resources, management resources, and manufacturing resources.

Logistics resources encompass the necessary assets for receiving and delivering materials and products to various participants in the supply chain. Management resources, on the other hand, are utilized to coordinate and supervise the entire product fulfillment process. Lastly, manufacturing resources consist of the available product manufacturing machines.

To efficiently organize the product fulfillment process, a product precedence diagram is developed from the crowdsourcing contract. This diagram contains manufacturing tasks, denoted as M_{t_k}, as illustrated in Fig. 6.1c. These manufacturing tasks are then decomposed and regrouped based on the manufacturing specifications and requirements. By mapping the tasks from the product precedence diagram with the product fulfillment resources, the product fulfillment process is established, as depicted in the figure.

Within this product fulfillment process, each manufacturing task is allocated to a designated manufacturing provider, while logistics tasks are generated based on the precedence and physical locations of the manufacturing providers. These logistics tasks are subsequently distributed to logistics providers, thus establishing an effective and coordinated product fulfillment process. Overall, the resource layer plays a vital role in the platform-driven crowdsourced manufacturing system, ensuring optimal utilization of resources and seamless coordination of tasks among decision agents to achieve successful product fulfillment.

The product fulfillment process involves two key components: logistics routes and planning outflow during task execution. Logistics routes facilitate the transportation of products and materials between physical locations, carried out by entities of logistics resources. This process ensures the smooth movement of items within the supply chain. On the other hand, the planning outflow focuses on part manufacturing and assembly, moving products through various machines among different manufacturing providers.

Throughout the execution of both logistics routes and planning outflow, real-time stream product fulfillment data is collected using smart sensing IoT networks. This data provides valuable insights into the ongoing processes, allowing local project managers, logistics providers, and manufacturing providers to closely monitor the working conditions and progress.

Additionally, the stream data is uploaded to a centralized space for supervision and management by distant participants. This enables remote stakeholders to have visibility and control over the product fulfillment process, promoting collaboration and efficient coordination within the platform-driven crowdsourced manufacturing system.

6.4 Blockchain-Based Smart Contracting Services

Smart contracting services in manufacturing require robust information management and secure data exchange between multiple stakeholders. As illustrated in Fig. 6.2, the proposed architecture integrates three complementary components that work together to enable secure and efficient contracting: a blockchain network, a distributed database based on Interplanetary File System (IPFS) technology, and a function portal. The IPFS-based distributed storage system acts as an intermediary layer, bridging the function portal and blockchain network. This arrangement offers two key advantages: First, it provides efficient storage for large volumes of raw operational data uploaded through the function portal. Second, it maintains organized repositories of processed data extracted from the blockchain network, making information readily accessible to stakeholders for analysis and decision-making.

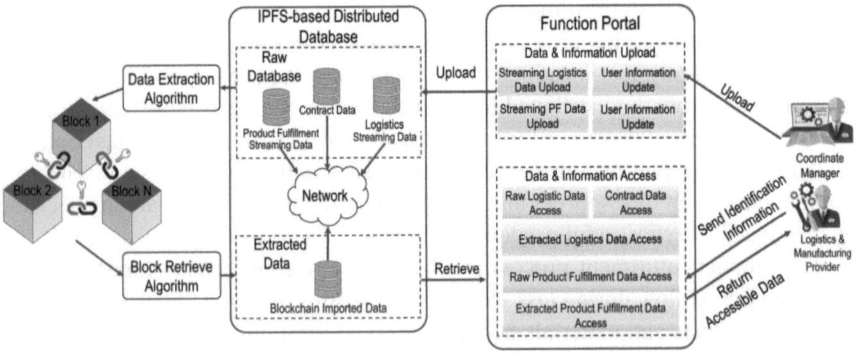

Fig. 6.2 System architecture of smart contracting management

6.4.1 IPFS-Based Distributed Storage System

IPFS is significant for the smart contracting service which operates on a distributed network rather than traditional client-server architecture. In this system, when the crowdsourcing participants upload files, IPFS generates a unique cryptographic hash address known as a content identifier (CID). This CID functions as a digital fingerprint that uniquely identifies each uploaded file and ensures its authenticity. When files are modified, IPFS generates new CIDs, creating an automatic version control mechanism that makes any changes readily detectable. IPFS delivers high-performance distributed storage while maintaining data security (Zheng, Li, et al., 2018). The proposed architecture employs IPFS to manage two distinct categories of data in crowdsourcing manufacturing tasks.

The first category is raw data which encompasses product fulfillment streaming data captured in real-time, logistics streaming data from IoT-enabled networks, and contract data containing task specifications and guidelines. The other category is extracted data which represents processed information derived from raw data using specialized algorithms. Each service provider operates as an independent node in the crowdsourcing network, contributing to the network by uploading their local streaming data. The coordinate manager oversees the contract data, which includes manufacturing specifications, logistics requirements, precedence relationships, and execution protocols.

To transform raw data into actionable insights, data extraction algorithms are employed to identify and process critical information about task execution status. This extracted data flows to the blockchain network through secure transactions. A specialized block data retrieval algorithm then organizes this information within the blockchain and creates corresponding records in IPFS. This systematic approach enables the function portal to efficiently manage product fulfillment information while reducing data management complexity.

6.4.2 Function Portal

The function portal serves as the central interface for managing task execution status and information tracking. It provides comprehensive services for both logistics and manufacturing providers, operating under the oversight of a coordinating manager. The portal's functionality is organized into two main fields:

1. ***Data and Information Upload Field:*** The data and information upload field comprises four essential service functions. The first one is known as "User Information Update". This function manages critical organizational data including provider names, contact details, and locations. The information is stored in contract data using IPFS under the coordinate manager's supervision. While the coordinate manager maintains full database access, logistics and manufacturing providers receive read-only permissions. The CID system enables all participants to monitor changes in contract data.

The second service function is called "Streaming Logistics Data Upload," and its purpose is to enables logistics providers to seamlessly transfer streaming data to IPFS through the function portal. The system establishes direct connections between local databases and IPFS, automatically generating and tracking CIDs for each data upload. Update intervals are determined through collaborative agreements between logistics providers and the coordinate manager, ensuring efficient data synchronization.

The third service function, "Contract Information Update," serves the coordinate manager's needs. Reserved for coordinate manager use, this function facilitates contract modifications based on stakeholder feedback and negotiations. IPFS maintains comprehensive records of these updates, ensuring transparency throughout the contract lifecycle.

The fourth service function, "Streaming Product Fulfillment Data Upload," creates secure pathways between manufacturers' local databases and IPFS for raw product fulfillment data. The system automatically updates CIDs when local data changes, creating an efficient tracking mechanism. Update frequencies are established through mutual agreement between manufacturing providers and the coordinate manager.

By employing these distinct service functions and utilizing the power of IPFS and CID, the data and information upload field achieves a robust and reliable system for managing and tracking essential data and updates from various participants.

2. ***Data and Information Access Field:*** This component provides sophisticated data access services that support the supply chain network's information needs. It enables downstream participants to access upstream task execution data while maintaining appropriate security controls. The system implements role-based access permissions that vary based on task requirements and participant roles. Five core services facilitate this functionality:

   "Contract Data Access" provides permission-based access to contract information, execution guidelines, and product specifications. The CID system enables both current and historical data retrieval.

   "Raw Logistics Data Access" enables authorized participants to view logistics data through the function portal. The IPFS-based storage system maintains comprehensive CID records for tracking both current and historical execution status.

   "Extracted Logistics Data Access" delivers processed logistics data from the blockchain-imported database. Participants can retrieve execution status information using blockchain transaction hashes.

   "Raw Product Fulfillment Data Access" mirrors the logistics data access functionality for manufacturing data, providing CID-based access to all authorized product fulfillment information.

   "Extracted Product Fulfillment Data Access" complements raw data access by providing streamlined access to processed manufacturing data through transaction hashes.

6.4.3 Blockchain Network

The blockchain network receives transactions containing two main types of extracted data: "User Information" and "Task Execution Status."

"User Information" represents the first type of data stored in the blockchain. The system simplifies complex provider information into essential components: user types (identifying logistics or manufacturing providers) and user IDs (provider identifiers). When users update their information through the portal, the system uploads detailed data to IPFS and creates corresponding blockchain records. Each user receives a blockchain network account, with their address mapped to their unique ID.

The second type of data stored in the blockchain is "Product Fulfillment Data." This data is transacted to the blockchain network after users update the raw product fulfillment data in IPFS. The extracted message reflects the task execution status, including details such as the update time of the file in IPFS, the start and end times of collecting local product fulfillment data, the physical location of the task execution, the executing task's name, and the number of finished tasks. In this extracted data, the status is tracked at larger time intervals for better efficiency. Additionally, multiple manufacturing processes are simplified into a binary state, indicating either a completed task or an incomplete one. A label is added to this extracted data to identify it as product fulfillment data. Similar to User Information, the extracted data is transacted to the blockchain network using the same account as the one used for sending the extracted user information.

The third type of extracted data is "Logistics Data." Similar to product fulfillment data, raw logistics data stored in IPFS is also extracted using a similar mechanism. The data extraction algorithm simplifies status changes and enlarges the time intervals for tracking. Complex transportation processes are also simplified to improve the data representation. The blockchain network receives the logistics data transaction after the user updates the raw logistics data in IPFS.

Overall, the transactions sent to the blockchain network contain simplified user information and extracted task execution status, enabling efficient tracking and collaboration among participants in the supply chain network.

6.5 IoT-Enabled Distributed Manufacturing Data Streaming Management

IoT technology plays a crucial role in the platform-driven crowdsourced manufacturing ecosystem, granting access to resources and entities through a ubiquitous network. This technology is a fundamental requirement for achieving real-time task execution in this advanced manufacturing landscape. Embracing real-time task monitoring not only propels traditional manufacturing towards greater efficiency and intelligence but also paves the way for a decentralized crowdsourcing

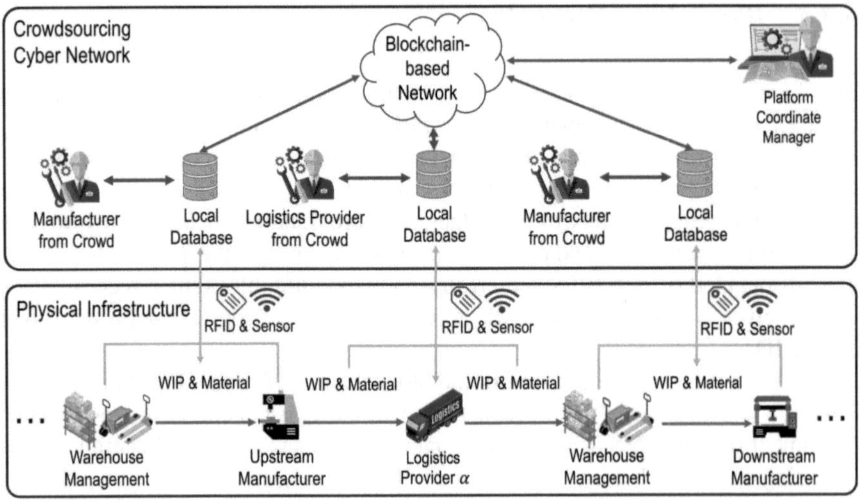

Fig. 6.3 IoT-enabled distributed data streaming for cyber-physical supply chain information sharing

manufacturing network, facilitated by a cyber-physical production system (Lee et al., 2019).

In Fig. 6.3, we present an IoT-enabled cyber-physical system architecture designed specifically for platform-driven crowdsourced manufacturing. This architecture seamlessly integrates physical infrastructure at the lower level with a crowdsourcing cyber network at the higher level, enabled by intelligent sensors. By leveraging technologies such as RFID tags and barcodes, the system can effectively trace product fulfillment and logistics service resources through a manufacturing crowdsourcing process. Moreover, the synthesis of IoT-streamed product fulfillment data offers valuable insights into the status of intra- and inter-manufacturer work-in-progress and material flows.

The initial tracing data is uploaded to a local database, granting local manufacturers the ability to monitor and control production in real-time. Additionally, copies of this raw tracing data are shared and extracted, transferring them to a blockchain-based network. Once integrated into the blockchain, any updates to the tracing data can be easily tracked and accessed by relevant stakeholders, including the coordinate manager and other service providers.

Through this comprehensive approach, IoT technology empowers platform-driven crowdsourced manufacturing with real-time capabilities, fostering an intelligent and interconnected ecosystem that enhances efficiency and transparency across the supply chain.

Figure 6.4 illustrates the operational process of the information service system. Initially, the raw product fulfillment data is collected from the local network and updated at specific intervals determined locally. Simultaneously, the streaming data file in the local storage undergoes a separate time interval before being copied and

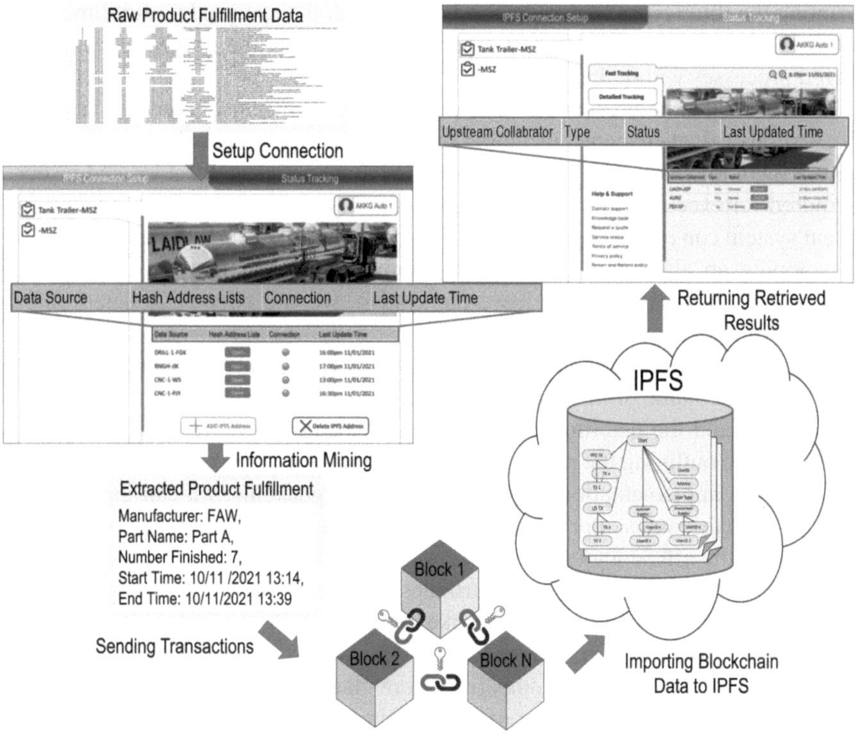

Fig. 6.4 Operation flows of integrated information services

transferred to the IPFS. This time interval is collaboratively determined by both the coordinate manager and the local service provider, considering the specific task's requirements and prevailing conditions. The connection setup interface plays a crucial role in establishing the link between the local storage and the IPFS, with manufacturing providers and logistics providers taking the lead. Throughout the connection establishment activity in this interface, the coordinate manager closely monitors the process.

Moreover, the connection setup interface provides essential services for users to inspect and manage the connection status of the local database. For instance, users can review the hash address list of an updated raw data file and check the timestamp of its last update. In the proposed system, the coordinate manager, manufacturing provider, and logistics provider employ the connection setup interface to share contract, raw product fulfillment data, and raw logistics data, respectively. This seamless information service system facilitates efficient data sharing and collaboration among stakeholders, leading to improved coordination and streamlined operations in the platform-driven crowdsourced manufacturing network.

Information extraction occurs once the raw data has been updated, and this extracted data serves to summarize the task execution status based on the raw data.

Among the three types of raw data to be extracted, the raw product fulfillment data contains the most comprehensive information, enabling the tracing of unsynchronized processes within data files.

To elaborate, the process within the raw product fulfillment data can be defined by several critical elements, including the time when the process begins or ends, the entity ID of the interacting entity involved in the process, and the allocated resource required for executing the process. Each entity entering the defined product fulfillment system can only exit the system upon completing the final designated step. By developing an algorithm to count the number of entities that have successfully undergone the last process to leave the system, we can effectively determine the number of finished tasks.

The extracted data from the raw product fulfillment data includes the start and end times, the name of the manufacturer responsible for updating the raw data, the task name, and the determined number of finished tasks. Figure 6.5 presents the pseudo-code outlining the product fulfillment data extraction algorithms.

With the implementation of these extraction algorithms, we can gain valuable insights into the status of product fulfillment tasks, enhancing decision-making, and optimizing the efficiency of the platform-driven crowdsourced manufacturing network. The extracted information contributes significantly to streamlining operations and promoting coordination among stakeholders.

The extracted data is transmitted to the blockchain network, offering improved feedback to remote viewers compared to raw data. The reduced data size ensures compatibility with a wide range of blockchain technologies, making the proposed information service system adaptable and efficient. To enhance data management

```
Input: 1. Name of the task.
       2. Table that record when does a process happen in the system. The process is defined as: an entity occupies
a resource for executing a certain action and one entity requires several processes for leaving the system.
Output: Extracted data that summarize the overall action exaction results.
begin
   Acquire the time when the first process happens in the table as t_start;
   Acquire the time when the last process ends in the table as t_end;
   Identify the last process associate with entities for leaving the system p_l;
   L ← {};
   Foreach row in table:
      Obtain the EntityID in each row and save as I;
       If I not in L:
          Append I in the list L;
       end
   end
   N ← length of L;
   return t_start;
   return t_end;
   return name of task;
   return N;
end
```

Fig. 6.5 Pseudo-code of product fulfilment data extraction

```
struct User{
    string userID;
    string address;
    string usertype;
    list PFDtx;
    list LDtx;
    list upstream;
    list downstream;
    function addPFD(PFD) {
        self.PFDtx.append(PFD);
    }
    function addLD(LD) {
        self.PFDtx.append(LD);
    }
    function addupstream(US) {
        self.upstream.append(US);
    }
    function adddownstream(DS) {
        self.downstream.append(DS);
    }
}
Connect to Blockchain network
for i in range(1 to number of block ) {
  list tx_list ← list of transactions in block i;
  for c_tx in tx_list {
    string extracted_data ← string stored in c_tx
    if extracted_data contains user information{
      string c_uid ← userID stored in extracted_data;
      User globals()c_uid;   //create a User struct instance named with string in c_uid
      globals()c_uid.address ← sender's account address;
      globals()c_uid.usertype ← usertype information in extracted_data;
      adddownstream(string that contains downstream userID);
      adduptream(string that contains upstream userID);
    }
    elseif extracted_data contains product fulfillment data{
      addPFD(c_tx.address);  //append the current transaction hash to PFDtx field
    }
    elseif extracted_data contains user logistic data{
      addLD(c_tx.address);  //append the current transaction hash to LDtx field
    }
  end
```

Fig. 6.6 Pseudo-code of structuralizing blockchain information

services, the data within the blockchain network is structured and stored in the IPFS. The algorithm responsible for this data structuring is outlined in Fig. 6.6.

The algorithm begins by defining a struct containing essential fields such as user ID, account address, user type, lists of sent extracted product fulfillment data, logistics data, and adjacent upstream or downstream service providers' user IDs. Upon connecting to the blockchain network, the algorithm traverses each transaction within every block. For transactions containing user information, the algorithm creates an instance of the user struct and names it after the sender's user ID. Furthermore, the sender's address and information about adjacent service providers are updated accordingly. Similarly, for transactions containing product fulfillment data, the algorithm adds the transaction hash to the struct instance named after the sender's user ID. The same process is followed for transactions containing logistics data. Once all blocks have been processed, a list is generated to store all defined struct instances, which is then exported to IPFS for secure storage.

When accessing extracted data through the status tracking interface, the structuralized data acts as a catalog, providing the cyber location of accessible information to the user. Upon sending an access request through the status tracking interface, the corresponding struct instance named after the user ID is retrieved and reviewed by the access management algorithm. The algorithm then searches for struct instances named after user IDs that belong to upstream service providers from the list. Once the struct instances of these service providers are obtained, all transaction hashes sent by them are returned to the user. As a result, users can view the extracted data in the blockchain using these transaction hashes.

By employing this structured data approach, the information service system ensures efficient data access, enabling better collaboration and coordination among stakeholders within the platform-driven crowdsourced manufacturing network.

6.6 Case Study of Tank-Trailer Manufacturing Crowdsourcing and Supply Chain Operations

The proposed blockchain-based information service system is verified through a case study on tank-trailer crowdsourced manufacturing. This case study showcases the system's performance and feasibility in fulfilling a product with the collaboration of five crowdsourced manufacturers and two logistics providers. It highlights the functions of smart contracts, distributed information management, and product fulfillment data streaming, all powered by IoT technology. The case study is divided into four parts: the description of the crowdsourcing case, smart contract generation, information management for task execution, and product fulfillment streaming data services.

Throughout the case study's development process, various environments and tools are utilized, each serving a specific purpose. These environments and tools are categorized accordingly in Table 6.1. The case study is based on the Ethereum blockchain network, chosen for its robust developer support and ability to create "Smart Contracts" that serve as digital twins of traditional contracts, ensuring secure transaction management. To facilitate the case study, a blockchain test network based on Ethereum is established using Truffle and Ganache-cli. Truffle, a development framework for Ethereum, provides essential services such as smart contract management and test network management. Leveraging Truffle, a personalized Ethereum test network is created, and users can adjust network properties, including block size, block mining logic, and transaction costs. Ganache-cli, acting as an auxiliary tool for Truffle in this case study, accesses the Ethereum test

Table 6.1 Development environments and tools employed in the case study

Usage purpose	Tools and environment
Building blockchain test network	Ubuntu Linux 21.10, 24 processors, 32 GB RAM
	Ganache-cli
	Truffle
Simulating PF data	Simio
Developing web-based raw PF data upload interface	IPFS
	JavaScript, CSS
	React.js
	Web3.js
	Meta Mask
	Solidity
Developing blockchain information importer	Web3.py
	IPFS

network built by Truffle and displays blockchain information through a user-friendly graphical interface, serving as a demonstration tool for the test Ethereum network.

The use of Ethereum and the associated tools empower the case study, allowing for comprehensive testing and evaluation of the blockchain-based information service system. This ensures that the proposed system is efficient, secure, and well-suited for the complexities of tank-trailer crowdsourced manufacturing and similar decentralized manufacturing networks.

This case study utilizes Discrete Event Simulation (DES) software to generate simulated product fulfillment data, allowing for the verification of the real-time streaming function. DES simulations are commonly employed to assess design performance, particularly in the late development stages. Simio, a widely used software, is applied for this purpose. Simio offers model-to-model transition capabilities, allowing simulations to derive and export various outputs, including the trace of the simulation model as product fulfillment data.

A web-based raw product fulfillment data upload interface is a critical component of this case study. This interface facilitates the transfer of user-uploaded files to IPFS through a web browser. Additionally, it processes the raw product fulfillment data for extraction and forwards the extracted task execution status to the blockchain network. IPFS serves as the foundation tool for providing distributed file storage service, enabling remote access for users. For web UI development, the React JavaScript library is the primary tool employed.

To interact with an Ethereum node using HTTP, IPC, and WebSocket, the Web3 JavaScript library is utilized in this case study. MetaMask, a cryptocurrency wallet integrated with web browsers, offers account management services, allowing users to confirm or reject blockchain transactions. Smart contracts written in Solidity execute blockchain commands, enabling seamless interaction with the blockchain network.

When utilizing the web-based interface, the Web3 library establishes the connection between IPFS and the test blockchain network, while MetaMask provides account management services for users during the confirmation or rejection of blockchain transactions. Smart contracts play a crucial role in executing blockchain commands and interacting with the blockchain network effectively.

Furthermore, a blockchain information importer algorithm is implemented as a critical element in this case study. As mentioned earlier, this algorithm categorizes and structures transaction information before sending the structured data to IPFS. To assist in this development process, the Web3 Python library is employed, allowing users to retrieve information from the blockchain test network.

Overall, the combination of DES simulations, a web-based interface, Web3 library, MetaMask, and smart contracts enables efficient and reliable verification of the proposed information service system in the context of tank-trailer crowdsourced manufacturing.

1. ***Description of crowdsourcing case:*** Figure 6.7 illustrates the genetic product process structure of the tank-trailer product variants in this case study. Through crowdsourcing task decomposition and derivation, five specific tasks are generated. To execute these tasks, a total of five product fulfillment providers and two logistics providers are carefully selected to form a collaborative crowdsourcing supply network.

In the depicted scenario of the tank-trailer crowdsourcing supply network, the five manufacturers serve as the product fulfillment providers, each playing a crucial role in contributing to the overall product fulfillment process. Additionally, the

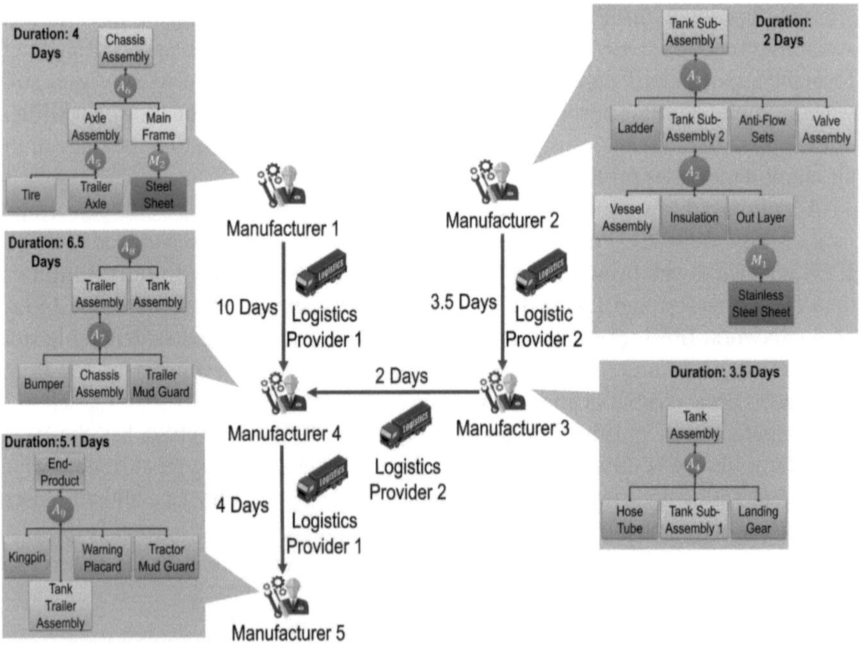

Fig. 6.7 Crowdsourcing supply network for tank trailer manufacturing

logistics providers play a pivotal role in transporting the finished sub-assemblies between the product fulfillment providers, facilitating the smooth flow of operations within the network.

This collaborative setup demonstrates the efficiency and effectiveness of the proposed platform-driven crowdsourced manufacturing approach, where multiple stakeholders work in synergy to accomplish the product fulfillment tasks in an agile and decentralized manner. The interconnected supply network ensures seamless coordination and optimized resource utilization, resulting in enhanced productivity and a streamlined manufacturing process.

2. ***Smart contracts generation case:*** Figure 6.8 depicts the deployment of the smart contract "Storage" to the blockchain network after initializing the test network through Truffle. In the test network, the block size is defined by the "gas limit," set at 6721975. To simulate the proof of authority consensus mechanism, a new block is mined at specific time intervals. Notably, Block 0 is mined immediately after the deployment of the smart contract "Storage."

Upon deployment, the smart contract "Storage" is assigned the Ethereum address "0xF7955da8a54C23547c86f379047c852D57B54886," as indicated in the figure. This address serves as the destination for all information uploads to the blockchain network, taking the form of transactions sent to this specific address. The smart contract "Storage" functions as an essential portal, acting as an intermediary between users and the blockchain network. It verifies the identities of users and efficiently receives and manages the information transmitted by them.

By leveraging the capabilities of the smart contract "Storage," the proposed information service system ensures secure and reliable communication between users and the blockchain network, facilitating seamless data exchange and streamlined operations within the platform-driven crowdsourced manufacturing network.

The smart contract "Storage" serves a vital role in the case study, containing three functions: user information upload, product fulfillment data upload, and logistics data upload. However, it only provides partial service and requires the assistance of other components to offer a complete solution. Following initialization, the coordinate manager registers information of both product fulfillment providers and

Fig. 6.8 Smart contract and its storage block

logistics providers to the blockchain network using the user information upload service. For each user, the coordinate manager enters three essential pieces of information: Ethereum account address, user ID, and user type. To ensure security and access control, the Ethereum account addresses of the coordinate managers are stored in the smart contract for verification. If the transaction sender's address does not match the stored addresses, non-coordinate manager users cannot access the user information upload function. In collaboration with the web-based interface and MetaMask, the smart contract works as the backend for direct interaction with the coordinate manager. This combined effort allows the smart contract to handle the information uploads effectively.

In the case study, the coordinate manager uploads information for five product fulfillment providers and two logistics providers to the blockchain network. This mapping process is crucial for the other two functions to operate effectively and is sent to the blockchain through the second transaction. By effectively coordinating with other components, the smart contract "Storage" ensures secure and efficient management of user information, facilitating smooth operations within the platform-driven crowdsourced manufacturing network.

The Product fulfillment data upload function comprises two main parts. The first part involves verifying the sender's information. When a product fulfillment provider uploads its raw streaming data in the product fulfillment web-based interface, the function utilizes the provider's account address as a key to retrieve the corresponding user ID mapped with the address. If the user ID contains the letter "M," the function grants permission to the user to send product fulfillment transactions to the blockchain. This verification process is a simplified example, but similar algorithms could be applied to incorporate other elements for more robust identity verification.

Similarly, the Logistics fulfillment data upload function operates with a similar algorithm. However, in this case, the letter "L" is used to check the types of users, as the user IDs of the two logistics providers are "L1" and "L2" in the case study. By utilizing these verification mechanisms, the information service system ensures that only authorized users, based on their user IDs and types, can upload product fulfillment data and logistics data to the blockchain network. These streamlined and efficient verification processes add an extra layer of security and integrity to the platform-driven crowdsourced manufacturing network, enabling smooth and reliable data uploads and interactions.

3. ***Information management for task execution:*** The information management in this case study involves handling both raw streaming data stored in the IPFS and extracted data stored in the blockchain. The coordinate manager is responsible for uploading information related to crowdsourced tasks, task specifications, and the supply chain precedence file to IPFS. The uploaded files are given a unique hash address called CID, which acts as a pointer to the shared file accessible to all participants. To access the file, participants can simply add the hash address after "https://ipfs.io/ipfs/" and browse the HTTP address in their web browser. As files stored in IPFS are immutable, participants can always access the shared

file using its hash address. If necessary, the coordinate manager can upload an updated file to IPFS; however, the updated hash address must be shared with other participants.

For product fulfillment providers and logistics providers, the process of uploading raw data to IPFS is integrated with the web-based interface. After uploading the file to IPFS, the information is extracted from the file and sent to the blockchain network, generating a corresponding CID. Consequently, the CID of the raw data and the extracted task execution information are stored together in the blockchain. To enhance versatility, the information stored in the blockchain is categorized as a json file, which is also stored in IPFS and accessible through the web-based interface.

4. ***Product fulfillment streaming data execution services:*** In this section, we will elaborate on the detailed process of raw product fulfillment data upload and extraction. To generate raw product fulfillment data, we utilize the Simio simulation model. Figure 6.9 provides an overview of the layout of five manufacturers in the Simio model, where an entity progresses through several servers, simulating the machining process.

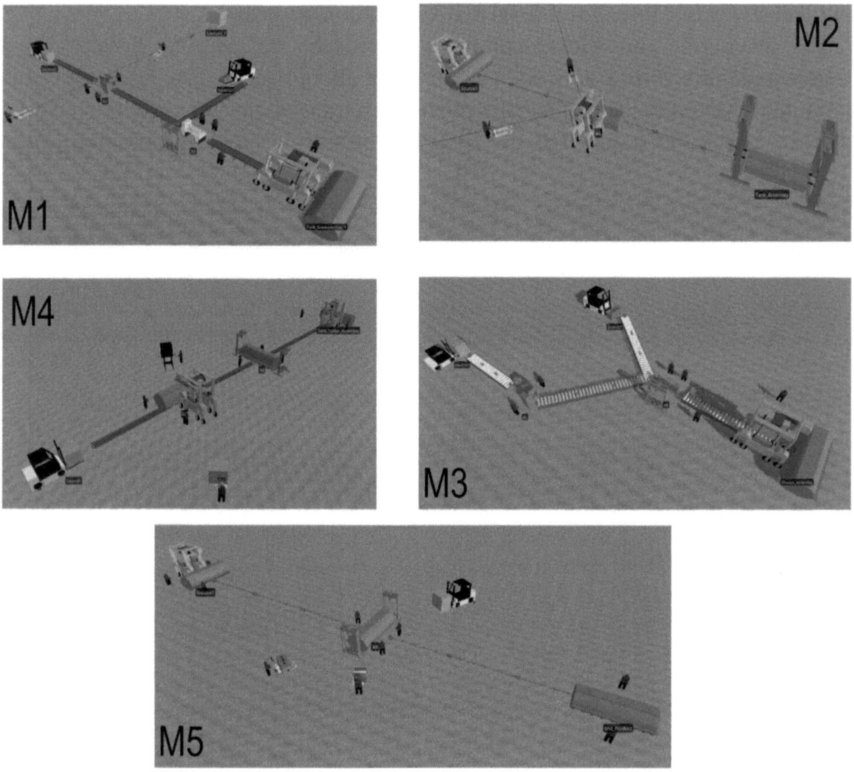

Fig. 6.9 Simio simulation model of manufacturers

In each Simio model, an entity is considered complete once it reaches the sink object at the end of the assembly line. The model trace data in CSV format, generated by Simio simulation, is used as the raw product fulfillment data. As a DES software, the model trace data records every event occurring during the simulation in chronological order.

To simulate the acquisition of raw data at specific time intervals, the simulation is paused multiple times for each model, and model trace data is generated at each pause. This approach allows us to capture the necessary information and construct the raw product fulfillment data in a manner that reflects the real-time task execution status. By employing the Simio model and the corresponding model trace data, we can effectively generate realistic and dynamic raw product fulfillment data for further processing and analysis within the platform-driven crowdsourced manufacturing network.

In the case study, all five manufacturers and two logistics providers utilize the web-based interface for uploading their data. Transactions that are sent within the same time interval are grouped together and included in the same block. Figure 6.10 provides a visual representation of a block mined in the test network, where each block contains various types of transactions from different users. The information contained in the block is converted into a Json file format and stored in IPFS, making it easily accessible for retrieval and verification purposes.

By leveraging this blockchain-based approach, the platform-driven crowdsourced manufacturing network ensures that all uploaded data is securely and immutably stored in the blockchain, providing a transparent and auditable record of the task execution status. The use of IPFS for storing block information further

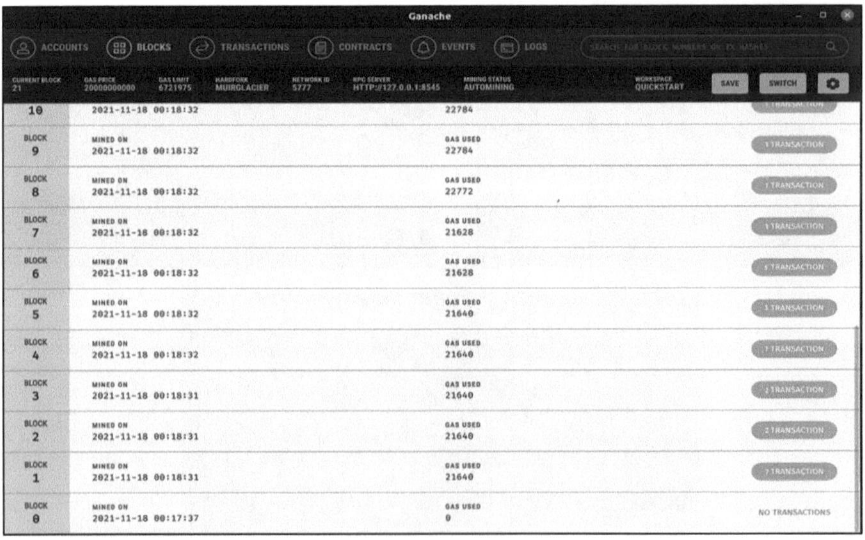

Fig. 6.10 Block mining for transactions from different users

enhances the efficiency and accessibility of the data, allowing all participants to easily access and review the pertinent details of the task execution process.

6.7 Concluding Remarks

The blockchain-based information service system presented in this chapter advances the field of crowdsourced manufacturing through its novel three-layer architecture. The integration of contracting, cyber, and resource layers enables secure smart contracting while facilitating efficient data management and stakeholder collaboration. The case study implementation demonstrates the architecture's capability to handle complex manufacturing scenarios while maintaining data security and transparency. This system represents a significant step forward in platform-driven manufacturing by providing a robust framework for information sharing and smart contracting. The successful validation through practical application confirms that blockchain technology can effectively address key challenges in crowdsourced manufacturing, particularly in areas of security, traceability, and trust establishment among distributed stakeholders. These findings contribute to both theoretical understanding and practical implementation of blockchain-enabled manufacturing services, offering valuable insights for researchers and practitioners in the field.

This study presents several limitations that open avenues for future research. First, the current implementation focuses exclusively on scenarios with predefined design parameters, where manufacturers participate only through bidding processes. This approach overlooks the potential benefits of concurrent engineering, where open innovators could collaborate directly with manufacturers during product development phases. Additionally, the system's evaluation of manufacturer qualifications could be more comprehensive, particularly in complex scenarios involving multiple stakeholder requirements.

Future research should address these limitations through several key directions. A critical area for investigation is the evolutionary dynamics of manufacturing capabilities and product families. This includes studying how manufacturer competencies develop through collaborative relationships and how product families adapt to align with evolving manufacturing strengths. Research should also examine the co-evolution of product families and manufacturing services in response to market demands and technological advancement. Furthermore, the robustness of the proposed methods requires more extensive evaluation across diverse criteria and complex manufacturer profiles. Future studies should verify the system's effectiveness in handling varied task execution parameters, including negotiations involving different due dates, revenues, and unit tardiness penalty costs. Such investigations would enhance the system's applicability across a broader spectrum of manufacturing scenarios while advancing our understanding of blockchain-enabled information services in crowdsourced manufacturing.

References

Agostinho, C., Ducq, Y., Zacharewicz, G., Sarraipa, J., Lampathaki, F., Poler, R., & Jardim-Goncalves, R. (2016). Towards a sustainable interoperability in networked enterprise information systems: Trends of knowledge and model-driven technology. *Computers in Industry, 79*, 64–76.

Bahga, A., & Madisetti, V. K. (2016). Blockchain platform for industrial internet of things. *Journal of Software Engineering and Applications, 9*, 533–546.

Büyüközkan, G., & Göçer, F. (2018). Digital supply chain: Literature review and a proposed framework for future research. *Computers in Industry, 97*, 157–177.

Chen, S., Cai, X., Wang, X., Liu, A., Lu, Q., Xu, X., & Tao, F. (2022). Blockchain applications in PLM towards smart manufacturing. *The International Journal of Advanced Manufacturing Technology, 118*, 2669–2683.

Chida, T., Kaihara, T., Fujii, N., & Kokuryo, D. (2019). Stability analysis on resource matching in crowdsourced manufacturing. *Procedia CIRP, 81*, 405–410.

Crosby, M., Pattanayak, P., Verma, S., & Kalyanaraman, V. (2016). Blockchain technology: Beyond bitcoin. *Applied Innovation, 2*, 71.

Ding, K., Jiang, P., Leng, J., & Cao, W. (2016). Modeling and analyzing of an enterprise relationship network in the context of social manufacturing. *Proceedings of the Institution of Mechanical Engineers, Part B: Journal of Engineering Manufacture, 230*, 752–769.

Du, M., Ma, X., Zhang, Z., Wang, X., & Chen, Q. (2017). A review on consensus algorithm of blockchain. In *2017 IEEE international conference on systems, man, and cybernetics (SMC)* (pp. 2567–2572). IEEE.

Giannakis, M. (2011). Management of service supply chains with a service-oriented reference model: the case of management consulting. *Supply Chain Management: An International Journal, 16*(5), 346–361.

Gong, X. (2021). *Platform-driven crowdsourced manufacturing for manufacturing as a service.* Ph.D Dissertation, Georgia Institute of Technology, Atlanta, GA.

Gong, X., Jiao, R., Jariwala, A., & Morkos, B. (2021). Crowdsourced manufacturing cloud platform and intelligent cognitive assistants for delivery of manufacturing as a service: fundamental issues and outlook. *The International Journal of Advanced Manufacturing Technology*, 1–11.

Grabowska, S. (2020). Smart factories in the age of Industry 4.0. *Management Systems in Production Engineering, 28*(2), 90–96.

Guo, W., & Jiang, P. (2019). Manufacturing service order allocation in the context of social manufacturing based on Stackelberg game. *Proceedings of the Institution of Mechanical Engineers, Part B: Journal of Engineering Manufacture, 233*, 1890–1901.

Hamalainen, M., & Karjalainen, J. (2017). Social manufacturing: When the maker movement meets interfirm production networks. *Business Horizons, 60*, 795–805.

Huddar, Y., Kumatagi, P., & Latte, M. (2017). Digital supply chain management—A review. *IARJSET, 4*, 34–37. https://doi.org/10.17148/IARJSET/NCDMETE.2017.10

Iddris, F. (2018). Digital supply chain: survey of the literature. *International Journal of Business Research and Management, 9*(1), 47–61.

Ivanov, D., & Dolgui, A. (2021). A digital supply chain twin for managing the disruption risks and resilience in the era of Industry 4.0. *Production Planning & Control, 32*(9), 775–788.

Ivanov, D., Dolgui, A., & Sokolov, B. (2022). Cloud supply chain: Integrating industry 4.0 and digital platforms in the "supply chain-as-a-service". *Transportation Research Part E: Logistics and Transportation Review, 160*, 102676.

Jiang, P., & Ding, K. (2018). Analysis of personalized production organizing and operating mechanism in a social manufacturing environment. *Proceedings of the Institution of Mechanical Engineers, Part B: Journal of Engineering Manufacture, 232*, 2670–2676.

Jiang, P., Leng, J., Ding, K., Gu, P., & Koren, Y. (2016). Social manufacturing as a sustainable paradigm for mass individualization. *Proceedings of the Institution of Mechanical Engineers, Part B: Journal of Engineering Manufacture, 230*, 1961–1968.

Jiao, R., Gong, X., Song, M., Wang, S., & Helo, P. (2024). An X-as-a-service (XAAS) reference model for smart manufacturing systems. In *Advances in mechatronics and mechanical engineering (AMME) book series* (pp. 47–67). IGI Global.

Jiao, R., Luo, J., Malmqvist, J., & Summers, J. (2022). New design: opportunities for engineering design in an era of digital transformation. *Journal of Engineering Design, 33*(10), 685–690.

Kadar, B., Egri, P., Pedone, G., & Chida, T. (2018). Smart, simulation-based resource sharing in federated production networks. *CIRP Annals, 67*, 503–506.

Kouhizadeh, M., & Sarkis, J. (2018). Blockchain practices, potentials, and perspectives in greening supply chains. *Sustainability, 10*, 3652.

Kusiak, A. (2018). Smart manufacturing. *International Journal of Production Research, 56*, 508–517.

Kusiak, A. (2019). Service manufacturing: Basic concepts and technologies. *Journal of Manufacturing Systems, 52*, 198–204.

Kusiak, A. (2020). *Service manufacturing= Process-as-a-service+ manufacturing operations-as-a-service*. Springer.

Lee, J., Azamfar, M., & Singh, J. (2019). A blockchain enabled cyber-physical system architecture for Industry 4.0 manufacturing systems. *Manufacturing Letters, 20*, 34–39.

Leng, J., Jiang, P., Xu, K., Liu, Q., Zhao, J. L., Bian, Y., & Shi, R. (2019). Makerchain: A blockchain with chemical signature for self-organizing process in social manufacturing. *Journal of Cleaner Production, 234*, 767–778.

Leng, J., Jiang, P., & Zheng, M. (2017). Outsourcer–supplier coordination for parts machining outsourcing under social manufacturing. *Proceedings of the Institution of Mechanical Engineers, Part B: Journal of Engineering Manufacture, 231*, 1078–1090.

Leukel, J., Kirn, S., & Schlegel, T. (2011). Supply chain as a service: a cloud perspective on supply chain systems. *IEEE Systems Journal, 5*(1), 16–27.

Li, Z., Barenji, A. V., & Huang, G. Q. (2018). Toward a blockchain cloud manufacturing system as a peer to peer distributed network platform. *Robotics and computer-integrated manufacturing, 54*, 133–144.

Li, Z., Wang, W., Liu, G., Liu, L., He, J., & Huang, G. (2018). Toward open manufacturing. *Industrial Management & Data Systems, 118*(1), 303–320.

Maas, S., Hartmann, E., & Herb, S. (2014). Supply chain services from a service-dominant perspective: a content analysis. *International Journal of Physical Distribution & Logistics Management, 44*(1/2), 58–79.

Nakamoto, S. (2008). Bitcoin: A peer-to-peer electronic cash system. *Decentralized Business Review*, 21260.

Qu, T., Lei, S., Wang, Z., Nie, D., Chen, X., & Huang, G. Q. (2016). IoT-based realtime production logistics synchronization system under smart cloud manufacturing. *The International Journal of Advanced Manufacturing Technology, 84*, 147–164.

Song, M., Gong, X., Du, J., Lu, T., & Jiao, R. J. (2024). Population dynamics modeling of crowdsourcing as an evolutionary Cooperation-Competition game for fulfillment capacity balancing and optimization of smart manufacturing services. *Computers & Industrial Engineering, 197*, 110572.

Tan, C., Hu, S. J., Chung, H., Barton, K., Piya, C., Ramani, K., & Banu, M. (2017). Product personalization enabled by assembly architecture and cyber physical systems. *CIRP Annals, 66*, 33–36.

Tao, F., Zhang, L., Venkatesh, V., Luo, Y., & Cheng, Y. (2011). Cloud manufacturing: a computing and service-oriented manufacturing model. *Proceedings of the Institution of Mechanical Engineers, Part B: Journal of Engineering Manufacture, 225*, 1969–1976.

Tao, F., Zuo, Y., Da, X., & L. & Zhang, L. (2014). IoT-based intelligent perception and access of manufacturing resource toward cloud manufacturing. *IEEE Transactions on Industrial Informatics, 10*, 1547–1557.

Tian, F. (2016). An agri-food supply chain traceability system for China based on RFID & block-chain technology. In *2016 13th international conference on service systems and service management (ICSSSM)* (pp. 1–6). IEEE.

Weber, I., Xu, X., Riveret, R., Governatori, G., Ponomarev, A., & Mendling, J. (2016). Untrusted business process monitoring and execution using blockchain. In *International Conference on Business Process Management* (pp. 329–347). Springer.

WHU. (2015). Digital supply chains: increasingly critical for competitive edge, European Kearney/WHU Logistics Study 2015. https://www.kearney.com/article/-/insights/digital-supply-chains-increasingly-critical-for-competitive-edge

Xu, X., Pautasso, C., Zhu, L., Gramoli, V., Ponomarev, A., Tran, A. B., & Chen, S. (2016). The blockchain as a software connector. In *2016 13th working IEEE/IFIP conference on software architecture (WICSA)* (pp. 182–191). IEEE.

Xue, L., Zhang, C., Ling, H., & Zhao, X. (2013). Risk mitigation in supply chain digitization: System modularity and information technology governance. *Journal of Management Information Systems, 30*(1), 325–352.

Yu, W., Chavez, R., Liu, Q., & Cadden, T. (2023). Examining the effects of digital supply chain practices on supply chain viability and operational performance: a practice-based view. *IEEE Transactions on Engineering Management, 71*, 10413–10426.

Zhang, J., Brintrup, A., Calinescu, A., Kosasih, E., & Sharma, A. (2021). Supply chain digital twin framework design: an approach of supply chain operations reference model and system of systems. *arXiv Preprint*. https://doi.org/10.48550/arXiv.2107.09485

Zheng, Q., Li, Y., Chen, P., & Dong, X. (2018). An innovative IPFS-based storage model for blockchain. In *2018 IEEE/WIC/ACM international conference on web intelligence (WI)* (pp. 704–708). IEEE.

Zheng, Z., Xie, S., Dai, H.-N., Chen, X., & Wang, H. (2018). Blockchain challenges and opportunities: A survey. *International Journal of Web and Grid Services, 14*, 352–375.

Chapter 7
The Role of Blockchain in Food Supply Chains: Enhancing Traceability

Elif Dursun and Yavuz Gunalay

Abstract The digital transformation era has ushered in groundbreaking advancements in supply chain management, with blockchain technology emerging as a promising solution to address issues like food safety, traceability, and transparency in the global food supply chain. This chapter explores the feasibility and effectiveness of blockchain technology in enhancing traceability within food supply chains, specifically focusing on Kashar cheese production. By leveraging the Hyperledger platform and employing an exploratory research methodology, the study aims to design a conceptual blockchain model tailored to address the unique challenges and requirements of the cheese supply chain, promising to revolutionize stakeholder interactions, information sharing, and collaboration. The dairy foods industry plays a crucial role in global food supply chains, with Kashar cheese facing traceability challenges despite its economic significance. Further research is needed to address limitations and challenges of the proposed blockchain model, including practical testing, coverage expansion, adoption challenges, and exploring alternative blockchain platforms for scalability, applicability, and sustainability in food supply chains.

Keywords Blockchain technology · Traceability · Food supply chain · Hyperledger

---

E. Dursun (✉) · Y. Gunalay
Department of Logistics Management, Faculty of Economics, Administrative and Social Sciences, Bahçeşehir University, Istanbul, Turkey
e-mail: elif.dursun@bau.edu.tr

Y. Khojasteh, S. M. M. Jahan Abadi (eds.), *Emerging Technologies in Supply Chains*, International Series in Operations Research & Management Science 373, https://doi.org/10.1007/978-3-032-01218-0_7

7.1 Introduction

Globalization of food supply chains has manifold benefits, yet it has introduced a heightened risk of food-borne diseases. This critical issue raises the concern of whether the food we consume is safe, a question that has become increasingly difficult to answer. Tracking the origin of the food on our plates is like assembling a complex, thousand-piece puzzle. Even if food is locally processed, one cannot be certain that all inputs such as seeds and pesticides are locally sourced. Companies continuously strive to reduce costs and maximize profits, often seeking cost-cutting measures throughout the supply chain.

These developments are also influenced by consumers. Businesses are forced to adapt their operating procedures in response to the growing consumer demand for rapid, low-cost, and high-quality food goods. Managing these processes is difficult in the food industry, which is rapidly growing to satisfy increasing consumer needs. From suppliers of raw materials to final customers and back, food supply chains entail a complex web of information exchange and transactions. For every organization, handling these complexities without any problems is essential. Even with technological improvements, contemporary supply chains sometimes don't offer total traceability and transparency.

Providing dependable product delivery, avoiding risks and contamination, enabling real-time information exchange, guaranteeing information security and authenticity, automating information sharing, and creating extensive traceability systems within a frequently centralized management framework are the main concerns in food supply chains. Furthermore, the lack of trustworthy information and bad decisions brought on by information security problems erode customer confidence.

Due to these intricacies, companies are compelled to look for alternate ways to give customers and other supply chain participants reliable information. Promising opportunities are presented by contemporary technical developments like artificial intelligence (AI), the Internet of Things (IoT), near-field communication (NFC), quick response (QR) codes, and blockchain technology (BCT). In particular, the distributed, decentralized, secure, and consensus-based nature of blockchain technology makes it unique. Blockchain's ledger features, which were first made popular by cryptocurrencies in the financial industry, are finding strong uses across a number of industries.

The tracking and documentation of a product and all related data from its point of origin to its point of consumption and back is known as supply chain traceability. To find anomalies and unforeseen occurrences in the supply chain, effective traceability is essential. Enterprises use various instruments and approaches to monitor data and merchandise movement in their supply chains; yet these endeavors frequently lack uniform structures bolstered by robust legal support. BCT offers a crucial answer to this centralized environment's traceability problem. The distributed-decentralized structure of blockchain, along with its consensus process, immutable data record, transparent and accessible network, proof mechanism, and automated smart contracts, make it an innovative tool for improving traceability in food supply chains.

The traceability of food production would be greatly increased by using blockchain, which is crucial for businesses as well as for the population's general health and sustainability. Food safety is a critical matter that requires more attention from businesses and consumers, as seen by recent outbreaks of food-borne illnesses. The continued occurrence of tainted food cases implies that current traceability methods are inadequate. Therefore, it is imperative to investigate alternate approaches to food chain traceability using developing technology.

The primary objective of this exploration is to determine the requirements of an effective traceability system and to develop a conceptual blockchain model that benefits both businesses and consumers. The central research question addressed is: "Will blockchain technology be an alternative traceability system for food supply chains?"

Given the exploratory nature of the topic, this chapter focuses on developing a thematic model for food chain traceability on top of blockchain system. Beforehand, qualitative analysis of reports through secondary research is conducted. A thorough review and analysis of pilot studies from related literature are undertaken to formulate a framework for determining the requirements of a traceability system. Following this, the conceptual framework of the blockchain model is designed.

To concretize this conceptual framework, an example of a Kashar cheese supply chain is used. This application aims to illustrate how blockchain technology would be effectively integrated to improve traceability, ensure food safety, and restore consumer trust. Through blockchain's decentralized and transparent nature, it is possible to create a more efficient and trustworthy food supply chain system, capable of addressing the current challenges and improving overall food safety standards.

In summary, while the worldwide food supply chains pose significant challenges to food safety, advanced technologies, particularly blockchain, offer promising solutions. By utilizing blockchain's decentralized ledger system, businesses can achieve greater transparency, security, and efficiency in tracking food products from origin to consumption. This enhanced traceability system not only helps in identifying and mitigating risks but also in making sure of uniqueness and safety of food products, thereby fostering a healthier and more trustworthy food supply ecosystem.

7.2 Concepts and Related Literature

The review begins by defining blockchain terminology and discussing various blockchain technology systems. Next, it covers the significance of the supply chain in the food industry and examines current studies on supply chain management (SCM). Finally, it reviews existing traceability systems in the food industry, identifying the benefits and challenges of building supply chain traceability systems, setting the foundation for a blockchain traceability model.

7.2.1 Blockchain Technology (BCT)

Blockchain is essentially a digital ledger of transactions that are distributed through a network of nodes. Defined by Gupta (2017) as "a shared, distributed ledger that facilitates the process of recording transactions and tracking assets in a business network," blockchain technology incorporates peer-to-peer networking, cryptography, distributed databases, consensus mechanisms, and immutability, significantly influencing business strategies and relationships. Bashir (2017) adds that blockchain is "cryptographically secure, append-only, and updateable only via consensus among peers." There are three network types: Centralized, decentralized, and distributed. Centralized networks are vulnerable due to their single point of failure structure. In contrast, distributed systems like blockchains mitigate this vulnerability by distributing data across all nodes. Every node possesses a copy of the validated data, making it extremely difficult to alter.

7.2.1.1 Blockchain Terminology

Cryptography is fundamental to blockchain, ensuring data confidentiality, integrity, authentication, and non-repudiation. Julius Caesar's Cipher was an early form, evolving to modern cryptography which David Chaum and Nick Szabo advanced significantly. Cryptographic techniques like hashing play a vital role in blockchain (Gates, 2017).

Hash Functions and Hashing It enables conversion of data of any size into a fixed-size string, ensuring data integrity and unique identification. Popular hash function families include MD (Message Digest) and SHA (Secure Hash Algorithm), with applications in digital signatures and data integrity protocols (Laurence, 2017).

Peer-to-Peer (P2P) Networks enable decentralized activities without a central authority, essential for blockchain. Variants include pure, hierarchical, and hybrid P2P networks, each with unique characteristics and applications, such as Spotify and BitTorrent for hybrid systems (Li, 2017).

Consensus Mechanisms ensure that data added to the ledger is validated by the network, reinforcing data integrity. Different alternatives include Proof of Work (PoW), Proof of Stake (PoS), and Practical Byzantine Fault Tolerance (PBFT). ***Proof of Work (PoW)*** requires computational power to validate transactions. Miners solve complex problems to add blocks, discouraging fraudulent behavior due to high costs. However, it's energy-intensive and slow. ***Proof of Stake (PoS)*** relies on the number of coins a participant holds. Validators, chosen based on stake and other factors, validate transactions, making the process faster and eco-friendlier. Ethereum's Casper protocol and Peercoin are notable examples. ***Practical Byzantine Fault Tolerance (PBFT)*** focuses on maintaining consensus even if some nodes act maliciously. It doesn't rely on mining or rewards but ensures system integrity through node agreement on transactions (Morabito, 2017; Singhal et al., 2018; Usta & Doğantekin, 2018).

7.2.1.2 Blockchain Types and Their Applications in Various Industries

Blockchain networks exhibit a range of access and operational methodologies tailored to fulfill diverse requirements such as contractual actualization, data storage, and monetary transactions. The key classification pivots around access rights:

Public and Permissionless Blockchains: Accessible to any participant for transaction reading, writing, and creation. This model epitomizes decentralization and inclusivity. ***Private and Permissionless Blockchains:*** Allow unrestricted writing access; however, reading and transaction initialization are confined to selected entities, offering a blend of openness and controlled dissemination. ***Public and Permissioned Blockchains:*** Restricts writing and validation rights to a select group, while allowing public visibility of the blockchain, balancing transparency with selective participation. ***Private and Permissioned Blockchains:*** Highly restrictive, limiting transactional interactions (reading, writing, creating) exclusively to authorized participants, ensuring stringent privacy and access control (Drescher, 2017). ***Consortium Blockchains:*** Bambara and Allen (2018) introduce a new category, the consortium blockchain, which caters to a predefined set of validating nodes, exemplifying a "partially decentralized system." This model often serves consortia of organizations where consensus and transaction validation are managed collaboratively.

Sector-Specific Blockchain Applications

Legal Industry: Blockchain technology, particularly through smart contracts, is poised to revolutionize legal practice by automating and securing contractual agreements. Legal professionals might evolve into technologists, ensuring precise dispute resolution and agreement fulfillment. The projected growth in global legal services, compounded by the inefficiencies in document management, positions blockchain as a vital tool for enhancing accessibility, transparency, efficiency, and reducing operational costs (Bambara & Allen, 2018). ***Insurance Sector:*** The insurance industry stands to gain significantly from blockchain through enhanced product and market strategies driven by efficiency, cost savings, and fraud mitigation. Smart contracts can facilitate parametric (index-based) products, streamline claim handling, and bolster anti-money laundering (AML) and know-your-customer (KYC) procedures. The integration of IoT and blockchain could address the oracle problem, thereby automating insurance claims effectively (Usta & Doğantekin, 2018). ***Financial Services:*** Blockchain, the progenitor of decentralized financial systems (DeFi), offers an alternative to traditional finance by eliminating intermediaries. Innovations in commercial financing, trade finance, and cross-border transactions are evidence of blockchain's disruptive potential. DeFi facilitates decentralized currencies, payment services, and fundraising, ushering in a new era of financial inclusivity and borderless transactions (Chen & Bellavitis, 2020). ***Supply Chain and Logistics:*** Supply chain management might undergo revolutionary changes as a result of blockchain technology's increased traceability, transparency, and cost effectiveness. Many times, modern technologies like RFID and ERP are unable to deliver dependable and safe solutions. By ensuring improved monitoring and provenance of items from manufacture to delivery, blockchain's immutable ledger and smart contracts may reduce the risk of counterfeiting and inefficiencies. By enabling real-time visibility, blockchain's capacity to foster participation may also greatly increase customer happiness and confidence (Consensys, 2020).

Prominent Blockchain Platforms

- ***Bitcoin:*** The pioneering blockchain platform, facilitates decentralized P2P payment transactions, removing the need for financial intermediaries. This platform is a Proof of Work (PoW) mechanism to secure and validate transactions, ensuring transparency and reducing transactional inefficiencies.
- ***Ethereum:*** Ethereum extends beyond simple transactions, offering a robust environment for decentralized applications (dApps) and smart contracts. The transition to a Proof of Stake (PoS) mechanism in Ethereum 2.0 aims to enhance network scalability and performance, making it a versatile platform for businesses reliant on contractual agreements.
- ***Hyperledger:*** An open-source collaborative platform provided by the Linux Foundation, focuses on developing modular and interoperable blockchain frameworks tailored for enterprises. It supports various industry-specific projects, offering high security and flexibility through its diverse frameworks such as Hyperledger Fabric and tools like Hyperledger Composer.
- ***Corda:*** Corda, developed by R3, caters specifically to enterprise needs by facilitating streamlined and secure business transactions through pre-agreed smart contract logic and unique features like notary pools. Corda emphasizes interoperability and cost reduction, positioning itself as a critical infrastructure for business agreements and data processing (Brown, 2018; Nakamoto, 2008; Takyar, 2024; The Linux Foundation, 2025).

The way blockchain technology is developing shows how revolutionary it may be in many different industries by improving security, efficiency, and transparency. Blockchain architectures are intentionally modified to satisfy a range of corporate and organizational objectives, from private, permissioned systems that provide strict control to public, permissionless networks that enable broad involvement. Together, these focused uses and strong platforms point to a big, varied influence of blockchain in the future on a variety of industries, from supply chain management to legal and financial services.

7.2.2 Blockchain in Supply Chain

Blockchain technology has the ability to completely transform supply chains in a number of industries because of its decentralized, transparent, and secure features. Blockchain technology has the potential to improve the traceability, security, integration, and transparency of traditional supply chain systems, thereby mitigating several inefficiencies and difficulties. The automotive, pharmaceutical, culinary, agricultural, and energy sectors are important application areas. A number of academic and practical studies have investigated blockchain's feasibility and benefits in the supply chain context. Vayadande et al. (2024), in their paper, uses smart contracts and Ethereum for agriculture traceability by using QR codes. Toader et al. (2024) aims to find what are the effecting factors for intention to use and the actual usage of blockchain-driven agri-food supply chain platforms by utilizing the Unified Theory of Acceptance and Use of Technology (UTAUT) including data from 175 participants from agri-food companies across eight European countries. Another paper by Safeer and Pulvento (2024) builds a solution that uses blockchain technology to enhance transparency, traceability and operational efficiency of Italian tomato supply chain by integrating Solidity, Remix IDE, MetaMask wallet, and Sepolia Testnet. Bär et al. (2023) proposes a system called Digital Product Passport for dairy products, which enables all stakeholders to trace products and processes in real-time. Prasad and Chidambaranathan (2024) aspires to implement blockchain technology in Dairy Supply Chain (DSC) in the Telangana region, seeking to illuminate the current state of milk collection procedures and the consequential impact of integrating Information and Communication Technology (ICT) tools. Tse et al. (2017) propose a decentralized food supply chain model for China to improve information security and traceability. Sternberg and Baruffaldi (2018) examine the disruptive potential of blockchain in supply chains and highlight implementation challenges. Casado-Vara et al. (2018) suggest a multi-agent system utilizing smart contracts for order monitoring in agricultural supply chains. Benton et al. (2018) emphasize blockchain's role in enhancing traceability and operational efficiency, citing IBM and Maersk's successful pilot of Hyperledger. Thiruchelvam et al. (2017) propose a blockchain model to enhance traceability, sustainability, and market access in the Burundian coffee sector. Santonino III et al. (2018) and Abeyratne and Monfared (2016) explore blockchain's application for tracking and integrating documentation in aviation. Hypothetical scenarios suggest blockchain can significantly enhance data accuracy, decision-making, and fraud prevention in manufacturing. Casino et al. (2020) find blockchain adoption enhances trust, efficiency, and quality but raises scalability and data management issues in dairy sector. Queiroz and Fosso Wamba (2019) conduct empirical studies in India and the USA, revealing significant barriers to blockchain adoption, particularly in emerging markets, due to infrastructural and technical constraints.

7.2.2.1 Early Adopters and Case Studies

Several major corporations have begun implementing blockchain technology in their supply chains. ***IBM Food Trust:*** Enhances food chain transparency, safety, and efficiency by using Hyperledger Fabric. Notable pilot studies include tracking avocados for Walmart, reducing verification time from days to seconds, and ensuring the quality and safety of beef logistics. The joint offering—called iFoodDS Trace Exchange with IBM Food Trust—will offer businesses in the food industry and their suppliers a comprehensive, scalable, and secure traceability solution (Civitillo, 2023). ***Oracle Blockchain Solutions:*** Provides a SaaS platform based on

Hyperledger Fabric for end-to-end visibility and risk aversion in supply chains. Case studies include temperature tracking for food delivery, trust-building in beer brewing, and equipment tracking for reverse logistics (Oracle, 2021). ***Provenance:*** Offers transparency solutions for the food and beverage industry, enabling consumers and businesses to trace the journey and impact of products. Notable projects include tracking fish for Princess Group and preventing fraud in the Italian tomato supply chain for Napolina (Provenance, 2025). ***Fetch.ai:*** Utilizes AI and blockchain to create decentralized digital economy infrastructures. Applications include smart city projects in Germany, decentralized commodity exchanges, and intelligent supply chains for optimization and efficiency (Fetch.ai, 2020).

7.2.2.2 Advantages of Blockchain in Supply Chains

Main advantages include as follows: ***Cost Efficiency:*** Reduction in documentation and audit-related costs, enhanced payment tracking, and improved inventory management. ***Speed and Trust:*** Rapid access to information and transactions, and increased trust through decentralized and immutable records. ***Sustainability:*** Encourages the use of local materials, reduces environmental impact, and motivates renewable energy consumption. ***Security and Flexibility:*** Encryption and consensus mechanisms ensure secure and reliable data sharing, and the decentralized ledger fosters flexible, transparent operations (Kshetri, 2018).

7.2.2.3 Challenges and Limitations

As every technological transition has challenges, blockchain technology also contains hardships. In short, these can be listed as technical limitations, scalability issues, efficiency and maintenance issues, technology adaption problems, lack of regulative instruments, lack of qualified people in implementation, finding the right platform to deploy, and excessive computing power needed for the technology to operate (Agarwal, 2018; Min, 2019; Perboli et al., 2018).

7.2.2.4 Food Supply Chains and Safety

The food sector has received heightened attention due to its critical impact on human health and environmental sustainability. Blockchain can address issues of traceability and safety, helping ensure food quality, reduce waste, and prevent health-related problems. Studies and case applications reveal blockchain's potential for providing transparent, traceable, and reliable food supply chains. International and national bodies like WHO, Codex Alimentarius, and the European Commission are pivotal in developing and enforcing food safety standards. Blockchain technology can support these regulatory frameworks by offering secure and transparent methods for monitoring and verifying compliance. Academic research supports blockchain's transformative potential in various supply chain sectors. Studies highlight the importance of integrating emergent technologies like IoT, RFID, and AI with blockchain to enhance supply chain traceability, safety, and efficiency. In summary, blockchain technology has shown significant promise for improving various dimensions of supply chains. These encompass increased traceability, enhanced security, better integration, and heightened transparency. The continued academic and practical exploration of blockchain applications in sectors such as automotive, pharmaceuticals, food, agriculture, and energy will likely yield innovative solutions that address traditional supply chain inefficiencies and pave the way for more robust and resilient future supply chains (Yele and Litoriya, 2024; European Commission, 2000; FAO and WHO, 2019; Statista, 2024a; World Healt Organization, 2024).

7.2.3 Traceability

Scientia potentia est.—"Knowledge is power," underscores the timeless vitality of knowledge in our lives, both socially and in business. Traceability is the practice of tracing knowledge across various aspects, imperative for individuals, corporations, and governments. The significance lies in understanding the origins and impacts of what we consume, a necessity heightened by the complex global supply chains of today. The practice of traceability mitigates unconscious production and consumption, providing a profound impact on managing side effects. From a business standpoint, traceability is crucial as it offers insights into products, processes, financial flows, and inventory management. Modern consumers demand transparency, challenging businesses to ensure reliable information about product origins and processes. This practice extends beyond mere logistics, embedding itself as a cornerstone of trust and safety in various sectors, particularly food.

7.2.3.1 Food Traceability and Its Necessity

Ensuring food traceability becomes essential as globalization complicates the food supply chain. Back in the good old days, people farmed their own food and knew where it came from. These days, components for goods come from all over the world, thus keeping track of "What, where, when, and how" in food manufacturing calls for a sophisticated record system. Because of this intricacy, several definitions and standards have been created to make tracing easier. Many definitions define food traceability as the capacity to track food through all phases of processing, distribution, and manufacturing while guaranteeing its authenticity and safety.

Identification, connection, recording, gathering, storing, and verifying information are all part of a traceability system along the food supply chain. It strikes a balance between organizational and management resources, established processes, regulatory frameworks, and norms. Systems can be paper-based, which is appropriate for smaller businesses, or electronic-based, which is favored in bigger, internationally integrated businesses.

A well-devised traceability system ensures effective internal and external data management, enabling companies to track materials and processes within and outside their organization. Several models and frameworks, such as those developed by Moe (1998), Olsen and Borit (2013), and Bosona and Gebresenbet (2013), emphasize the integration of traceable resource units (TRUs), identification keys, process and product characteristics, accessible mechanisms, and advanced technologies like Wireless Sensor Networks (WSN) and the Internet of Things (IoT).

7.2.3.2 Practical Implications and Benefits in Food Chains

The food industry has a greater need for reliable traceability systems since it is more likely to experience well-known outbreaks and public health accidents. For the purpose of preserving customer confidence and lessening the impact of food safety violations, these systems provide accountability and transparency from the point of production to the point of consumption. Companies must comply with strict traceability regulations enforced by regulatory authorities, including the European Food Safety Authority (EFSA) in Europe, and keep comprehensive records covering all stages of the food supply chain. Main steps to implementing traceability systems include ***Pre-Planning:*** Establishing clear, achievable objectives based on available resources and technology. ***Data Management:*** Employing systems that can comprehensively track and store data across supply chains. ***Advanced Technologies:*** Utilizing technologies like IoT, blockchain, and RFID to enhance traceability. It basically aims to achieve food safety by readily identifying and isolating problems, establishment of reliable and transparent food histories to build trust, and achieving efficiency gains in inventory, quality management, and communication. Table 7.1 summarizes some blockchain applications for food traceability.

7.2.3.3 Key Issues, Regulations, and Standards

Food traceability faces significant hurdles classified into technical, managerial, and environmental categories. Issues like high investment costs, prioritization of other business areas, and lack of expertise and comprehensive knowledge can impede adopting robust traceability systems. The food industry's unique challenges, such as diverse product characteristics, multi-stakeholder environments, and the necessity to prevent cross-contamination, add to the complexity. Factors to impact traceability systems can be listed as: ***Technological Complexity:*** High initial investments, particularly for small companies. ***Regulatory Compliance:*** Adapting to evolving legislative frameworks at national and international levels. ***Industry Specifications:*** Tailoring systems to diverse food types and processes.

Table 7.1 Summary of proposed blockchain applications for food traceability in literature

Reference	Objective	Contribution	BCT platform
Rodríguez-Molina et al. (2024)	Introducing a cost-effective Software-as-a-Service (SaaS) solution that leverages Internet of Things (IoT) technology and blockchain—specifically, the VeChain network—to provide real-time data on crop conditions and weather forecasts	The proposed solution enables farmers to make informed decisions by offering actionable insights and monitoring crops growth and general progress over time, thus promoting more efficient and sustainable farming practices	The VeChain network
Das et al. (2024)	To illustrate the diverse ways in which blockchain technology can be applied in the agriculture sector, additionally, it will resolve current problems	Blockchain technology presents a viable means of promoting a more reliable, superior, safer, and environmentally sustainable agri-food system in the future	Agri-Chain on the Ethereum blockchain
Sakthivel et al. (2024)	To investigate how a technological blockchain architecture addresses supply chain challenges in precision Agriculture to take up challenges related to transparency traceability, and security	A robust registration phase, efficient authentication mechanisms, and optimized data management strategies are the key components of the proposed architecture	The Hyperledger network
Singh Thakur et al. (2024)	To tackle issues in the grain supply chain, a blockchain-based solution is proposed	Combination of non-fungible token (NFT), An Internet of Things-based circuit, sensors, and GPS module is used that unites farmers, warehouses, government central and state agencies, transporters, and food corporations	The Ethereum network
Venkateswara Reddy et al. (2025)	Presenting a Green Blockchain-based traceability technology that is used as part of the EU-funded "Typicalp" project to monitor the Fontina PDO cheese supply chain	The whole production chain has been digitalized thanks to proven traceability technology, which gives Fontina consortium operators and end users instantaneous access to irreversible data	The Algorand Blockchain
Lalitha and Sasikala (2024)	To propose a Chain of Things (CoT) based traceability system (CoT-TS) to alleviate the shortcomings of the prevailing traceability system in agriculture	The proposed CoT-TS fetches the benefits of transparent, tamper-proof, and incontrovertible transactions integrated with decentralized monitoring and tracing operations	NA

(continued)

Table 7.1 (continued)

Reference	Objective	Contribution	BCT platform
Feng et al. (2024)	To design and implement a tilapia-blockchain IoT traceability system (T-BITS) based on Hyperledger Fabric	The T-BITS system is more effective to capture and trace the critical ambient and quality information for tilapia cold chain	Hyperledger Fabric
Valarmathi et al. (2024)	To introduce AgriBlockIoT, a decentralized blockchain-based traceability system designed for managing supply chains in the agri-food sector	Enhanced traceability within this system helps minimize wastage, detect suspicious activities, and streamline auditing and record-keeping in the food supply chain	Ethereum and Hyperledger Sawtooth

Legal and Regulatory Frameworks

International standards include ISO and HACCP. ISO Standards (ISO 9001:2000 and ISO22000:2005) focus on quality and food safety management, incorporating traceability as a core component while HACCP (Hazard Analysis and Critical Control Points) provides guidelines for tracking hazardous materials within food products. Regional legislations, on the other hand, can be listed as:

- ***European Union:*** Mandates strict traceability compliance through EFSA regulations like EU No 1169/2011 (European Parliament, CEU, 2021), General Food Law (GFL), requiring detailed labeling and traceability.
- ***United States:*** Enforces guidelines via documents like "Traceability for Food Marketing and Food Safety," stressing similar compliance.
- ***Turkiye:*** Aligning with EU standards, Turkiye's regulations emphasize comprehensive traceability systems through legislative acts like Law No: 5996 (İstanbul İl Tarım ve Orman Müdürlüğü, 2010).

7.2.3.4 Benefits and Advances in Traceability Technology

Putting in place a strong traceability system has several advantages, from improved regulatory compliance to increased operational efficiencies. Traceability systems enable (Queiroz et al., 2019; Tian, 2017):

- Accurate order fulfillment.
- Reduced inventory costs.
- Mitigation of information asymmetry.
- Swift responses to unforeseen supply chain disruptions.
- Enhanced customer and stakeholder satisfaction.
- Improved organizational management.

Additionally, according to Moe (1998), some specific benefits include:

- Optimization and control of production processes and product quality.

- Rapid problem resolution and minimization of costs.
- Compliance with audit and certification requirements.
- Effective recall management, preserving brand integrity.
- Competitive differentiation through transparency and consumer trust.

7.2.3.5 Challenges in Traceability

Despite its benefits, implementing traceability systems in food supply chains is fraught with challenges, such as those identified by Kamilaris et al. (2019) and Pearson et al. (2019):

- ***Complex Supply Chains:*** Diverse product characteristics, such as bulk vs. packaged goods, complicate the traceability. Also, fresh products demand more rigorous tracking due to perishability.
- ***Stakeholder Involvement:*** If all supply chain actors do not participate, the system's efficacy is compromised. Small entities often find it prohibitive due to high entry costs.
- ***Data Management:*** Managing extensive data across multiple stages and entities is difficult. Cross-contamination and preserving the integrity of information across batches further exacerbate these challenges.
- ***Perception and Knowledge:*** In some sectors, traceability is seen as burdensome or unnecessary. A lack of thorough understanding and prioritization can hinder widespread adoption.

7.2.3.6 Future Directions in Food Chain Traceability

Initiatives for food traceability are trending in the direction of a more comprehensive strategy using cutting-edge technology and collaborative frameworks. Important factors influencing the landscape of the future include: ***IoT and blockchain***: Blockchain technology promises to improve supply chain data accuracy and security when it is integrated with IoT devices. Blockchain maintains the data's integrity and immutability while IoT delivers real-time logistical and environmental data. ***Smart Contracts***: Supply chain operations may run more smoothly and dependably by automating transaction and verification procedures within the blockchain.

Harmonizing international standards is essential to streamlining international food supply chains. To create norms that are accepted by everybody, regional coalitions and international organizations like ISO must work together. To stay up with global norms and technological breakthroughs, nations constantly update and improve their legal frameworks. Ensuring strong traceability frameworks requires a broad and flexible regulatory framework. It is critical that scholarly and applied studies on food traceability continue. Focus areas include ***Economic Viability***: Looking at ways to use cutting-edge traceability systems at a reasonable cost for smaller supply chain partners. ***Data management*** is the process of creating complex yet intuitive systems that enable the smooth integration and administration of data amongst several stakeholders in supply chains. ***Customer Engagement***: Improving customer interfaces to deliver traceable information that is understandable and easily available, fostering transparency and confidence. ***Industry Cooperation***: Creating networks of suppliers, producers, regulators, and technology providers who work well together can help to create a food supply chain that is more responsive and robust. Industry consortiums and public-private partnerships can spearhead group initiatives for standardized traceability procedures.

Traceability is a complex topic with administrative, legal, and technical components that is involved in food supply chains. Robust traceability systems are vital, as underscored by the growing consumer consciousness and more stringent regulatory mandates. Notwithstanding some challenges, the use of revolutionary technologies like blockchain and IoT has the potential to alter food traceability practices. The food industry may benefit from these advancements by being more dependable, transparent, and efficient; these attributes will ultimately enhance public health and attract more consumers. Long-term research, collaborative efforts, and adaptable legal frameworks are essential to enhancing the efficiency of traceability programs and their widespread adoption throughout the food chain.

7.3 A Blockchain Based Traceability Model Development for Dairy Foods Sector

This chapter outlines a methodical approach to building a blockchain-based traceability model for the dairy food industry, with a particular emphasis on the supply chain for Kashar cheese. First, the dairy sector is examined in detail to give a complete background, and then the Kashar cheese supply chain is thoroughly examined. The paper goes into detail on the elements of the blockchain traceability model that are relevant to the supply chain participants and product phases. Presented is a prototype supply chain flow scenario for the production of cheese using the suggested blockchain technology.

7.3.1 Methodological Overview

The research aims to build a conceptual framework for implementing a blockchain-based traceability mechanism in food supply chains. Given the imperative nature of food traceability in mitigating foodborne diseases, the study underscores the importance of reliable, secure, and transparent information regarding the food's history throughout the supply chain. Traceability ensures quality and hygiene control, beneficial to producers, auditors, and consumers. Kashar cheese was selected due to its nutritional significance, encapsulating high protein and calcium content. The model development began with an analysis of a Kashar cheese manufacturing company's supply chain flow, utilizing a generic supply chain diagram previously established.

The initial phase of creating the blockchain model involved requirement identification and platform selection, guided by criteria such as security, authenticity, and comprehensive traceability of all materials, transactions, and relevant entities. These requirements culminated in the selection of the Hyperledger platform, deemed suitable due to its feature set despite its limitations concerning smart contract capabilities.

Subsequent steps entail identifying process actors to delineate their roles and synergies within the blockchain platform, followed by the determination of required and generated data at each supply chain stage. Data requirements were aligned with the Critical Tracking Events (CTEs) and Key Data Elements (KDEs) frameworks as updated by the Institute of Food Technologists (IFT). Following component specification, a conceptual design phase elucidates the relationships among design elements categorized into actors, assets, transactions, and inquiries, ensuring proper security and access controls.

7.3.2 Sector Information

The dairy foods sector is vital due to its substantial contribution to nutrition, with significant global production and market dynamics. In 2024, global milk production reached 906 million tons, with notable contributions from India, Brazil, the EU, Russia, USA, Australia, Argentina, the UK, and New Zealand. The EU-27 led production, followed by US, India, and China (Statista, 2024b). Challenges in Africa included adverse weather and high operational costs. Global dairy trade demonstrated a rising trend, with Europe as the largest exporter and Asia as the largest importer. The global market for dairy foods was valued at US $487.4 billion in 2023, projected to grow to US $629.3 billion by 2030. (Foreign Agricultural Service/ USDA, 2024) Turkiye's collected cow milk and drinking milk in 2024 stood at approximately 23 million metric tonnes, with milk, yoghurt, and cheese remaining the primary products. Cheese production summed up to 800,000 metric tonnes per year (770 from cow's milk, 30 from sheep, goat, buffalo, and mixed milk). Through strategic initiatives such as the '2024–2028 Livestock Roadmap,' Turkiye is poised to overcome existing challenges and solidify its standing in the global dairy market (Milkypedia, 2025; Voluntary Report, 2024).

The Deloitte report (Deloitte, 2017) highlights shifting consumer demands towards healthier diets and the increasing need for transparency and safety in the dairy food supply chain. Key challenges include price volatility, evolving consumer expectations for product safety and quality, sustainability, and adaptation to new technologies. Rising distrust in brands underscores the necessity for robust traceability systems to provide detailed product information and enhance consumer confidence.

7.3.3 *Kashar Cheese Supply Chain Flow*

The chapter proposes a blockchain traceability model tailored to a generic Kashar cheese supply chain of a company previously lacking an effective traceability mechanism, despite using Enterprise Resource Planning (ERP) software (Ediz, 2014). With a significant production capacity and extensive operational infrastructure, the company processed various dairy products. The journey of milk from procurement, through stages of checking, pasteurization, and conversion with other ingredients, to the final product's packaging and distribution, is meticulously mapped out. Each process stage includes assigning unique party IDs to facilitate traceability within the blockchain system (Fig. 7.1).

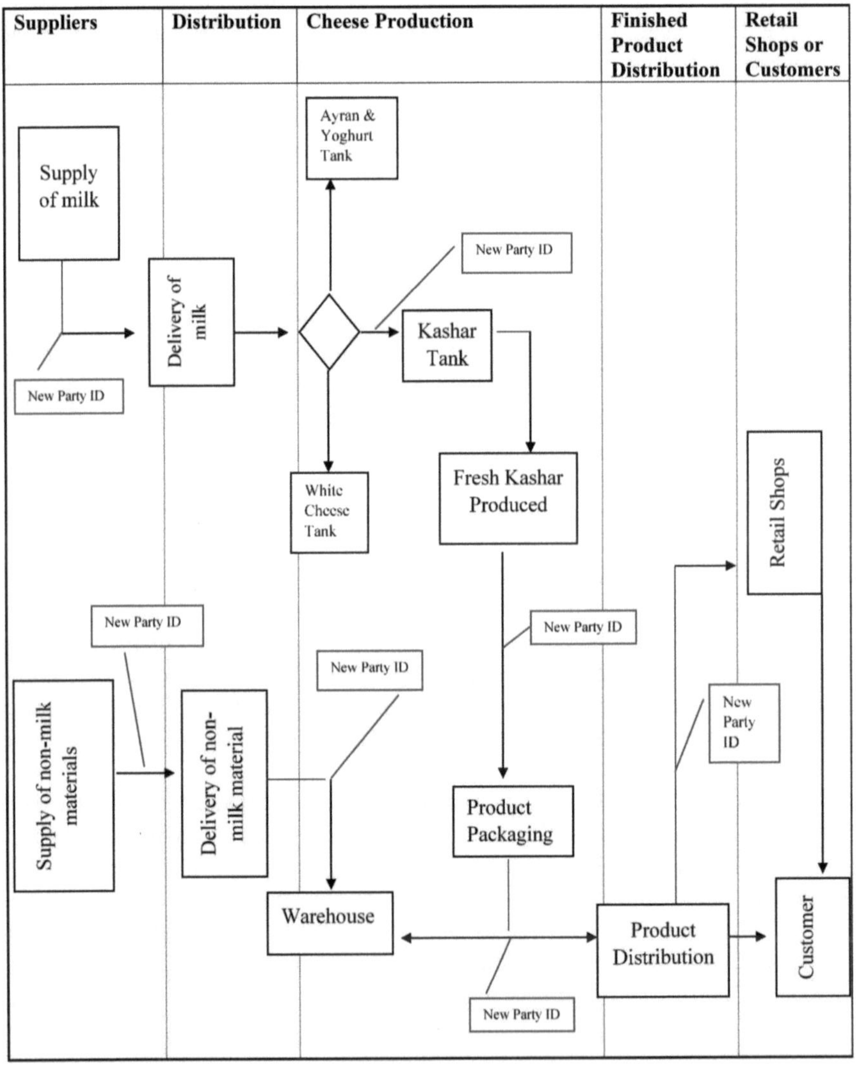

Fig. 7.1 Kashar cheese supply chain flow diagram

7.3.4 *Conceptual Blockchain Traceability Model*

The study outlines the creation of a blockchain-based traceability model for the Kashar cheese supply chain. Essential requirements from literature include information security, authenticity, and traceability throughout all production and distribution stages, achieved through synchronized networks and smart contracts. The Hyperledger framework is chosen due to its alignment with these requirements. However, the evolving nature of blockchain technology and the emergence of platforms like Ethereum 2.0 indicate potential shifts in suitability and preference for blockchain solutions, highlighting the competitive landscape and the necessity for continuous adaptation. Figure 7.2 shows the important differences between both platforms with respect to various aspects.

After selecting the blockchain platform and conducting a requirement analysis, the next stage is to identify the participants in the various processes. In order to accurately specify the roles and skills of the players who will be incorporated on the platform, this is necessary. Figure 7.3 depicts the actors involved in the Kashar cheese manufacturing flow.

Figure 7.4 displays the actors and their roles on the platform. Suppliers are responsible for raw material supply, information supply, and transactional data supply. Manufacturers, on the other hand, have more roles from record keeping of raw material to asset data keeping. External bodies like regulatory authorities or nongovernmental organizations (NGOs) are responsible for auditing for validations of the procedures. Figure 7.5 displays the regarding activities to fulfill these roles. Please see Fig. 7.5 for details.

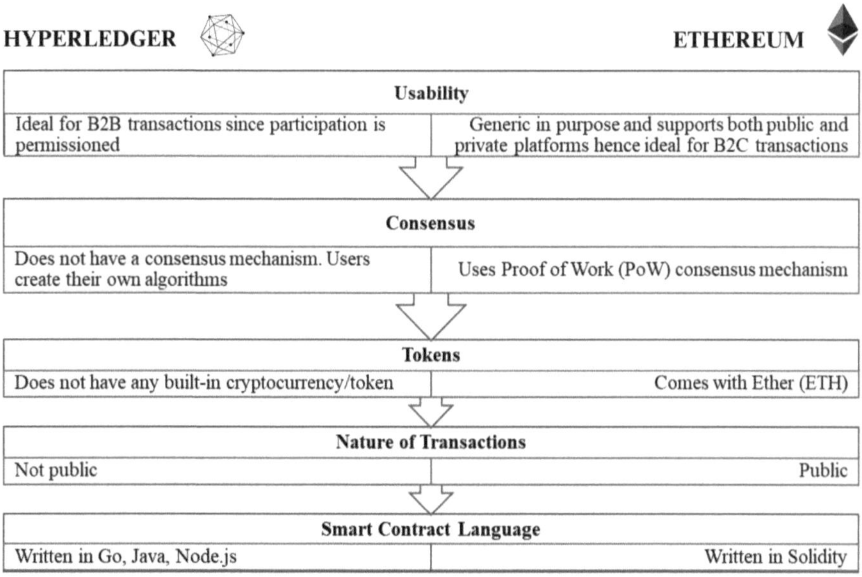

Fig. 7.2 Ethereum vs. Hyperledger (Anwar, 2021)

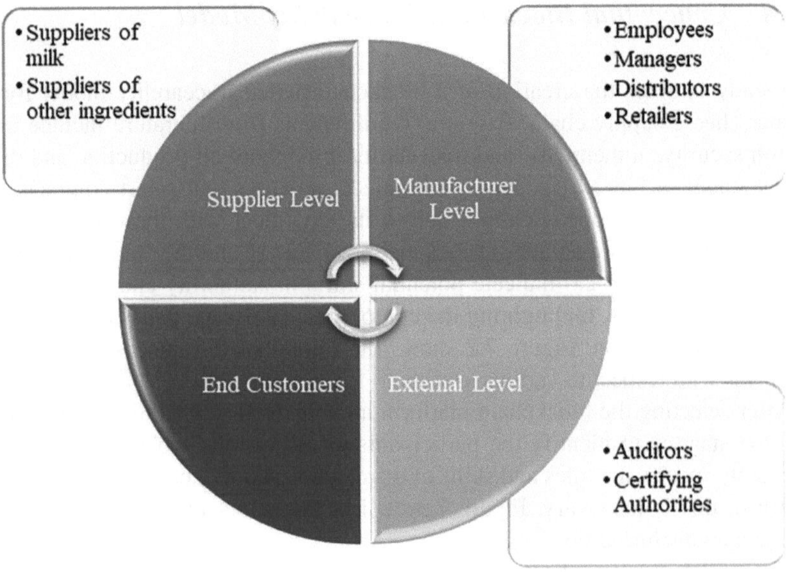

Fig. 7.3 Kashar cheese manufacturing actors and levels of actors

Identifying the necessary and generated data entries throughout the chain is the third phase. The data created by suppliers at the time the firm purchases the raw materials used in production is where this all begins. The Institute of Food Technologists' (IFT) updated Critical Tracking Events (CTEs), and Key Data Elements (KDEs) framework is consulted when determining the necessary data. Refer to Table 7.2 for further details.

Figures 7.6, 7.7, and 7.8 show the details of each level actors data generation. Figure 7.6, e.g., includes the supplier data that should be integrated into the platform. It is very essential to get these data from supplier to provide a transparent chain of flow during other levels from manufacturing to the end customers.

7.3.5 A Brief Scenario on the Proposed Blockchain Platform

The process begins with the registration of actors, assets, smart contracts, and transaction details on the platform using an admin panel on the Hyperledger network. The initial transaction is made by the supplier of raw materials, such as milk, packaging materials, and other ingredients necessary for producing Kashar cheese. The supplied data is logged into the system, triggering a contractual agreement. Raw materials are converted into assets, each tagged with a supplier ID. Suppliers input all required information into the Hyperledger platform via an access key. Certifications are created or verified as needed and uploaded onto the system.

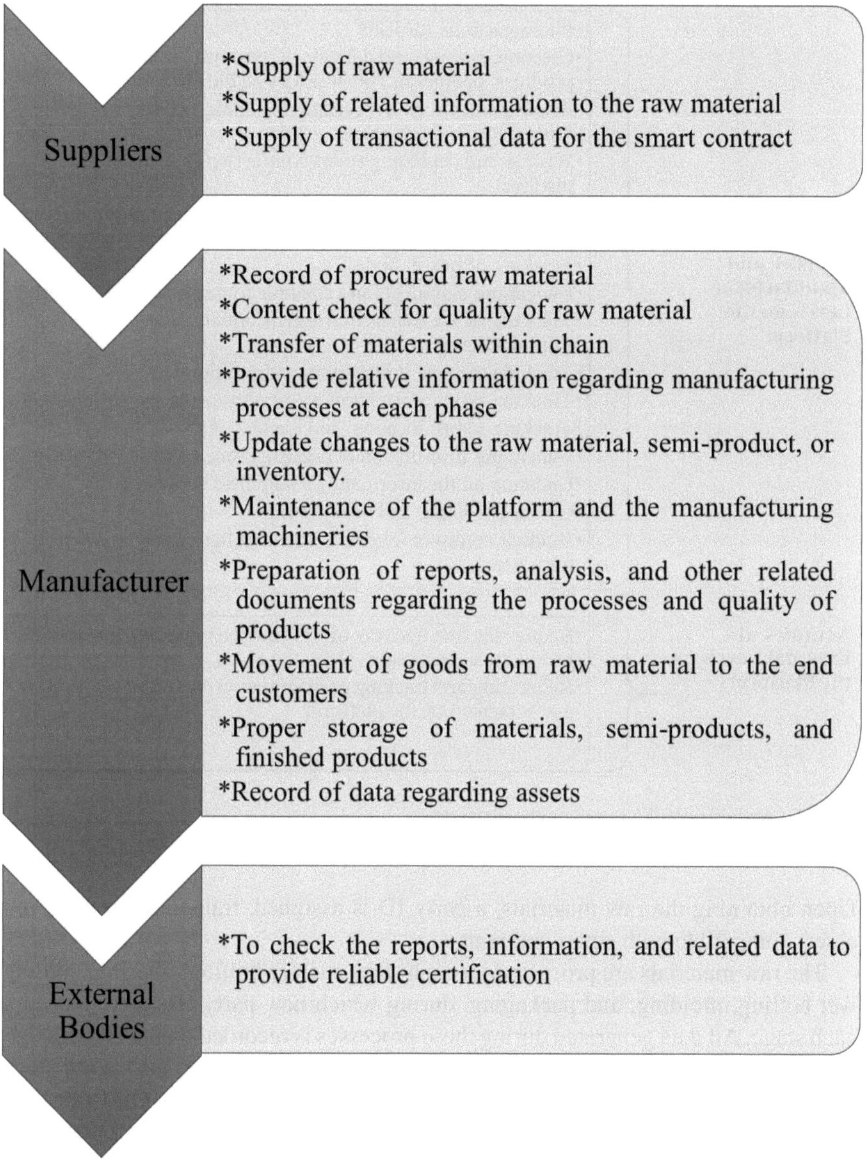

Fig. 7.4 Actors and their roles

Periodic external audits and certifications are carried out by regulatory bodies to ensure the manufacturer meets necessary conditions for producing Kashar cheese, such as proper equipment, trained employees, suitable temperature and location conditions, hygiene standards, nutritional requirements, and accurate labeling.

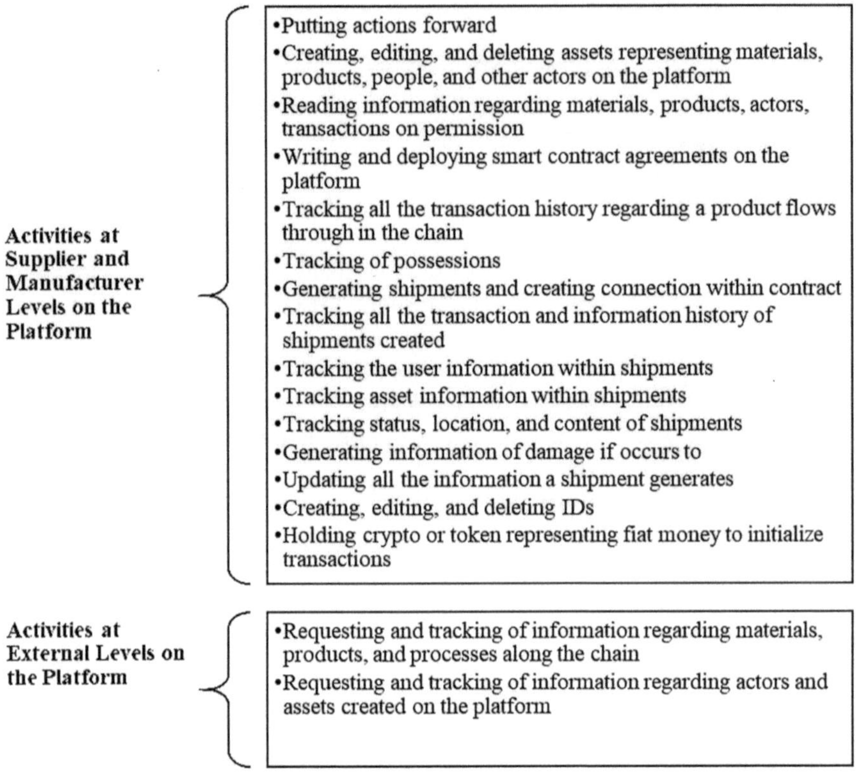

Fig. 7.5 Activities of the actors on the platform

Upon obtaining the raw materials, a party ID is assigned, transforming them into assets compatible with smart contracts.

The raw materials are processed through steps such as boiling, filtering, curding, wet boiling, molding, and packaging, during which new party IDs are attached at each stage. All data generated during these processes is recorded on the Hyperledger platform, with each party ID immutably documented. During the packaging stage, essential data for the end customer is recorded on the package using QR Code technology. This code can be scanned by mobile phones to direct the customer to the application platform, where they can access detailed product information according to permissions.

After the final product stage, distribution to warehouses or retailers is tracked, with transportation and storage conditions, distributor IDs, and timestamps recorded on the platform. Figure 7.9 illustrates the Hyperledger architecture for the cheese manufacturing company.

Table 7.2 IFT suggested key data elements (KDEs) for capture and recordkeeping at critical tracking events (CTEs) (Zhang & Bhatt, 2014)

	Critical tracking events (CTEs)					
	Transportation (exchange of goods)		Transformation (creation/ manipulation of products)		Depletion (exit from system)	
	Shipping	Receiving	Input	Output	Consumption	Disposal
Currently required KDEs						
Event Owner (firm submitting information)	X	X	X	X	X	X
Date/Time	X	X	X	X	X	X
Event Location	X	X	X	X	X	X
Trading Partner[a]	X	X	X			
Item (the good)	X	X	X	X	X	X
Lot/Batch/Serial Nr.	X	X	X	X	X	X
Quantity	X	X	X	X	X	X
Unit of measure	X	X	X	X	X	X
Linking KDEs						
Activity Type (for example, Purchase Order, Bill of Lading, Work Order)	X	X	X	X		
Activity ID (number associated with PO, BOL, Work Order)	X	X	X	X		
Transfer type[b]	X	X				
Transfer number[b]	X	X				
Lot/Batch Relevant Date[c]	X	X	X	X	X	X
Carrier ID	X	X				
Trailer number	X	X				

[a]In the event of a Shipping CTE, the trading partner is the immediate subsequent recipient of the shipment. In the event of a Receiving CTE, the trading partner is the immediate previous supplier of the product. In the event of a Transformation CTE, the trading partner is the supplier of the input into the Transformation

[b]If the Activity Type and ID are not linked to a particular shipment of a product (that is, a purchase order that is fulfilled by multiple shipments over time), then the Transfer Type and ID are used to indicate the particular shipments that are linked to the Activity Type and ID

[c]If there is a different Lot/Batch designation on a consumer-level product, such as a "best-by" date, it must link to the manufacturer-assigned lot number

Supplier Level Data

1. Supplier information (company ID, location, brief history, contact information, production capacity, certifications, employee profile, etc.)
2. Information regarding milk as major raw material (Milking time, employees' information responsible from milking, location of milking, animal ID, milking cup ID, milk quantity, milk contents – pH, fat, water, intensity, heath, dry material, antibiotic data, etc.-, party ID of milk, and movement of party ID)
3. Information regarding other ingredients
4. Information regarding supplementary raw material (packaging material ID, quantity, quality, movement of material, etc.)

Fig. 7.6 Actors' data generated on the platform: supplier level

Manufacturer Level Data

1. Procurement of raw material (supplier information, raw material information – party ID, quantity, quality, location, movement, responsible employee -, time stamp)
2. Content quality check (content measures – pH, fat, water, antibiotic levels -, time stamp of quality check, suitability report)
3. Raw material movement (record and tracking of party ID of collected milk through transfer to the tanks, mixing, boiling, drum filter, curding, wet boiling, molding, packaging, storing, and finally distribution processes, reports, extra notes, certificates for production, etc.)
4. Raw material location (e.g., in tank no 1, in boiling room 1, etc.)
5. Employee ID (name, contact info, position, assigned roles, access permissions, etc.)
6. Machinery ID (description, functionality, time bought, maintenance requirements, etc.)
7. Process descriptions (each process Kashar cheese pass through should be described in detail to provide automation through smart contracts)
8. Relationships between processes (order of flows should be defined precisely)
9. Distributor ID (company ID, location, contact, brief information, capacity, asset requirements – e.g., cold chain transportation -, certificates, etc.)
10. Inventory information (record of items, semi-products, finished products, etc. by party IDs to enable proper track of inventory)
11. Actor related information (user data, category, access date, access keys, permissions, etc.)
12. Retailer ID (company ID, location, contact, brief information, demand data, demand fulfillment, etc.)
13. Semi-product information (time stamp, quantity, quality, whey produced, waste produced, location, movement, etc.)
14. Finished product information (time stamp, quality, shape, quantity, packaging, labeling, storage, distribution, etc.)

Fig. 7.7 Actors' data generated on the platform: manufacturer level

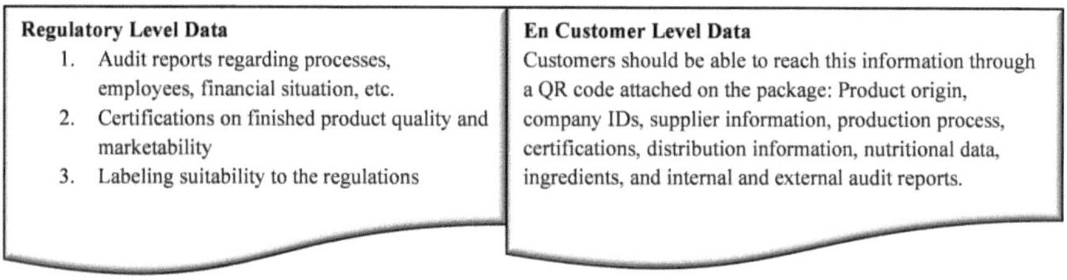

Fig. 7.8 Actors' data generated on the platform: regulatory and end customer level

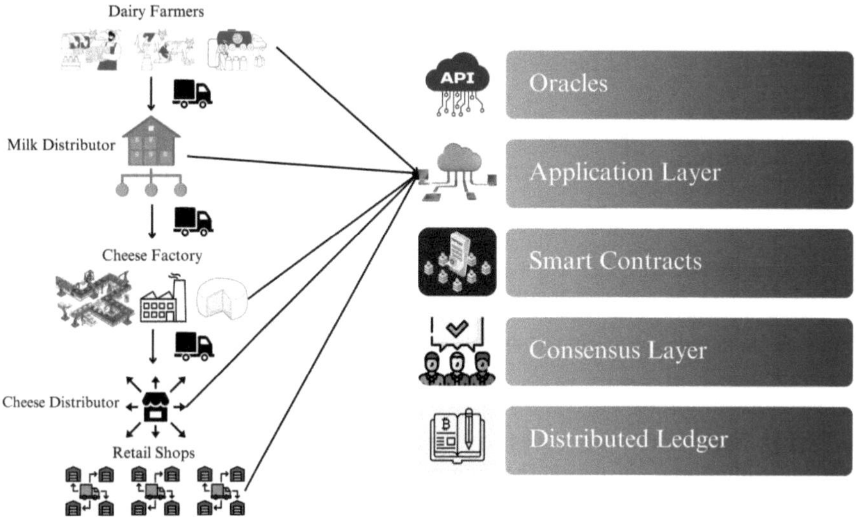

Fig. 7.9 Hyperledger architecture for cheese production

7.3.6 *Conceptual Model Design*

After defining the model components, we proceed to the conceptual design, detailing design elements and their relationships. The design elements are categorized into actors, assets, transactions, and inquiries. Actors were previously defined, with the admin added as the main controller of the platform. Assets represent products and materials on the platform, encompassing products, product movements, and contracts establishing conditions. Transactions reflect the activities of actors on the platform, recorded on the ledger and irreversible. Inquiries are commands to retrieve specific data, such as asset owners or locations.

The next step involves setting permissions to ensure the model's security and accessibility. This defines who can view and perform actions on the platform and how data related to actors, transactions, and assets are managed. The admin can create, edit, delete, read, and make inquiries on the platform. Regulatory actors,

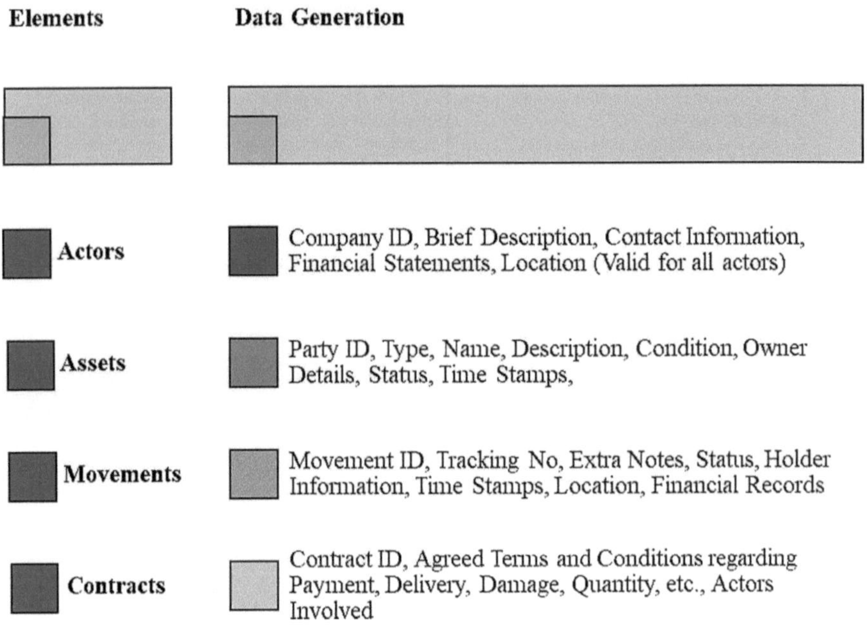

Fig. 7.10 Architecture elements and their data generation

however, have read-only access to all recorded data to conduct accurate audits and certifications, ensuring the company's authenticity and safety. Refer to Figs. 7.10 and 7.11 for detailed information on transactions, permissions, and recorded data.

7.4 Evaluation of the Model

The proposed blockchain model demonstrates a robust framework for addressing traceability and compliance challenges in Kashar cheese production. By integrating Hyperledger's permissioned network architecture, the model establishes a secure, immutable ledger for tracking raw materials (e.g., milk, packaging) from supplier to end consumer. Three core strengths emerge as below:

1. **Automated Compliance:** Smart contracts enforce predefined conditions (e.g., hygiene standards, temperature controls), reducing human error and ensuring adherence to regulatory requirements during audits.
2. **Granular Traceability:** Party IDs and QR codes enable stakeholders to trace product history at every stage (e.g., boiling, packaging), while restricting data access via permission tiers ensures confidentiality.

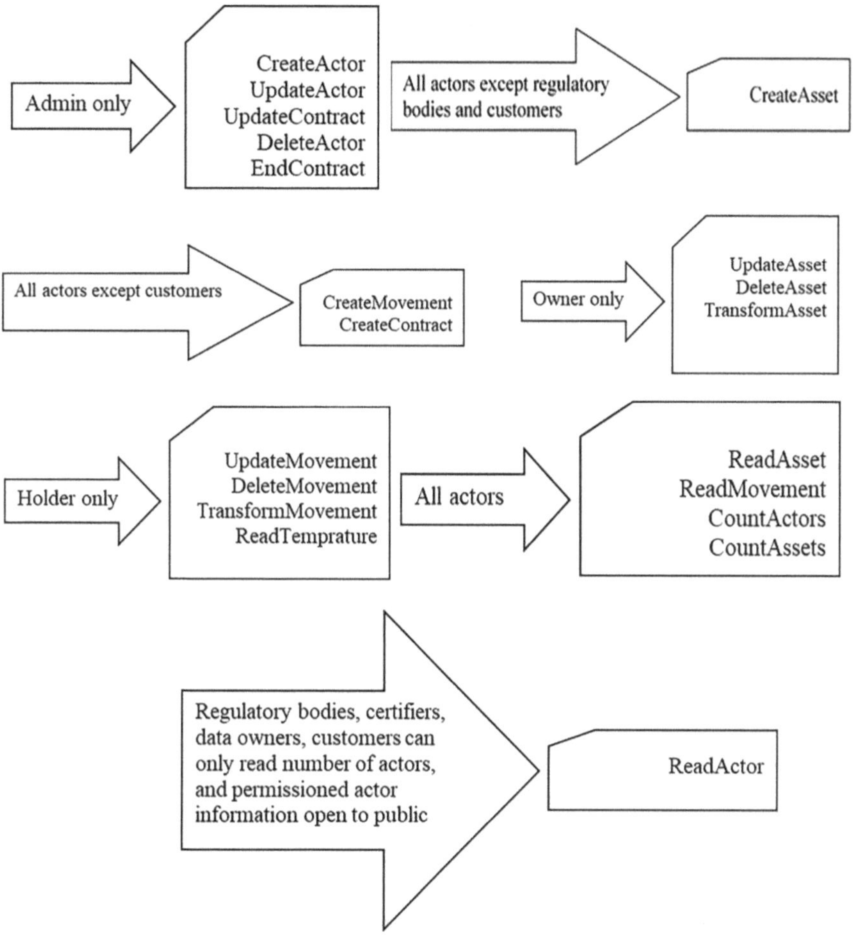

Fig. 7.11 Transactions and permissions

3. **Stakeholder Accountability:** Immutable records of transactions, certifications, and distributor IDs create a transparent chain of custody, preventing fraudulent practices such as mislabeling or adulteration.

Critical Analysis

While the model holds transformative potential, its success hinges on addressing adoption barriers like technical training, interoperability, and cost-benefit trade off. In brief explanation, these mean:

1. **Technical Training:** Small-scale suppliers may require subsidized training to navigate Hyperledger's interface.
2. **Interoperability:** Integration with legacy systems (e.g., ERP software) remains untested, necessitating pilot studies.

3. **Cost-Benefit Trade-offs:** Initial set-up costs (e.g., QR code infrastructure) must be justified by measurable return on investment (ROI), such as reduced recall expenses or premium pricing for traceable products.

Limitations, however, uphold some consideration. The model's reliance on consistent data input by suppliers and manufacturers introduces risks of incomplete or inaccurate entries. Additionally, the administrative complexity of managing permissions (Fig. 7.11) may hinder scalability, particularly for smaller producers lacking technical expertise.

7.4.1 Expected Results

Implementation of the blockchain platform is anticipated to generate the following outcomes:

1. **Enhanced Consumer Trust:** QR code-enabled access to production data (e.g., nutritional content, hygiene certifications) empowers consumers to verify product authenticity, aligning with growing demand for food transparency.
2. **Streamlined Audits:** Regulatory bodies' read-only access to real-time data (Fig. 7.11) reduces audit durations by 30–50%, as manual record-checking is replaced with automated verification.
3. **Reduced Operational Costs:** Smart contracts minimize disputes among stakeholders by automating contractual obligations (e.g., payment upon delivery confirmation), cutting administrative overhead by an estimated 15–20%.

These results align with findings from similar blockchain applications in dairy supply chains (e.g., IBM Food Trust), though the Hyperledger-based model's focus on permissioned access offers distinct advantages for safeguarding proprietary processes.

7.4.2 Practical Implications

The proposed model has significant implications regarding regulatory aspects for automated audits, brand reputation by transparency for end-consumers, and operational efficiency by reducing unnecessary processes and enhancing dispute resolution processes. Technically, it provides data privacy, scalability, reduced costs, and energy efficiency. Below, we can classify implications for literature and industry.

1. **Academic Implications**

   This chapter contributes to blockchain literature by:

(a) Proposing a novel conceptual framework for dairy supply chains, integrating Hyperledger's permissioned structure with dynamic asset tagging (e.g., party IDs).
(b) Highlighting underexplored challenges, such as balancing data transparency with commercial confidentiality in permission settings (Fig. 7.10).
(c) Providing a methodological blueprint for adapting blockchain architectures to artisanal food production, a sector often overlooked in favor of large-scale agriculture.

2. **Industrial Implications**
   For the dairy industry, the model offers actionable pathways to:

(a) **Mitigate Food Safety Risks:** Real-time monitoring of storage conditions (e.g., temperature during transportation) reduces spoilage and contamination incidents.
(b) **Strengthen Regulatory Compliance:** Automated certification uploads and audit trails simplify adherence to EU and USDA standards, critical for export-oriented producers like Kashar manufacturers.
(c) **Foster Collaboration:** Shared ledger access incentivizes suppliers, manufacturers, and distributors to align practices, reducing bottlenecks and fostering long-term partnerships.

7.5 Conclusions

This chapter presents a conceptual blockchain model designed to enhance traceability in the Kashar cheese supply chain, built upon a rigorous examination of existing literature and technological insights into food chain traceability and blockchain applications. The primary objective was to explore the viability of blockchain technology as an alternative system for food supply chain traceability. To address this objective, we developed a blockchain framework using the Hyperledger platform tailored to a dairy supply chain system. This study distinguishes itself from preceding research by leveraging emerging technologies and offering a novel approach to existing traceability systems.

The study indicates that insufficient legal frameworks and a deficiency of standardized procedures exist for building traceability systems in food supply chains. The essential elements of a successful traceability system are not well defined, and the roles and duties of the many parties involved in the supply chain—from suppliers to final consumers—are not well defined. Complete traceability requires careful data capture and synchronization throughout a network accessible to all authorized users. But theoretical ideas frequently deviate from actual situations. Sophisticated software solutions may not be enough to shield businesses from problems with data synchronization, network communication efficiency, and dependability. These problems are made worse by the lack of precise instructions for creating efficient traceability systems; this is especially problematic given the circumstances of our nation. The bullwhip effect, which results from insufficient process and inventory traceability, further negatively impacts cost-profit ratios.

For the sake of society's welfare, food supply chains must be secure and traceable as inconsistencies can have a negative impact on public health as well as the business reputations of suppliers, manufacturers, and retailers. A strong traceability system enables prompt intervention to prevent minor problems from becoming major ones. A thorough traceability system that identifies all activities and actors makes it easier to handle foodborne outbreaks and product recalls effectively. Gaining and keeping the trust of customers is still quite difficult. It takes effort to establish a trustworthy connection, but maintaining one in the face of potential food scandals is even more difficult. Consumer confidence may be strengthened faster and more thoroughly with a transparent network that offers full traceability than with conventional techniques.

Since being made widely known by Bitcoin, blockchain technology has gained interest from a variety of industries outside of cryptocurrency. It is not a cure-all; it is still a nascent technology that has to be refined and put through a rigorous testing process in order to maximize its potential. However, it offers a decentralized, distributed future. Expertise, sustainability, application, and scalability are among the current constraints. In order to better satisfy the demands and expectations of companies and customers in the future, a greater knowledge of the characteristics and possible enhancements of new blockchain models will emerge as they are developed and tested. By putting up a concept that is meant to be evaluated, this study seeks to further the current development by improving knowledge of the usefulness of blockchain technology in supply chain management.

This research makes significant contributions to both academic literature and practical application. It critically examines the terminology and foundational concepts of blockchain technology, the terminology of traceability, previously designed major traceability models, as well as the benefits and drawbacks of these systems from the perspective of food supply chains. The study delineates the essential requirements for a functional traceability system, identifies and explains the roles and actions of system participants, defines the transactions to be monitored, and compares various platforms based on their features.

The proposed model serves as a foundational framework to understand the requirements, stakeholders, internal and external production factors, distribution processes, and consumer perspectives within the supply chain. The conceptual model offers insights into the origins of finished products, the comprehensive tracking of processes, and the network structure of a food manufacturing entity. It enhances transparency by providing traceability within permissible limits. The proposed blockchain model partially mitigates food safety concerns through its immutable and distributed platform, fostering a trustworthy network among stakeholders. By consolidating all participants under a distributed network, the asymmetry of information is addressed, enabling real-time communication and enhancing decision-making practices. This leads to better demand and inventory forecasts, cost and market price analysis, and innovation opportunities.

7.5.1 Limitations and Further Research Suggestions

A major limitation of this study is the lack of practical testing for the model design, which would provide more concrete implications for businesses, given the limited number of existing model designs and pilot studies. Additionally, the complex multi-stakeholder structure of food chains restricts the model's scope to the Kashar cheese supply chain. Expanding the model to cover all product segments and incorporating international trade considerations would offer new perspectives and implementation opportunities.

The determination of system requirements is also limited to existing literature and case studies from pioneering companies. A comprehensive analysis involving multiple case studies and sectoral questionnaires would yield deeper insights into the sector-specific and systematic needs of food supply chains. The current maturity level of blockchain technology presents another constraint, as the options and features available on platforms are still evolving. Future advancements in blockchain platforms and features will inevitably influence and transform traceability model designs.

Blockchain-based systems necessitate full participation from all actors in the supply chain, a challenging task despite the numerous benefits promised. Small and medium-sized enterprises (SMEs) face significant hurdles due to the initial high costs associated with building a blockchain network. Therefore, future research should focus on conducting simulations or real-case testing of the model, delving into model requirements through quantitative analysis from all stakeholder perspectives, expanding the data generation and smart contract aspects of the model, evaluating alternative blockchain platforms such as Oracle, Ambrosus, Ethereum 2.0, assessing model performance against derived requirements, and comparing existing traceability software with the proposed blockchain model post-testing.

In synthesizing the key findings and contributions of this chapter, it is evident that the integration of blockchain technology in food supply chains holds substantial promise for enhancing traceability, transparency, and trust. The conceptual model proposed serves as a crucial steppingstone toward realizing this potential, although the complexity of practical implementation and the multi-faceted nature of food supply chains pose significant challenges.

The lack of standardized frameworks and clear legislative guidelines highlights a critical gap that necessitates further research and policy development. The emphasis on the need for synchronization and data integrity across the network underscores the technical and operational hurdles that must be addressed. The potential benefits of blockchain—real-time communication, enhanced decision-making, improved forecasting, and trust-building—are compelling, yet they are contingent upon the resolution of scalability, cost, and participation barriers.

References

Abeyratne, S. A., & Monfared, R. P. (2016). Blockchain ready manufacturing supply chain using distributed ledger. *International Journal of Research in Engineering and Technology, 5*(9), 1–10. https://doi.org/10.15623/ijret.2016.0509001

Agarwal, S. (2018). *Blockchain technology in supply chain and logistics*. Thesis: S.M. in engineering and management, Massachusetts Institute of Technology, System Design and Management Program. Retrieved January 1, 2025, from https://www.researchgate.net/publication/328333423_Blockchain_technology_in_supply_chain_and_logistics

Anwar, H. (2021). *Hyperledger vs Ethereum: A clash of two emerging technologies*. Retrieved from https://101blockchains.com/hyperledger-vs-ethereum-2/

Bambara, J. J., & Allen, P. R. (2018). *Blockchain: A practical guide to developing business, law, and technology solutions*. McGraw-Hill Education.

Bär, M., Colombo, A., Torres, J., Fernandez, E., Rico, M., & Caliusco, M. (2023). An Industry 4.0-compliant digital product passport approach for realising dairy product traceability. In *IECON 2023—49th annual conference of the IEEE Industrial Electronics Society* (pp. 1–9). IEEE. https://doi.org/10.1109/IECON51785.2023.10312481

Bashir, I. (2017). *Mastering blockchain: Distributed ledgers, decentralization and smart contracts explained.* L. Subramanian (Ed.). Packt.

Benton, M. C., Radziwill, N. M., Purritano, A. W., & Gerhart, C. J. (2018). Blockchain for supply chain: Improving transparency and efficiency simultaneously. *Software Quality Professional, 20*(3), 28.

Bosona, T., & Gebresenbet, G. (2013). Food traceability as an integral part of logistics management in food and agricultural supply chain. *Food Control, 33*(1), 32–48. https://doi.org/10.1016/j.foodcont.2013.02.004

Brown, R. G. (2018, May). *The Corda platform: An introduction.* Retrieved January 1, 2025, from https://www.r3.com/wp-content/uploads/2018/05/corda-platform-whitepaper.pdf.

Casado-Vara, R., Prieto, J., De la Prieta, F., & Corchado, J. M. (2018). How blockchain improves the supply chain: case study alimentary supply chain. *Procedia Computer Science, 134*, 393–398. https://doi.org/10.1016/j.procs.2018.07.193

Casino, F., Kanakaris, V., Dasaklis, T. K., Moschuris, S., Stachtiaris, S., Pagon, M., & Rachaniotis, N. P. (2020). Blockchain-based food supply chain traceability: a case study in the dairy sector. *International Journal of Production Research, 59*(19), 1–13. https://doi.org/10.1080/00207543.2020.1789238

Chen, Y., & Bellavitis, C. (2020). Blockchain disruption and decentralized finance: The rise ofdecentralized business models. *Journal of Business Venturing Insights, 13*, e00151. https://doi.org/10.1016/j.jbvi.2019.e00151

Civitillo, W. (2023). *iFoodDS and IBM forge new path to food safety with IBM Food Trust™*. Retrieved January 1, 2025, from https://www.ibm.com/blog/ifoodds-and-ibm-forge-new-path-to-food-safety-with-ibm-food-trust/

Consensys. (2020). *Blockchain in supply chain management*. Retrieved January 1, 2025, from https://consensys.net/blockchain-use-cases/supply-chain-management/

Das, P., Jain, C., Singh, A., & Singh, M. (2024). Agri-chain: A blockchain-empowered smart solution for agricultural industry. In M. Sumithra, M. Sathyamoorthy, M. Manikandan, R. Dhanaraj, & M. Ouaissa (Eds.), *Computational intelligence in internet of agricultural things part of the book series: Studies in computational intelligence (SCI)* (Vol. 1170, pp. 221–245). Springer. https://doi.org/10.1007/978-3-031-67450-1_9

Deloitte. (2017). *Continous interconnected supply chain—Using blockchain and Internet of Things in supply chain traceability*. Delloitte. Retrieved January 1, 2025, from https://www2.deloitte.com/content/dam/Deloitte/lu/Documents/technology/lu-blockchain-internet-things-supply-chain-traceability.pdf

Drescher, D. (2017). *Blockchain basics: A non-technical introduction in 25 steps* (1st ed.). Apress Media. https://doi.org/10.1007/978-1-4842-2604-9

Ediz, Ç. (2014). *Süreç Bazlı Dinamik Ürün Ağaçlarıyla Ürün İzlenebilirliği Modeli*. Ph.D. thesis. T.C. Sakarya Üniversitesi Sosyal Bilimler Enstitüsü. Retrieved January 1, 2025, from https://www.academia.edu/103087483/S%C3%BCre%C3%A7_bazl%C4%B1_dinamik_%C3%BCr%C3%BCn_a%C4%9Fa%C3%A7lar%C4%B1yla_%C3%BCr%C3%BCn_izlenebilirli%C4%9Fi_modeli

European Commission. (2000). *Wite paper on food safety*. Retrieved January 2, 2025, from https://eur-lex.europa.eu/legal-content/EN/ALL/?uri=CELEX:51999DC0719

European Parliament, CEU. (2021). *Labelling of foodstuffs*. Retrieved January 1, 2025, from Regulation (EU) No 1169/2011: https://eur-lex.europa.eu/legal-content/EN/LSU/?uri=CELEX:32011R1169

FAO and WHO. (2019). *Codex Alimentarius Commission—Procedural manual twenty-seventh*. Retrieved January 1, 2025, from openknowledge.fao.org: http://www.fao.org/3/ca2329en/CA2329EN.pdf

Feng, H., Fan, J., Ji, Y., Glamuzina, B., & Ma, R. (2024). Reliable quality traceability for tilapia cold chain using blockchain and machine learning techniques. *Journal of Food Process Engineering, 47*(12), e70016. https://doi.org/10.1111/jfpe.70016

Fetch.ai. (2020). Building connections—How Fetch.ai is enabling smart cities. *Medium*. Retrieved January 1, 2025, from https://medium.com/fetch-ai/building-connections-how-fetch-ai-is-enabling-smart-cities-f257b3de7a54

Foreign Agricultural Service/USDA. (2024, July). *Dairy: world markets and trade*. Retrieved January 1, 2025, from https://downloads.usda.library.cornell.edu/usda-esmis/files/5t34sj56t/3r076m345/zs2601365/dairy.pdf

Gates, M. (2017). *Blockchain: Ultimate guide to understanding blockchain, bitcoin, cryptocurrencies, smart contracts and the future of money* (1st ed.). Wise Fox.

Gupta, M. (2017). *Blockchain for dummies* (IBM Limited Edition ed.). Wiley.

İstanbul İl Tarım ve Orman Müdürlüğü. (2010). *5996 SAYILI KANUN*. Retrieved January 1, 2025, from https://istanbul.tarimorman.gov.tr/Menu/72/5996-Sayili-Kanun: https://istanbul.tarimorman.gov.tr/Belgeler/SolMenu/RESMIGAZETE2018/5996SAYILIKANUN.pdf

Kamilaris, A., Fonts, A., & Prenafeta-Boldύ, F. X. (2019). The rise of blockchain technology in agriculture and food supply chains. *Trends in Food Science & Technology, 91*, 640–652. https://doi.org/10.1016/j.tifs.2019.07.034

Kshetri, N. (2018). Blockchain's roles in meeting key supply chain management objectives. *International Journal of Information Management, 39*, 80–89. https://doi.org/10.1016/j.ijinfomgt.2017.12.005

Lalitha, R., & Sasikala, T. (2024). Enhancing product traceability in agricultural supply chain using chain of things technology. In *2024 5th international conference on data intelligence and cognitive informatics (ICDICI)* (pp. 804–811). IEEE. https://doi.org/10.1109/ICDICI62993.2024.10810817

Laurence, T. (2017). *Blockchain for dummies*. Wiley.

Li, Z. (2017). *A hybrid peer-to-peer framework for supply chain visibility*. Master of Science in Electric and Computer Engineering Thesis. Purdue University.

Milkypedia. (2025, January 23). Turkey: A comprehensive overview of the dairy market. *dairynews.today*. Retrieved January 1, 2025, from https://dairynews.today/milkypedia/country/tr/

Min, H. (2019). Blockchain technology for enhancing supply chain resilience. *Business Horizons, 62*(1), 35–45. https://doi.org/10.1016/j.bushor.2018.08.012

Moe, T. (1998). Perspectives on traceability in food manufacture. *Trends in Food Science & Technology, 9*(5), 211–214. https://doi.org/10.1016/S0924-2244(98)00037-5

Morabito, V. (2017). *Business innovation through blockchain: The B³ perspective*. Springer. https://doi.org/10.1007/978-3-319-48478-5

Nakamoto, S. (2008). *Bitcoin: A peer-to-peer electronic cash system*. Retrieved January 1, 2025, from https://bitcoin.org/bitcoin.pdf

Olsen, P., & Borit, M. (2013). How to define traceability. *Trends in Food Science & Technology, 29*(2), 142–150. https://doi.org/10.1016/j.tifs.2012.10.003

Oracle. (2021). *Oracle intelligent track and trace.* Retrieved January 1, 2025, from https://www.oracle.com/a/ocom/docs/applications/scm/ds-intelligent-track-and-trace.pdf

Pearson, S., May, D., Leontidis, G., Swainson, M., Brewer, S., Bidaut, L., & Zisman, A. (2019). Are distributed ledger technologies the panacea for food traceability? *Global Food Security, 20*, 145–149. https://doi.org/10.1016/j.gfs.2019.02.002

Perboli, G., Musso, S., & Rosano, M. (2018). Blockchain in logistics and supply chain: A lean approach for designing real-world use cases. *IEEE Access, 6*, 62018–62028. https://doi.org/10.1109/ACCESS.2018.2875782

Prasad, R., & Chidambaranathan, S. (2024). Digitizing traceability in the dairy supply chain: A case study of the Telangana region. In *2024 International conference on intelligent systems for cybersecurity (ISCS)* (pp. 1–6). IEEE. https://doi.org/10.1109/ISCS61804.2024.10581117

Provenance. (2025). *Case studies*. Retrieved January 1, 2025, from https://www.provenance.org/case-studies

Queiroz, M., Telles, R., & Bonilla, S. H. (2019). Blockchain and supply chain management integration a systematic review of the literature. *Supply Chain Management, 25*(2), 241–254. https://doi.org/10.1108/SCM-03-2018-0143

Queiroz, M. M., & Fosso Wamba, S. (2019). Blockchain adoption challenges in supply chain: An empirical investigation of the main drivers in India and the USA. *International Journal of Information Management, 46*, 70–82. Retrieved from https://doi.org/10.1016/j.ijinfomgt.2018.11.021

Rodríguez-Molina, J., Urtecho, R., Martínez-Núñez, M., & Garbajosa, J. (2024). Enhancing agriculture through IoT and blockchain: A VeChain-enhanced sustainable development approach for small-scale agricultural exploitations. *IEEE Access, 12*, 179962–179980. https://doi.org/10.1109/ACCESS.2024.3506510

Safeer, S., & Pulvento, C. (2024). Blockchain-backed sustainable management of Italian tomato processing industry. *Agriculture, 14*(7), 1120. https://doi.org/10.3390/agriculture14071120

Sakthivel, V., Periyaswamy, P., Lee, J.-W., & Prabu, P. (2024). Enhancing transparency and trust in agrifood supply chains through novel blockchain-based architecture. *Korean Society for Internet Information (KSII) Transactions on Internet and Information Systems, 18*(7), 1968–1985. https://doi.org/10.3837/tiis.2024.07.013

Santonino, M., III, Koursaris, C. M., & Williams, M. J. (2018). Modernizing the supply chain of Airbus by integrating RFID and blockchain processes. *International Journal of Aviation, Aeronautics, and Aerospace, 5*(4), Article 4. https://doi.org/10.15394/ijaaa.2018.1265

Singh Thakur, K., Ahuja, R., & Singh, R. (2024). IoT-GChain: Internet of things-assisted secure and tractable grain supply chain framework leveraging blockchain. *Electronics Special Issue Security and Privacy for Modern Wireless Communication Systems, 2nd Edition, 13*(18), 3740. https://doi.org/10.3390/electronics13183740

Singhal, B., Dhameja, G., & Panda, P. S. (2018). *Beginning blockchain: A beginner's guide to building blockchain solutions* (1st ed.). Apress Media. https://doi.org/10.1007/978-1-4842-3444-0

Statista. (2024a). *Food: Market data & analysis.* Statista, Junior Analyst. Statista. Retrieved January 2, 2025, from https://www.statista.com/study/55496/food-market-data-and-analysis/

Statista. (2024b, July). *Major producers of cow milk worldwide in 2024, by country.* Retrieved January 1, 2025, from https://www.statista.com/statistics/268191/cow-milk-production-worldwide-top-producers/

Sternberg, H., & Baruffaldi, G. (2018). Chains in chains—Logic and challenges of blockchains in supply chains. In *Hawaii international conference on system sciences, 51st* (pp. 3936–3943). HICSS. Retrieved from http://hdl.handle.net/10125/50382

Takyar, A. (2024). *Top blockchain platforms of 2024.* Retrieved January 1, 2025, from https://www.leewayhertz.com/blockchain-platforms-for-top-blockchain-companies/

The Linux Foundation. (2025). Projects. Retrieved January 1, 2025, from https://www.lfdecentralizedtrust.org/projects/tag/ledger-technology

Thiruchelvam, V., Mughisha, A. S., Shahpasand, M., & Bamiah, M. (2017). Blockchain-based technology in the coffee supply chain trade: Case of Burundi Coffee. *Journal of Telecommunication,*

Electronic and Computer Engineering, 10(3), 121–123. Retrieved from https://jtec.utem.edu.my/jtec/article/view/4722

Tian, F. (2017). A supply chain traceability system for food safety based on HACCP, blockchain & internet of things. In *2017 International conference on service systems and service management (ICSSSM)* (pp. 1–6). IEEE. https://doi.org/10.1109/ICSSSM.2017.7996119

Toader, D.-C., Rădulescu, C., & Toader, C. (2024). Investigating the adoption of blockchain technology in agri-food supply chains: Analysis of an extended UTAUT model. *Agriculture, 14*(4), 614. https://doi.org/10.3390/agriculture14040614

Tse, D., Zhang, B., Yang, Y., Cheng, C., & Mu, H. (2017). Blockchain application in food supply information security. In *2017 IEEE international conference on industrial engineering and engineering management (IEEM)* (p. 1357). IEEE.

Usta, A., & Doğantekin, S. (2018). *Blockchain 101 v.2* (2nd ed.). Bankalararası Kart Merkezi. Retrieved from https://www.bkm.com.tr/wp-content/uploads/2019/08/15082019_kitap.pdf

Valarmathi, V., Sakthi Mahendran, R., & Sabari Srinivas, H. (2024). Design and implementation of public distribution system using blockchain. In *2024 IEEE international conference on blockchain and distributed systems security (ICBDS)* (pp. 1–5). IEEE. https://doi.org/10.1109/ICBDS61829.2024.10837391

Vayadande, K., Jadhav, A., Patil, A., Lakde, V., Rathod, K., Nimbolkar, A., & Wadile, P. (2024). Agriculture supply chain using blockchain technology. In *2024 2nd international conference on computer, communication and control (IC4)* (pp. 1–7). IEEE. https://doi.org/10.1109/IC457434.2024.10486523

Venkateswara Reddy, K., Mahesh, G., & Suresh, B. (2025). Milk delivery: Enhancing transparency and fair payments for kenyan stallholder dairy farmers. *IPE Journal of Management, 14*(5), 83–90. Retrieved from https://apgcu.edu.in/pdf/mba-publications/2024-11.pdf

Voluntary Report. (2024, July). *2024-2028 livestock roadmap overview.* Retrieved January 1, 2025, from https://apps.fas.usda.gov/newgainapi/api/Report/DownloadReportByFileName?fileName=2024-2028%20Livestock%20Roadmap%20Overview_Ankara_Turkiye_TU2024-0030.

World Healt Organization. (2024). *Food safety*. Retrieved January 1, 2025, from https://www.who.int/news-room/fact-sheets/detail/food-safety

Yele, S., & Litoriya, R. (2024). Blockchain-based secure dining: Enhancing safety, transparency, and traceability in food consumption environment. *Blockchain: Research and Applications, 5*(2), 100187. https://doi.org/10.1016/j.bcra.2023.100187

Zhang, J., & Bhatt, T. (2014). A guidance document on the best practices in food traceability. *Comprehensive Reviews in Food Science and Food Safety, 13*(5), 1074–1103. https://doi.org/10.1111/1541-4337.12103

Chapter 8
Application of Blockchain in Reverse Supply Chain

Yanting Huang and Tianqin Shi

Abstract Reverse supply chains face various challenges such as incomplete end-of-life data, product untraceability, information asymmetry, and low perceived value of remanufactured products. Blockchain, as a promising new technology, can be used to store accurate and reliable information regarding manufacturers, suppliers, logistics, product life expectancy, reusability, and recycling. Data precision and completeness help manufacturers to reduce or avoid costly testing and time consuming reduces. Despite documented benefits and real-world examples of blockchain, limited research integrates blockchain with recycling strategies, advertising efforts, and information management in reverse supply chains. This chapter reviews these challenges and how blockchain help to address them. It specifically analyzes blockchain adoption under three recycling modes (the manufacturer collecting used products directly from customers, the manufacturer contracting a retailer for collection, and the manufacturer outsourcing the collection activities to a third-party recycler). Additionally, it explores interactions between blockchain adoption and advertising strategies, as well as information strategies.

Keywords Blockchain · Reverse supply chain · Recycling mode · Information strategies

Y. Huang
College of Management, Shenzhen University, Shenzhen, China

T. Shi (✉)
Lucas College and Graduate School of Business, San José State University, San Jose, CA, USA
e-mail: tianqin.shi@sjsu.edu

Y. Khojasteh, S. M. M. Jahan Abadi (eds.), *Emerging Technologies in Supply Chains*, International Series in Operations Research & Management Science 373, https://doi.org/10.1007/978-3-032-01218-0_8

8.1 Introduction

8.1.1 Background

With rapid economic development and shortening product life cycle, the amount of waste generated globally is increasing exponentially (Van Engeland et al., 2020). As shown in Fig. 8.1, the annual global production of e-waste (i.e., discarded electrical and electronic equipment) is approximately 50 million metric tons, and it is projected to reach 74.7 million metric tons by 2030 and 121 million metric tons by 2050 (Pongen et al., 2024). This drastic growth in e-waste generation poses significant challenges in managing sustainable waste management and transitioning towards a circular economy, as the proper collection, recycling, and disposal of these electronic products are essential for mitigating the environmental and public health impacts associated with the improper handling of such waste.

Failure to promptly handle discarded electronic waste can result in severe environmental pollution and pose risks to human health. Therefore, there is an urgent need to identify a suitable method for effectively recovering and processing electronic waste. The reverse supply chain emphasizes the potential for product returns, repairs, remanufacturing, and recycling during the product use phase, which can effectively contribute to addressing the global issue of excessive e-waste. Enterprises can also maximize the extraction of resource value and mitigate the growing volume

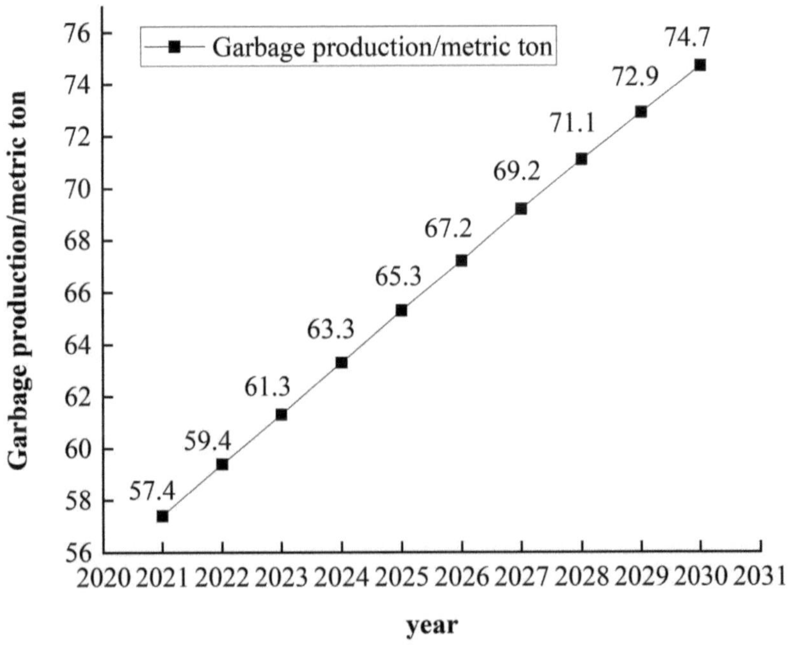

Fig. 8.1 Global annual production of electronic waste

of e-waste through the effective management and utilization of the reverse supply chain processes, thereby achieving more efficient resource utilization. In 2023, through effectively disposing of waste electronic materials, Apple recycled nearly 40,000 tons of electronic materials through continuous technological advances and 12.8 million refurbished devices and components were delivered to new users.[1]

8.1.2 Reverse Supply Chain

The reverse supply chain refers to the process in which products are returned from end consumers to manufacturers or suppliers. The process involves the return, remanufacturing, maintenance, recycling, and reuse of products, with the goal of reducing waste, maximizing products and materials reuse, and protecting the environment (Prahinski & Kocabasoglu, 2006). The fundamental structure of a reverse supply chain includes the following elements: product recycling, reverse logistics, remanufacturing or reprocessing, and final disposal (Fig. 8.2).

1. **Product recycling:** Product recycling encompasses that consumers return unwanted products to manufacturers or retailers for a variety of reasons, including quality issues, excess inventory, expired products, and other circumstances.
2. **Reverse logistics:** Reverse logistics encompasses the collection, transportation, and classification of recycled products, which refers to the process in which the

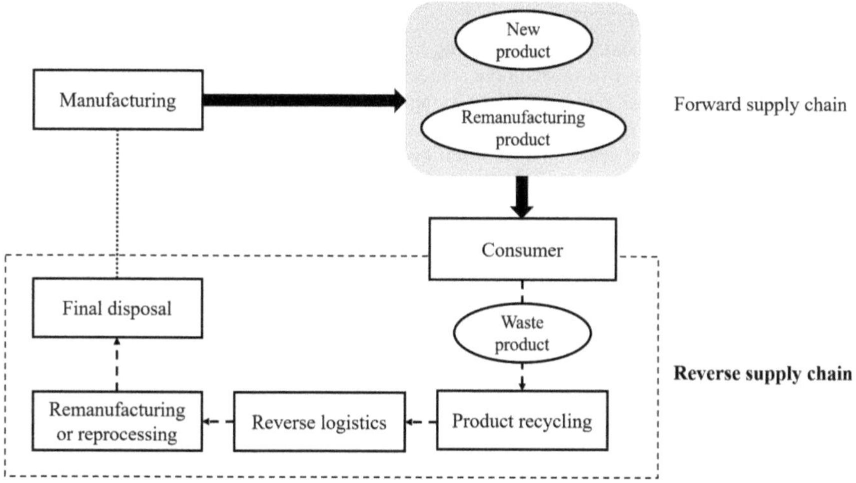

Fig. 8.2 Reverse supply chain structure

[1] https://www.apple.com.cn/shop/trade-in

merchant entrusts a third-party logistics company to deliver the delivered items from the user's designated location to the merchant's location.

3. **Remanufacturing and reprocessing process:** The remanufacturing and reprocessing process encompasses upgrading the product or restoring the raw materials for transfer to reprocessing operations. Such a process includes a wide range of activities, including testing, repairing, and upgrading recycled products, in order to achieve the secondary utilization of used materials.
4. **Final disposal:** The final disposal process requires environmentally friendly treatment of products that cannot be reused/recycled/remanufactured, taking into account factors such as feasibility, cost-effectiveness, and environmental friendliness, and selecting the most suitable treatment methods, such as material recycling and energy recovery (Van Engeland et al., 2020).

Reverse supply chains have been extensively utilized across various sectors, with its implementation differing among industries and business operations. Table 8.1 provides relevant examples.

The reverse supply chain is of vital importance to sustainable business practices and the development of the circular economy, as it involves the entire life cycle process of products, from production to recycling, reuse, and remanufacturing. The circular nature of the reverse supply chain enables the continuous recycling and reuse of resources, which is a key component of the circular economy model and promotes more sustainable and environmentally-friendly business operations.

Table 8.1 Reverse supply chain examples

Company	Reverse supply chain
Mercedes-Benz	By constructing battery recycling factories and employing an integrated mechanical-hydrometallurgical process, waste automotive lithium-ion batteries are recycled and reprocessed, achieving a battery material recovery rate as high as 96%. The recovered materials can be utilized to produce battery modules for 50,000 new vehicles, demonstrating a high level of efficiency and sustainability in the field of battery recycling management (Mercedes-Benz, 2024)
Hewlett-Packard (HP)	Through its Ink Cartridge Recycling Program, HP separates and extracts reusable components from the recycled cartridges for use in the production of new ink cartridges or other electronic products, showcasing a commitment to sustainability and circular economy practices within the management realm (HP, 2024)
H&M	H&M has initiated a global clothing recycling program in which used garments undergo a series of processes including cleaning, decomposition, filtration, carding, drawing, spinning, twisting, and weaving. Through these processes, the materials are either reused or redesigned. Additionally, H&M has launched an eco-friendly clothing line and an in-store rental service in Stockholm as a result of this initiative (H&M, 2024)
IKEA	Consumers can return unwanted old IKEA furniture to IKEA for recycling, encompassing various furniture types under the IKEA brand such as bookcases, dining tables, chairs, and more. The returned furniture must be relatively intact, without severe damage or deformation (Ikea, 2020)

8.2 Issues in Reverse Supply Chain

It is well known that although the recycling industry is constantly adapting and changing with economic development, there are still many unresolved issues such as information opacity, high supply chain security risks, and high costs, as defined in Table 8.2. These issues may make reverse supply chain operations difficult and increase the risk for supply chain members.

8.2.1 *Information Opacity*

In the reverse supply chain, information opacity refers to the state where information related to the product recovery process cannot be fully accessed, obtained, or understood by the relevant supply chain members. The lack of transparency of information in reverse supply chains makes it difficult for participants to trace the source and quality control of products, resulting in new or remanufactured products in the supply chain being affected by unknown sources, improper handling, or poor quality. In the process of recycling and reusing waste products, information gaps prevent the supply chain from ensuring the quality and compliance of recycled materials. This can result in recycled materials not meeting the expected standards, leading to environmental pollution, potential threats to the reputation of companies involved in recycling, and reduced consumer desire and confidence in recycling and purchasing remanufactured products. For example, in 2016, Samsung Electronics recalled its Galaxy Note 7 smartphone due to battery explosion issues. However, due to information gaps and communication issues among supply chain partners involved in the process of recycling mobile phones, tracking and tracing of recycled materials has become difficult. In the end, the necessity for multiple recalls had a detrimental effect on the reputation and recycling efficiency of Samsung Electronics.

8.2.2 *Data Security and Privacy*

The data security and privacy concerns in reverse supply chains encompass the personal identities and sensitive information of the various participants engaged in the product recovery activities, including manufacturers, retailers, recyclers, and

Table 8.2 The issues in the reverse supply chain

The main issues	Specific meanings
Information opacity	Information related to the product recovery process cannot be fully accessed, obtained, or understood by the relevant supply chain members
Data security and privacy	The personal identities and sensitive information of the various participants engaged in the product recovery activities
High costs	The main high costs involved in the reverse supply chain include the collecting costs, sorting costs, transporting costs, and remanufacturing costs

consumers. When consumers return used or discarded products for recycling, they are typically required to provide details about the product's usage history as well as their personal identification data. These personal or commercially sensitive information may then be shared among the members of the supply chain. If such information were to be compromised or leaked, it could lead to severe economic and reputational damages for both the enterprises and the individuals involved, thereby adversely impacting the collaborative relationships within the supply chain and undermining the overall efficiency of the reverse logistics operations.

8.2.3 High Costs

The main costs involved in the reverse supply chain include the costs of collecting, sorting, transporting, and remanufacturing discarded products. These costs are often significant and persistent.

1. **Product collection costs:** the costs of collecting used or discarded products from consumers or other sources. The reverse supply chain incurs additional costs due to the various processing steps required for product recovery, such as cleaning and treatment.
2. **Sorting costs**: the costs of inspecting, sorting, and categorizing the collected products based on their condition and potential for reuse or remanufacturing. The collection of used products necessitates specialized personnel to categorize, inspect, and evaluate the items, leading to significant labor-intensive classification costs.
3. **Transportation costs**: the costs of transporting the collected products to the appropriate facilities for further processing. The transportation costs primarily encompass the expenses associated with collecting and transporting the discarded or secondhand products to the recovery centers or remanufacturing facilities, including labor, vehicle, and fuel expenditures. For larger and heavier products, the transportation costs can be particularly high.
4. **Remanufacturing costs**: The costs of the remanufacturing process, which may involve disassembly, cleaning, repair, and reassembly of the product. Notably, remanufacturing costs constitute a significant portion of the overall reverse supply chain costs. This is because the remanufacturing process involves a thorough inspection, disassembly, cleaning, and repair of the recovered products, resulting in high process-related expenses. Furthermore, the remanufacturing or reprocessing operations require specialized equipment and technical personnel, incurring substantial capital investments.

In summary, the reverse supply chain faces various cost components, including collection, sorting, transportation, and remanufacturing, all of which contribute to the overall cost burden. The complexity and variability of these activities, as well as the need for specialized resources, make the management and optimization of these costs a critical challenge for the viability and profitability of reverse supply chain operations.

8.3 Blockchain-Driven Solutions

The term "blockchain" was first coined in the context of Bitcoin, with the earliest known use occurring in a 2008 white paper by Satoshi Nakamoto titled "Bitcoin: A Peer-to-Peer Electronic Cash System." In contrast to traditional ERP systems or RFID tags, blockchain combines distributed storage, peer-to-peer transmission, consensus mechanisms, cryptography, and other technologies to record transactions and information through continuously growing data blockchains, thereby ensuring data security and transparency. Blockchain can solve issues in the reverse supply chain through the distributed ledger technology, decentralization and smart contracts. In the following sections, we will elaborate in detail on how blockchain solves the current issues in reverse supply chains.

8.3.1 Blockchain Improving Information Transparency and Strengthens Information Sharing

Blockchain records all transactions and information in the supply chain on a shared ledger through the use of distributed ledger technology (DLT), which is a type of database technology characterized by a system of records distributed across different locations and maintained collectively by multiple participants. Simply put, this ledger is public and tamper-proof, and all participants can access and view the information in the ledger. This improves the transparency of the supply chain, allowing each participant to access real, accurate, and comprehensive information.

Research has shown that DLT provides strong information security for agricultural product supply chains, clothing and textile product supply chains, food supply chains, cross-border supply chains, diamond supply chains, etc. (Choi, 2023; Li et al., 2023; Niu et al., 2021, 2022). For EV battery recycling, blockchain can collect data from sensors for near real-time monitoring of products and processes. Some battery data such as location, timestamp, voltage, current, and temperature will be stored through distributed, immutable, and secure technology, ensuring distributed verification of incoming data, and relevant participants can also know all information about the storage process. From the extraction and production of waste battery raw materials to the use of new batteries in different fields, blockchain will track important information such as their recycling and remanufacturing. RecycleChain is a blockchain project in Singapore aimed at establishing a global waste recycling network. They utilized blockchain to achieve real-time recording and sharing of recycling information, enabling participants to more efficiently recycle and process waste.

8.3.2 Blockchain Ensuring Data and Privacy Security

Decentralization refers to the blockchain connecting data in the form of blocks, forming an immutable and continuously growing chain. Each block contains a certain number of transaction records, and the effectiveness and security of each block are ensured through cryptographic techniques and consensus algorithms. Due to the fact that blockchain data is stored on all nodes of the network and requires consensus to be modified, there is no single centralized control, achieving true decentralized management of data.

Moreover, the data blocks in blockchain are interconnected through cryptographic hash functions, and any tampering with the data blocks will affect the hash values of subsequent data blocks, thereby ensuring the integrity and security of transaction records. In addition, the transparency of blockchain means that any participant can view all transaction information on the ledger, but due to the use of encryption algorithms, the participant's identity can remain anonymous. This ensures that all information in the reverse supply chain using blockchain is transparent and visible to members of the supply chain. All members involved in the recycling process can view relevant information, including the users, sources, residual value, and attributes of the recycled products. This enhances trust among recycling participants and connects them together, enhancing their collaborative recycling capabilities.

Meanwhile, smart contracts in blockchain can also help reverse supply chains protect their data security and information privacy. A smart contract is a pre-set automation rule and condition that can automatically execute programs without third-party intervention, and only participants who meet specific conditions and join the contract can execute and access specific contract functions and data. In a reverse supply chain, the implementation of smart contracts ensures the privacy of data on discarded and remanufactured products, without the need for third-party platforms. Blockchain itself can act as a neutral and impartial intermediary to safeguard the information and rights of supply chain members

8.3.3 Blockchain Reducing Costs in Reverse Supply Chain

The application of blockchain technology can effectively cut costs, reduce energy pollution, and improve product quality (Lotfi et al., 2022; Xie et al., 2023). Meanwhile, by addressing the challenges of evaluating discarded products during the remanufacturing process, blockchain can help establish a quality evaluation system for recycled products and solve the problem of high evaluation costs for discarded products (Wang et al., 2023). Blockchain technology can build shared ledgers, enabling real-time data sharing and collaboration among different participants. The recycling and remanufacturing in the supply chain involve multiple links and stakeholders. Through the shared ledger of blockchain, it can achieve better collaborative work and information sharing, reduce information fragmentation and repetitive labor, and lower collaboration costs. At the same time, blockchain can implement incentive mechanisms to reward participants such as recyclers and remanufacturers through smart contracts, increase their enthusiasm and participation in recycling and remanufacturing, and reduce operating costs.

When blockchain technology is adopted in the supply chain of EV batteries, blockchain can collect information throughout the entire process of battery production, sales, use, scrapping, recycling, and utilization, which is conducive to battery residual energy detection, residual value evaluation, and reuse, which will reduce the cost of recycling (Li et al., 2020). The immutability and complete recording of product value transfer possessed by blockchain make the recording of batteries in the blockchain circulation process unique and accurate. Blockchain provides a basis for "point-to-point, sales one receives one" recycling. While controlling and supervising the battery collection process, blockchain can also effectively reduce the cost of secondary recycling for enterprises (Rufino Júnior et al., 2022).

Moreover, blockchain can create a virtual digital model that is exactly the same as the actual battery system, known as a digital twin. The digital twin is synchronized with the actual manufacturing system and records the lifecycle information of each EV battery, including data from various stages such as production, distribution, use, and recycling. This further demonstrates that digital twins can be used to evaluate battery degradation mechanisms, predict the performance of retired batteries, and reduce the operating costs of actual battery systems.

8.4 Blockchain Application in the Reverse Supply Chain

Companies involved in the reverse supply chain often face a multitude of challenges such as information transparency, data security, and high costs. However, blockchain technology can help address these issues (Hastig & Sodhi, 2020). Known as a secure, shared digital ledger, blockchain is decentralized, collaborative, anonymous, and traceable (Choi & Siqin, 2022; Moosavi et al., 2021). Research indicates that blockchain can consolidate data from various supply chain entities including production, sales, and recycling, thereby enhancing recycling efficiency and reducing costs in reverse supply chain (Babich & Hilary, 2018; Xie et al., 2023).

8.4.1 Blockchain Technology in Reverse Supply Chains Considering Recycling Modes

The current prevailing recycling models in reverse supply chains include manufacturer recycling, retailer recycling, and third-party recycler recycling. Blockchain technology has the potential to enhance operational efficiency and ensure the reliability and integrity of reverse supply chain processes across various recovery modes (Ma et al., 2022; Ma & Hu, 2022). In this subsection, we will examine the impacts of blockchain on supply chain members and the overall supply chain system in each recycling mode.

8.4.1.1 Blockchain in the Manufacturer Recycling Mode

The manufacturer recycling mode refers to the situation where manufacturers are responsible for recycling activities and wholesale products to the retailer, with the responsibility for recycling borne by the manufacturer. For example, both CATL (Contemporary Amperex Technology Co. Limited) and BYD (Build Your Dreams) have entered the recycling business as manufacturers of EV batteries and new energy vehicles.

In the manufacturer recycling mode, the retailer is unable to directly obtain from the manufacturer collection information such as the recycling quantity, location, and time of product. It makes it difficult for the retailer to accurately evaluate the value of product recycling, thereby affecting pricing and sales strategies for products.

However, through incorporating blockchain into the supply chain, manufacturer can update product recycling information in real-time on a blockchain platform, ensuring that retailers can accurately evaluate the product's recycling value, thereby reducing their decision-making and operational costs (Li et al., 2022; Rufino Júnior et al., 2022; Zhang et al., 2022). The blockchain platform will provide consumers with information regarding the source of EV batteries if the manufacturer and the retailer adopt blockchain in the retailing process. In terms of battery recycling, BYD and CATL have established subsidiaries, acquired recycling companies, and partnered with others to create a recycling network (Harper et al., 2019). Figure 8.3

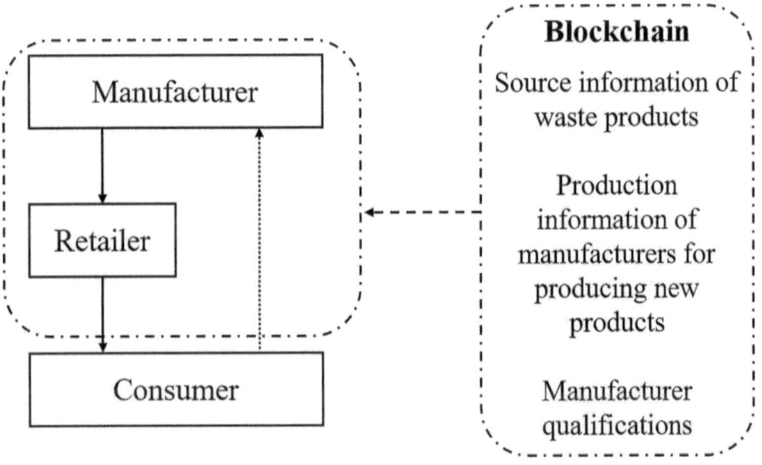

Fig. 8.3 Manufacturer recycling processes

shows the structure of the manufacturer recycling mode after adopting blockchain technology.

In the manufacturer recycling mode, blockchain will have an impact on the pricing strategy of the manufacturer. On the one hand, the use of blockchain by supply chain participants leads to an increase in wholesale prices (Niu et al., 2021). Irrespective of the scale, incorporating blockchain invariably introduces a unit cost, thereby reducing the manufacturer's profit margin. To compensate for their losses incurred from participating in the blockchain, manufacturers increase wholesale prices. On the other hand, the adoption cost of blockchain will affect the transfer prices of the manufacturer. If the costs are low, the adoption of blockchain can lead to a decrease in transfer prices; otherwise, the transfer prices will increase.

8.4.1.2 Blockchain in the Retailer Recycling Mode

The retailer recycling mode refers to the scenario where manufacturers entrust retailers (e.g., automobile dealers and repair shops) with the responsibility of recycling. Blockchain implementation can have an impact on the pricing strategy of retailers by increasing retail prices and sales.

In this mode, the entire supply chain lacks transparent recycling process information, and retailers need to invest more resources in data collection and analysis, which increases their operating costs. Manufacturers may also lack effective collaboration and communication mechanisms with retailers due to information issues.

The adoption of blockchain can help mitigate these problems. Blockchain provides an immutable record that tracks each stage of the recycling process, enabling both manufacturers and retailers to have a clear and accurate understanding of the recycling information at each step (Lotfi et al., 2022). Meanwhile, blockchain enables consumers to track the origin and destination of batteries, which increases consumer trust and thus expands market potential (Niu et al., 2022).

The vehicle owner can bring the unused vehicle to a 4S (Sales, Service, Spare parts and Surveys) store, where a professional team will conduct a vehicle condition assessment and assist the owner in completing the recycling process (Baum et al., 2022). Figure 8.4 shows the structure of the retailer recycling mode.

The application of blockchain will have a certain impact on retailer decision-making. It will lead to an increase in retail prices. Furthermore, the greater the cost of blockchain, the larger the increase in retail price; similarly, the stronger the market expansion effect due to blockchain, the greater the increase in retail prices. There are two main reasons for this: First, implementing blockchain increases supply chain costs, reducing the profit margin for retailers, who then tend to raise prices to compensate for these costs. Second, the market expansion effect enabled by joining the blockchain leads to an initial increase in market demand for retailers, who no longer rely on low pricing strategies to maintain sales quantity. With higher retail prices, the market sales volume remains substantial, thus incentivizing retailers to raise retail prices to gain more profit (Li et al., 2022).

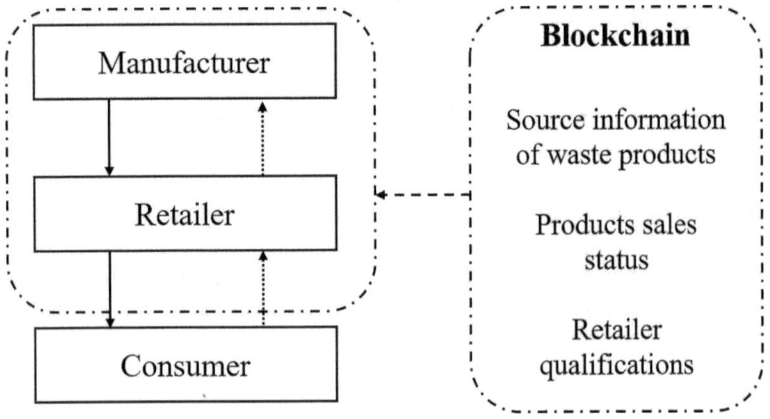

Fig. 8.4 Retailer recycling processes

Furthermore, the impact of blockchain on retail quantity is contingent upon the cost of implementing blockchain. Specifically, when the costs associated with blockchain are high, integrating this technology leads to a decrease in sales quantity; conversely, when the costs are low, it results in an increase in sales quantity. The underlying reasons are as follows: When blockchain costs are minimal, the retailer raises retail prices by a smaller margin, which in turn only slightly reduces sales quantity. The market expansion effects brought about by blockchain increase sales quantity, thus overall retail quantity increase. However, when blockchain costs are substantial, the increase in retail prices is significant, and the additional sales quantity generated by the market expansion effects of blockchain are insufficient to offset the decrease in sales quantity caused by the rise in prices. Therefore, overall retail quantity decreases.

8.4.1.3 Blockchain in the Recycler Recycling Mode

The recycler recycling mode refers to manufacturers or retailers entrusting third-party recycling agencies with the responsibility of recycling products (Fig. 8.5). In this mode, the processing techniques and quality standards of the third-party recyclers may not meet the requirements of manufacturers and retailers, which can lead to a decline in the quality and performance of the recycled products. This, in turn, can increase the costs of inspection and evaluation of the recycled products for both manufacturers and retailers.

Blockchain can improve the efficiency of residual energy detection, value assessment, and the reuse of batteries (Wang et al., 2023). Consequently, adopting blockchain reduces the recycling costs of batteries (Rufino Júnior et al., 2022). Companies such as GEM and Tinci Materials entrust third party businesses for battery recycling (Li et al., 2022).

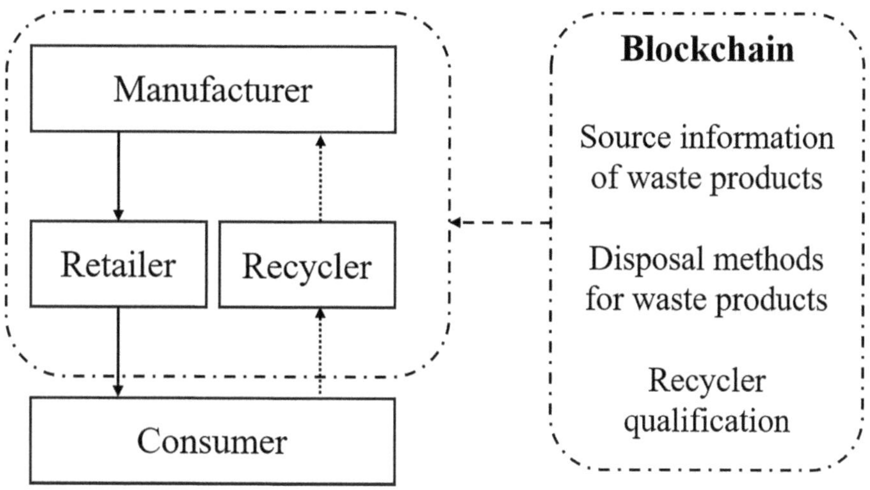

Fig. 8.5 Recycler recycling processes

The adoption cost of blockchain will affect the recycling price and volume. Specifically, low blockchain costs lead to an increase in both recycling prices and quantity. This is because when the costs of blockchain are lower than the cost savings (e.g., reduced inspection and evaluation costs), adopting blockchain reduces the recycler's overall operational costs. As a result, the manufacturer increases the transfer payment prices to extract higher profits from the recycler. Conversely, high blockchain costs results in a decrease in both recycling prices and quantity for the recycler.

8.4.1.4 Comparison of Blockchain in Three Recycling Modes

The impact of blockchain on supply chain members varies depending on the recycling mode. Here we will provide a detailed comparison of the prices, profits, and conditions for the adoption of blockchain by supply chain members in each recycling mode.

1. **The impact of blockchain on equilibrium strategies**

   The adoption of blockchain will affect the prices and sales of reverse supply chains. Regardless of the recycling mode, the adoption of blockchain is expected to enhance potential market demand, thereby stimulating an increase in sale and recycling prices as well as sales and recycling quantities. However, it is important to note that this can only be established when the cost of adopting blockchain is relatively low.

   Meanwhile, the impact of blockchain on transfer prices is smaller if there are two competing recyclers than if there is only one recycler. This is due to the fact that if only one of the two competing recyclers adopts blockchain, there will be a lesser impact of blockchain on recycling costs, hence reducing the manufacturer's ability to manipulate the transfer payment prices. Additionally, the change in recycling prices and quantity due to blockchain under the manufacturer recycling mode is greater than in other modes, with the manufacturer in this mode being more sensitive to blockchain costs. This is because the manufacturer is the direct beneficiary of using blockchain for recycling and bears the costs of blockchain in recycling. In the manufacturer mode, there is no transfer payment to be leveraged, making the manufacturer more sensitive to the costs of blockchain. Thus, in situations where the manufacturer is responsible for recycling, the impact of blockchain on recycling prices, recycling quantity, and hence the manufacturer's profits is greater.

Moreover, blockchain has the potential to further enhance the profits of supply chain members, contingent on the cost associated with blockchain. Regardless of the recycling mode, when the cost of blockchain adoption is low, the adoption of blockchain can potentially increase market demand and product prices, thereby further increasing the profits of supply chain members.

2. **The willingness of supply chain members to adopt blockchain technology**

   The willingness of various supply chain members to adopt blockchain is influenced by a variety of external factors. Overall, the willingness of the manufacturer, the retailer, and the recycler to adopt blockchain is largely influenced by blockchain costs, cost savings, and market expansion effects.

   In both the recycler and retailer recycling modes, it is evident that the manufacturer might adopt blockchain even when the cost-effectiveness of recycling through blockchain is low or negative. This is because of increased market potential leading to higher wholesale prices. If the cost of adopting blockchain is low or the market potential is high, blockchain adoption can lead to an increase in sales and hence the manufacturer' profits. When blockchain costs are high and the cost effects are significant, the manufacturer prefers to adopt the technology in the manufacturer mode. In such a mode, the manufacturers bear the recycling costs and the benefits from improved recycling efficiency are greater. When costs are low and cost effects are minimal, the manufacturer prefers to adopt the technology in the recycler mode because it does not bear the blockchain-related costs but can benefit from the market potential effect of blockchain.

   The retailer tends to adopt blockchain when blockchain costs are low and the market expansion effect is significant. In this scenario, blockchain facilitates an increase in retail quantity, which yields benefits for the retailer. However, in the retailer recycling mode, the retailer is inclined to adopt the technology in the recycler mode when both blockchain costs and market expansion effects are low. Note that the retailer bears the costs of blockchain recycling in the retailer mode, whereas these costs are borne by the recycler in the recycler recycling mode. This difference in cost bearer makes the retailer unwilling to adopt blockchain in the retailer-led mode.

   The recyclers' willingness to adopt blockchain is also influenced by blockchain costs and the cost-saving effects. When blockchain costs are low, the technology can lead to increased recycling quantity, thereby motivating the recyclers to adopt blockchain. However, as blockchain costs increase and the cost-saving effect diminishes, adopting blockchain may lead to an increase in recycling costs, at which point recyclers become unwilling to adopt the technology. If there are two competing recyclers in the supply chain, then their willingness to adopt blockchain is influenced by both the intensity of competition and the cost of blockchain. Assuming that recycler 1 adopts blockchain while recycler 2 does not, it is found that the less intense the competition, the more willing recycler 1 is to adopt blockchain; the lower the costs, the more likely recycler 1 will adopt the technology. If both recyclers adopt blockchain, then the greater the competitive coefficient, the more inclined they are to use blockchain. This is because when both recyclers can implement blockchain, competition between them intensifies. When the cost of blockchain is low, if recycler 2 employs the technology while recycler 1 does not, recycler 1 will lose competitive advantage when the competitive coefficient increases. Therefore, with a higher competitive coefficient, both recyclers are more willing to adopt blockchain, thereby raising the cost threshold for blockchain adoption to compete with each other.

8.4.2 Blockchain Technology in Reverse Supply Chains Considering Advertising Strategy

Reverse supply chain members face the challenge that consumers are skeptical of buying reused and remanufactured products due to their lack of understanding about the recycling process. However, advertising can address these issues by disseminating reverse supply chain information and sustainable concepts among consumers.

8.4.2.1 Advertising in Reverse Supply Chains

In this section, we will discuss the application of advertising in reverse supply chains and its impact on supply chain members and consumers.

1. **The introduction of advertising**

   Advertising refers to the dissemination of information and the promotion of products or services through various media channels, with the primary aim of arousing consumer interest, stimulating purchase desire, and ultimately facilitating transactions. Therefore, advertising plays a crucial role in the reverse supply chain in encouraging consumers participation in recycling activity. Below we will focus on the advertising strategies of the reverse supply chain in the context of blockchain technology.

   Both advertising content and advertising channels have an impact on the development of the reverse supply chain. Firstly, advertising content directly influences consumers' perceptions of recycled or remanufactured products. If advertising content can clearly convey the importance, process, and benefits of

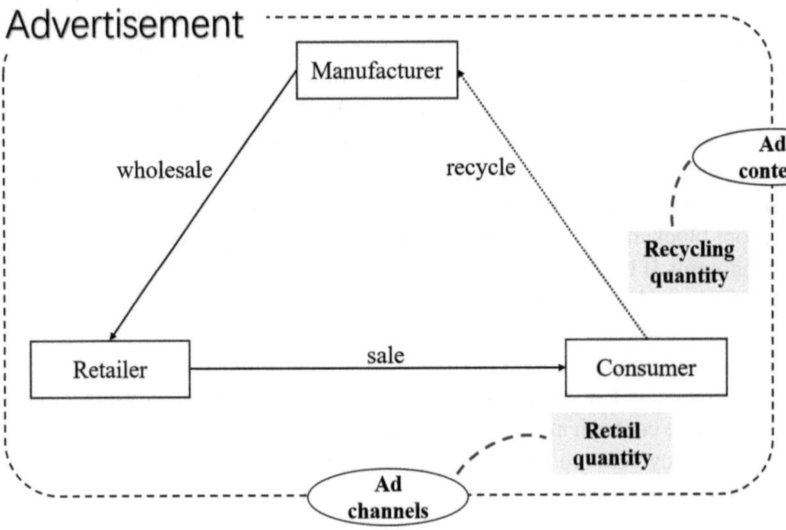

Fig. 8.6 Advertising in reverse supply chains

recycling and remanufactured products, it can enhance consumers' willingness to participate in recycling and sustainable purchasing, thereby increasing the volume of recycling and sales in the reverse supply chain. Secondly, the choice of advertising channels determines the number of potential consumers that can be reached. By carefully selecting media channels and targeting audiences, advertising can more effectively attract consumers who are likely to engage in product recycling, thus expanding the potential market of the reverse supply chain. In summary, through appropriate advertising content and channels, members of the reverse supply chain can devise better advertising strategies, expand the potential consumer market, and improve product recycling rates and sales (Fig. 8.6).

2. **The impact of advertising on reverse supply chains**

   The advertising strategy in reverse supply chains can encourage consumers to actively participate in recycling and increase the profits of supply chain members. In reverse supply chains, advertisements can stimulate consumers' interest in purchasing remanufactured products and increase their willingness to actively recycle waste products. Proper use of advertising can further encourage consumers to understand the benefits of recycling and reuse, actively participate in resource recycling and reuse, facilitate the recycling of waste products, and expand the market of remanufactured products. Previous studies have also found that advertising can indeed raise consumer awareness and motivate them to participate in the creation of green reverse supply chains (Jena et al., 2017; Johnson, 2013; Yue et al., 2006). Using advertising in the remanufacturing supply chain can also increase consumer trust in remanufactured products and attract consumers to return used items.

Moreover, the adoption of advertising can affect the decisions of members in the reverse supply chain (McAlister et al., 2007). Some studies suggest that the profits of retailers will increase as advertising effectiveness improves (Savaskan et al., 2004). Similarly, when profits increase, retailers will increase their advertising investment to obtain more profits (Hong et al., 2015; Sun et al., 2022).

8.4.2.2 The Advertising Strategies of the Reverse Supply Chain with Blockchain Technology

1. **The impact of blockchain on advertising**

   The adoption of blockchain technology can be beneficial to advertising. The decentralization and immutability of blockchain ensure the transparency and authenticity of advertising data and optimize the analysis and placement of advertising data, thus reducing the intermediary costs. Consumers can gain trust in advertisements by accessing authentic advertising data recorded on the blockchain. Through blockchain, supply chain participants can directly remit advertising fees to the respective entities, thereby decreasing intermediary costs and lowering advertising expenses (Fig. 8.7). For example, Ken Brook, co-founder and CEO of MetaX, has launched the first blockchain project for digital advertising supply chain, AdChain to build a transparent and secure advertising delivery ecosystem. Also, Japanese automaker Toyota has partnered with blockchain advertising analytics company Lucidity to reduce fraudulent behavior when purchasing digital advertising.

Blockchain technology can encourage advertising investments in reverse supply chains. Due to the fact that blockchain ensures the transparency, authenticity, and security of data, supply chain members are more likely to trust the reverse supply chain environment and willing to increase advertising investment to promote product sales and brand exposure (Li et al., 2019). Meanwhile, an increase in advertising investment can not only enhance the visibility of new products, remanufactured products, and brands, but also attract more consumers to make purchases, thereby achieving a closed loop supply chain and sustained growth (Liu et al., 2020).

2. **The impact of blockchain and advertising on equilibrium strategies**

   The combined use of blockchain and advertising in reverse supply chains can enhance the recovery rate of the supply chain. When the market effect of blockchain is significant, its adoption can lead to an increase in the recovery rate. As mentioned earlier, the adoption of blockchain can increase advertising investment, attracting more consumers to participate in recycling activities, and thereby further increasing the recovery rate of obsolete products.

   Furthermore, the combined use of blockchain and advertising can also lead to an increase in profits for supply chain members. This is because blockchain can help to increase market demand, reduce recycling costs, and improve the recovery rate, while advertising incentivizes consumers to purchase new products and express willingness to recycle old products. The dual impact of these factors may

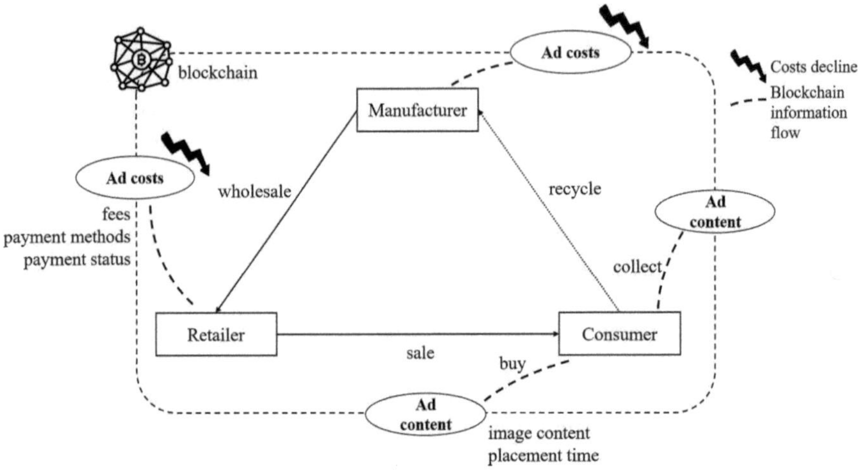

Fig. 8.7 Advertisement in blockchain technology

potentially increase the profits of supply chain members. Additionally, for manufacturers and recyclers, the extent to which blockchain impacts the recovery rate may influence their choice of recycling models. Other factors such as advertising costs, however, may exacerbate the costs for manufacturers and recyclers under certain recycling models, prompting them to select the most advantageous and optimal recycling model based on the degree of improvement in the recovery rate facilitated by blockchain.

3. **The impact of blockchain and advertising on recycling modes**

   In reverse supply chains, there are three prevalent recycling modes: manufacturer recycling, retailer recycling, and recycler recycling. The concurrent utilization of blockchain and advertising in reverse supply chains will influence the profits of supply chain members across various recycling modes, as well as their decisions on optimal recycling and advertising strategies.

   From the perspective of the manufacturer, as the promoting effect of blockchain on recycling rates gradually intensifies, the profit of the manufacturer in the retailer-recycling and manufacturer-advertising mode will gradually exceed that in the recycler-recycling mode and the manufacturer-recycling mode, and the growth rate will accelerate incrementally. This is due to the fact that in the retailer-recycling and manufacturer-advertising mode, the retailer responsible for recycling can save manufacturing costs, so that manufacturer does not need to invest significant resources in establishing recycling systems, which allows the retailer to purchase products from the manufacturer at a lower price. Furthermore, with the increasing impact of blockchain on recycling rate, the retailer has a greater opportunity to lower the price of products purchased from the manufacturer. On the other hand, the retailer responsible for recycling can benefit from recycling by establishing an environmentally-friendly and sustainable brand image, which can enhance the brand reputation and attract more consumers to purchase products. This expands the consumer market and indirectly increases the profits of the manufacturer.

For the retailer, the combined effects of advertising-driven recycling and blockchain-enabled recycling can increase the rate of profit growth across all recycling modes. As the blockchain-enabled recycling effect increases, the profits of the retailer in the retailer-recycling and manufacturer-advertising mode will first increase and then decrease as the advertising-driven recycling effect increases. In the retailer-recycling and manufacturer-advertising mode, as the advertising-driven recycling effect continues to grow, recycling costs will correspondingly increase. However, the revenues generated solely from recycling and advertising may not be sufficient to offset the gradually rising recycling costs, which could potentially result in a decline in the retailer's profit.

From the viewpoint of the recycler, when the blockchain's promotional effect on the recycling rate is relatively small, the recycler is always willing to take responsibility for recycling while leaving the advertising to the manufacturer. The reason is that the recycler can benefit more by focusing on the recycling operations itself, while leaving the advertising to the manufacturer who has the resources and expertise to effectively promote the recycling program. And as both the blockchain's promotional effect and the advertising's promotional effect on the recycling rate increase, the recycler may instead prefer to have the retailer take the responsibilities of advertising. Because the retailer often has a closer connection with consumers, the recycler will choose retailer advertising strategies. Meanwhile, as the promotion effect of advertising on recycling rate increases, no matter who is responsible for advertising investment, the profits of the recycler will first increase and then decrease due to the increase in advertising costs and new product costs. If advertising expenses exceed the growth rate of sales, the retailer's profit will decrease, leading to lower retail prices to maintain market demand.

8.4.3 *Blockchain Technology in Reverse Supply Chains Considering Information Strategies*

Faced with an increasingly complex market environment and consumer demands, supply chain members need to optimize information sharing strategies to ensure supply chain efficiency and attain higher profits. In this subsection, we will explore information sharing in reverse supply chains and evaluate the impact of blockchain on information strategies.

8.4.3.1 Information Strategies in the Reverse Supply Chain

In reverse supply chains, information sharing serves as a critical information strategy that enables supply chain members to adapt to the evolving market dynamics and consumer preferences. Information sharing among supply chain members is essential for enhancing coordination and collaboration in reverse supply chains, thus enabling the optimization of resource utilization and the mitigation of operational inefficiencies (Zheng et al., 2017). Supply chain members can leverage shared information to make informed decisions regarding product recovery, remanufacturing processes, and the identification of potential secondary markets (Meng et al., 2023; Sarkar & Guchhait, 2023). Effective information sharing and transparency within reverse supply chains can facilitate the identification of opportunities for value recovery from returned products and components. Apple actively promotes information disclosure and sharing in reverse supply chain management. They provide transparency through public environmental responsibility reports, showcasing progress in environmental initiatives and corporate social responsibility to consumers and stakeholders. Such information sharing not only enhances corporate transparency but also boosts consumer trust and brand loyalty.

8.4.3.2 Information Strategy of Reverse Supply Chain with Blockchain

As previously discussed, both blockchain and information sharing can bring benefits to reverse supply chains. When these two strategies are used simultaneously, they will further enhance the overall operational efficiency of reverse supply chains.

1. **The impact of blockchain and information strategies in reverse supply chain**
   Adopting blockchain and shares blockchain information in the reverse supply chain can reduce supply chain risks, optimize supply chain efficiency, and enhance cooperation among supply chain members:
   (a) **Reducing risks:** Through blockchain and information sharing, it is possible to monitor the flow and changes of products in the reverse supply chain in a real-time manner, thereby reducing risks and enhancing the safety of products.
   (b) **Optimizing efficiency:** The decentralized characteristics of blockchain and the accuracy of blockchain information can streamline processes, optimize resource allocation, improve efficiency, and reduce costs in reverse supply chains.
   (c) **Enhancing partnership:** The application of blockchain and blockchain information sharing can enhance the partnership between all parties in the reverse supply chain, establish trust mechanisms, promote information sharing and cooperation, and thus improve the overall business environment.

In summary, the simultaneous application of blockchain and information sharing may bring a more efficient, reliable, and secure management system to the reverse supply chain, which can help solve existing problems and improve overall operational effectiveness. Below we will analyze information sharing and blockchain adoption with three different scenarios.

2. **Three recycling modes in reverse supply chains with blockchain and information sharing**

   We consider a supply with one manufacturer, one retailer, and two competing recyclers (one authorized recycler and one unauthorized recycler) in the reverse supply chain and compare three different scenarios: (1) without blockchain and information sharing, (2) adopting blockchain without information sharing, and (3) adopting blockchain and sharing information.

   In the first scenario, supply chain members do not adopt blockchain. The manufacturer entrusts the authorized recycler and unauthorized recycler to be responsible for recycling, and then wholesales the remanufactured and new products to the retailer who then sells products to consumers. An authorized recycler typically holds legitimate business licenses, employs professional recycling personnel, and implements standard operating procedures. By November 2023, China's Ministry of Industry and Information Technology listed 156 companies officially permitted to recycle EV batteries in China (van Wyk, 2023). Smaller businesses that do not appear on this list are unauthorized recyclers. These entities often follow recycling processes that are not compliant with the established regulations. Figure 8.8a shows the structure of the reverse supply chain.

   In the second scenario, all supply chain members, except the unauthorized recycler, adopt blockchain and the manufacturer does not share blockchain information with the unauthorized recycler in this scenario. Figure 8.8b shows the reverse supply chain structure under this mode.

   In general, an unauthorized recycler has no access to blockchain information unless the manufacturer shares the access with the unauthorized recycler. We want to evaluate whether the manufacturer has an incentive to share blockchain information with the unauthorized recycler as illustrated in Fig. 8.8c.

8.4.3.3 The Impact of Information Strategy Under Blockchain on Reverse Supply Chains

In this section, we examine the effects of information sharing by the manufacturer to the unauthorized recycler on the equilibrium strategies and the willingness to adopt blockchain of supply chain members.

1. **The impact of information strategy on equilibrium strategies**

   Information strategies can affect supply chain prices and recycling quantities. Sharing blockchain information with the unauthorized recycler by the manufacturer will result in a lower transfer price, a higher unauthorized recovery price, and a higher unauthorized recovery quantity than no sharing. This is because sharing blockchain information with the unauthorized recycler will reduce the unauthorized recycler' recovery costs and allow the manufacturer to recover more battery waste from the unauthorized recycler at a lower transfer

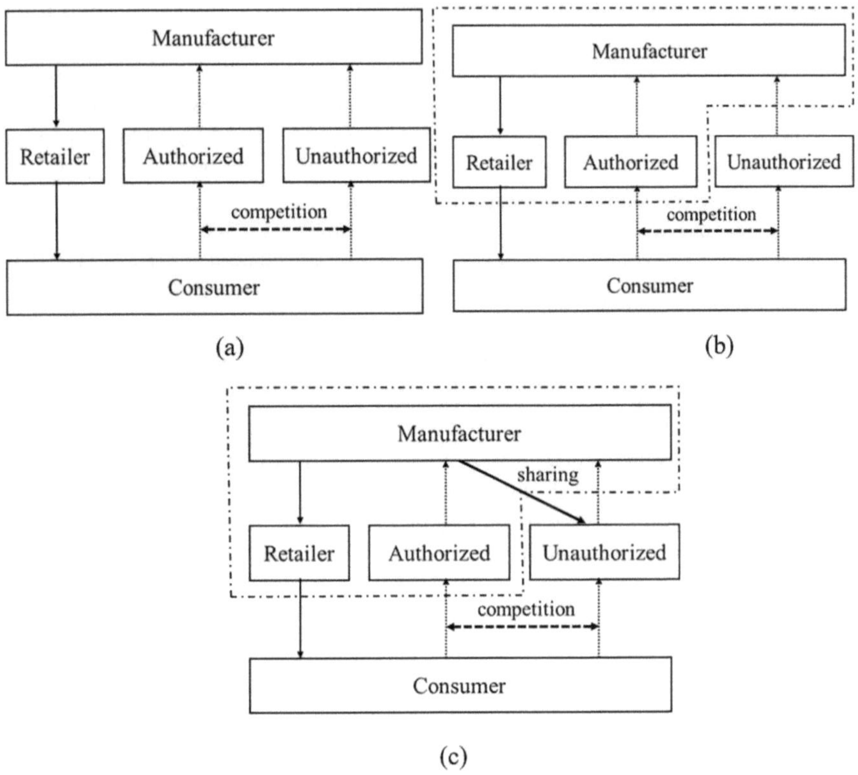

Fig. 8.8 Reverse supply chain structures with blockchain and information sharing

price. Moreover, the impact of information sharing on the recycling prices of the authorized recycler depends on the competition coefficient. Specifically, if the degree of competition intensity is low, then the recycling price of the authorized recycler decreases. When the degree of competition intensity is high, the authorized recycler has to set a higher recycling price in order to compete against the unauthorized recycler who increases the recycling prices with information sharing.

Sharing of blockchain information can increase the profits of some supply chain members. Sharing blockchain information increases the profit of the manufacturer and the unauthorized recycler but reduces the profit of the authorized recycler. This is because blockchain information sharing leads to a lower transfer price and a higher total recovery quantity and allows the unauthorized recycler to reduce recycling costs, increase recycling prices and quantity, and hence achieve a higher profit. However, information sharing does not impact the retail price, market share, and retailer quantity. This is because the information shared with the unauthorized recycler only affects the reverse supply chain, not the forward supply chain.

2. **The impact of information strategy on the willingness of supply chain members to adopt blockchain**

   The manufacturer has the incentive to share blockchain information with the unauthorized recycler because of the increased recycling volume and reduced transfer price. The willingness of the retailer to join the blockchain remains unchanged, as manufacturers' disclosure of information to the unauthorized recycler has no effect on the forward supply chain.

   However, when the manufacturer discloses information to the unauthorized recycler, the willingness of the authorized recycler to participate in the blockchain decreases. The reasons are as follows: If blockchain information is shared, the unauthorized recycler can reduce costs, resulting in a loss of the advantage for the authorized recycler joining the blockchain. The authorized recycler continues to incur higher recycling costs than the unauthorized recycler. Although blockchain can reduce the recycling costs of the authorized recycler, it has become less beneficial due to intensified competition from the unauthorized recycler. For example, GEM, a professional third-party recycling company, did not participate in the blockchain of Beili New Energy's battery recycling. Overall, such information sharing will reduce the willingness of some supply chain members to adopt blockchain and thus should be avoided.

8.5 Discussion

As previously mentioned, blockchain has been extensively adopted across various industries due to its exceptional data storage capabilities and transparency of information (Choi & Siqin, 2022). The research studies on the adoption of blockchain technology in reverse supply chain are growing rapidly, indicating that it is necessary and beneficial for supply chain members to adopt blockchain in reverse supply chain. The primary advantages of blockchain application in reverse supply chain include enhanced supply chain transparency and information traceability, as well as reduced operational management costs, which were also discussed in the preceding sections (Wang et al., 2024). For instance, Circulor, a UK-based blockchain technology company specializing in utilizing blockchain technology to track the flow of sustainable resources, collaborates with globally renowned electric vehicle manufacturers to incorporate blockchain technology in battery production and recycling. By tracing the entire battery lifecycle across both the production and recycling chains, it provides a highly transparent and traceable solution for the recycling and reuse of battery materials.

Despite the numerous benefits associated with the application of blockchain technology in reverse supply chain, it still faces technical limitations (Petratos & Faccia, 2023). Since blockchain employs distributed nodes and consensus algorithms to coordinate the execution of smart contracts and generate updated data, its operational speed is slower than that of centralized databases (Islam & Apu, 2024). This is because whenever a new block is added to the blockchain, it must perform all the processes executed by conventional databases, along with additional processes that impact its performance. Furthermore, each blockchain transaction must undergo digital signing and verification using a public or private key encryption scheme, which involves computationally complex and time-consuming signature verification.

Additionally, as a relatively new technology, blockchain has not been applied in reverse supply chain for an extended period, which leads to several ongoing challenges (Difrancesco et al., 2023). Firstly, during supply chain operations, not all enterprises are willing to upload all aspects of their operational and transactional data onto the blockchain, thereby weakening the fairness and efficiency of data information achieved (Chowdhury et al., 2023). Secondly, the security of transactions ensured by the immutability of blockchain is predicated on the authenticity of the transaction data. In other words, while blockchain can guarantee the reliability and immutability of data once it is uploaded, it cannot ensure the authenticity of the data. If the data source itself contains flaws or issues, it is unrealistic to expect blockchain to play a significant role. This is why blockchain technology is internationally recognized to potentially suffer from the "garbage in, garbage out" problem. Thirdly, blockchain requires relevant institutions to manage and control it to ensure that the system operates in a fair and transparent manner and that its applications are legal and compliant. This raises a question about who should control the blockchain and make decisions. Therefore, without the establishment of overall management norms and a supply chain system, as well as the concerted efforts of regulators, industries, and various stakeholders, the realization of blockchain value in reverse supply chain will remain elusive (Sharabati & Jreisat, 2024).

8.6 Conclusion and Future Research

In this chapter, we introduced the reverse supply chain, elaborated on the application of blockchain in the reverse supply chain, and further explored the joint effects of advertising strategies and blockchain on reverse supply chains. Current research suggests that adopting blockchain in reverse supply chains can improve supply chain efficiency and profits, reduce recycling process costs, and promote collaboration among supply chain members. However, it should be noted that in some cases, companies may not adopt blockchain during the recycling process. A higher adoption cost of blockchain can actually lead to a decrease in supply chain recycling efficiency, thereby reducing the profits of supply chain members. Therefore, when enterprises adopt blockchain in the recycling process, they still need to carefully evaluate the impact of various external factors on recycling to avoid excessive losses.

Meanwhile, our study found that blockchain adoption can increase advertising investment, reverse supply chain recovery rate, and supply chain profits, which echoes early relevant research. For instance, Kim et al. (2023) focuses on data transparency, consumer privacy, and advertising fraud. They find that blockchain-based advertising can leverage the transparency of data within the blockchain to ensure the data-driven characteristics of smart advertising for advertisers, intermediaries and publishers. It is believed that blockchain technology can enhance the advertising system, making it more trustworthy from consumer perspectives, which can foster increased investment in advertising to a certain extent. Plastic Bank, powered by its blockchain platform, fights against plastic pollution and poverty by paying plastic collectors in underprivileged community in a trustful way (Plastic Bank, 2024). The study by Markus and Buijs (2022) highlights the impacts of blockchain on various supply chain performance such as speed, quality, cost, dependability, and flexibility. However, different supply chain members may choose various recycling mode based on the impact of blockchain adoption and advertising strategies in improving recycling efficiency.

Finally, we also explored the issue of information strategy in the reverse supply chain with blockchain. It is found that if the manufacturer shares blockchain information with the unauthorized recycler, only the authorized recycler in the supply chain will incur a loss, and the level of competition between the two recyclers will affect the recycling price of authorized recycler.

Currently, many recycling enterprises have utilized blockchain to improve efficiency and secure recycling, but others are hesitant in implementing blockchain in recycling. Research on the application of blockchain in reverse supply chains currently mostly focuses on EV and EV battery recycling, with relatively little attention on other recyclable products. Future studies can examine other applications of blockchain in the reverse supply chain and explore the factors that facilitate the integration of blockchain in recycling and reuse of different recycled products.

References

Babich, V., & Hilary, G. (2018). *What OM researchers should know about blockchain technology* (Working paper). Georgetown University.

Baum, Z. J., Bird, R. E., Yu, X., & Ma, J. (2022). Lithium-ion battery recycling—Overview of techniques and trends. *ACS Energy Letters, 7*(2), 712–719.

Choi, T.-M. (2023). Supply chain financing using blockchain: impacts on supply chains selling fashionable products. *Annals of Operations Research, 331*(1), 393–415.

Choi, T.-M., & Siqin, T. (2022). Blockchain in logistics and production from Blockchain 1.0 to Blockchain 5.0: An intra-inter-organizational framework. *Transportation Research Part E-Logistics and Transportation Review, 160*, 102653.

Chowdhury, S., Rodriguez-Espindola, O., Dey, P., & Budhwar, P. (2023). Blockchain technology adoption for managing risks in operations and supply chain management: evidence from the UK. *Annals of Operations Research, 327*(1), 539–574.

Difrancesco, R. M., Meena, P., & Kumar, G. (2023). How blockchain technology improves sustainable supply chain processes: a practical guide. *Operations Management Research, 16*, 620–641.

Harper, G., Sommerville, R., Kendrick, E., Driscoll, L., Slater, P., Stolkin, R., Walton, A., Christensen, P., Heidrich, O., Lambert, S., & Abbott, A. (2019). Recycling lithium-ion batteries from electric vehicles. *Nature, 575*(7781), 75–86.

Hastig, G. M., & Sodhi, M. S. (2020). Blockchain for supply chain traceability: business requirements and critical success factors. *Production and Operations Management, 29*(4), 935–954.

H&M. (2024, March 27). H&M Group annual and sustainability report 2023. https://hmgroup.com/wp-content/uploads/2024/03/HM-Group-Annual-and-Sustainability-Report-2023.pdf

Hong, X., Xu, L., Du, P., & Wang, W. (2015). Joint advertising, pricing and collection decisions in a reverse supply chain. *International Journal of Production Economics, 167*, 12–22.

HP. (2024, July 1). How to recycle HP ink and toner cartridges: A complete guide. https://www.hp.com/us-en/shop/tech-takes/recycle-toner-cartridges

Ikea. (2020, October 13). Ikea to buy back used furniture in recycling push. https://www.bbc.com/news/business-54531619

Islam, S., & Apu, K. U. (2024). Decentralized vs. centralized database solutions in blockchain: Advantages, challenges, and use cases. *Global Mainstream Journal of Innovation, Engineering & Emerging Technology, 3*(4), 58–68.

Jena, S. K., Sarmah, S. P., & Sarin, S. C. (2017). Joint-advertising for collection of returned products in a reverse supply chain under uncertain environment. *Computers & Industrial Engineering, 113*, 305–322.

Johnson, J. P. (2013). Targeted advertising and advertising avoidance. *The RAND Journal of Economics, 44*(1), 128–144.

Kim, J., Lee, K. H., & Kim, J. (2023). Linking blockchain technology and digital advertising: How blockchain technology can enhance digital advertising to be more effective, efficient, and trustworthy. *Journal of Business Research, 160*, 113819.

Li, K., Lee, J. Y., & Gharehgozli, A. (2023). Blockchain in food supply chains: a literature review and synthesis analysis of platforms, benefits and challenges. *International Journal of Production Research, 61*(11), 3527–3546.

Li, M., Weng, J., Yang, A., Liu, J. N., & Lin, X. (2019). Toward blockchain-based fair and anonymous ad dissemination in vehicular networks. *IEEE Transactions on Vehicular Technology, 68*(11), 11248–11259.

Li, Q., Ji, H., & Huang, Y. (2022). The information leakage strategies of the supply chain under the block chain technology introduction. *Omega, 110*, 102616.

Li, X., Mu, D., Du, J., Cao, J., & Zhao, F. (2020). Game-based system dynamics simulation of deposit-refund scheme for electric vehicle battery recycling in China. *Resources, Conservation and Recycling, 157*, 104788.

Liu, D., Ni, J., Lin, X., & Shen, X. (2020). Transparent and accountable vehicular local advertising with practical blockchain designs. *IEEE Transactions on Vehicular Technology, 69*(12), 15694–15705.

Lotfi, R., Kargar, B., Gharehbaghi, A., Hazrati, H., Nazari, S., & Amra, M. (2022). Resource-constrained time–cost-quality-energy-environment tradeoff problem by considering blockchain technology, risk and robustness: a case study of healthcare project. *Environmental Science and Pollution Research, 29*(42), 63560–63576.

Ma, D., & Hu, J. (2022). The optimal combination between blockchain and sales format in an internet platform-based reverse supply chain. *International Journal of Production Economics, 254*, 108633.

Ma, D., Qin, H., & Hu, J. (2022). Achieving triple sustainability in reverse supply chain: The optimal combination of online platform sales format and blockchain-enabled recycling. *Computers Industrial Engineering, 174*, 108763.

Markus, S., & Buijs, P. (2022). Beyond the hype: how blockchain affects supply chain performance. *Supply Chain Management: An International Journal, 27*(7), 177–193.

McAlister, L., Srinivasan, R., & Kim, M. C. (2007). Advertising, research and development, and systematic risk of the firm. *Journal of Marketing, 71*(1), 35–48.

Meng, Q., Wan, L. J., & Zhao, Q. W. (2023). Uncertain cost information sharing strategy in a closed-loop supply chain. *International Journal of Technology Management, 92*(1-2), 95–111.

Mercedes-Benz. (2024, October 21). Own recycling factory to close the battery loop. https://group.mercedes-benz.com/company/news/recycling-factory-kuppenheim.html

Moosavi, J., Naeni, L. M., Fathollahi-Fard, A. M., & Fiore, U. (2021). Blockchain in supply chain management: A review, bibliometric, and network analysis. *Environmental Science and Pollution Research*. https://doi.org/10.1007/s11356-021-13094-3

Niu, B. Z., Dong, J., Dai, Z. P., & Jin, J. Y. (2022). Market expansion vs. intensified competition: Overseas supplier's adoption of blockchain in a cross-border agricultural supply chain. *Electronic Commerce Research and Applications, 51*, 101113.

Niu, B. Z., Dong, J., & Liu, Y. Q. (2021). Incentive alignment for blockchain adoption in medicine supply chains. *Transportation Research Part E, 152*, 102276.

Petratos, P. N., & Faccia, A. (2023). Fake news, misinformation, disinformation and supply chain risks and disruptions: risk management and resilience using blockchain. *Annals of Operations Research, 327*(2), 735–762.

Plastic Bank. (2024, May 5). Transparency and trust in ESG: The blockchain advantage. https://plasticbank.com/blog/transparency-and-trust-in-esg/

Pongen, I., Ray, P., & Govindan, K. (2024). Creating a sustainable closed-loop supply chain: An incentive-based contract with third-party E-waste collector. *Journal of Cleaner Production, 462*, 142351.

Prahinski, C., & Kocabasoglu, C. (2006). Empirical research opportunities in reverse supply chains. *Omega, 34*(6), 519–532.

Rufino Júnior, C. A., Sanseverino, E. R., Gallo, P., Koch, D., Schweiger, H. G., & Zanin, H. (2022). Blockchain review for battery supply chain monitoring and battery trading. *Renewable and Sustainable Energy Reviews, 157*, 112078.

Sarkar, B., & Guchhait, R. (2023). Ramification of information asymmetry on a green supply chain management with the cap-trade, service, and vendor-managed inventory strategies. *Electronic Commerce Research and Applications, 60*, 101274.

Savaskan, R. C., Bhattacharya, S., & Van Wassenhove, L. N. (2004). Reverse supply chain models with product remanufacturing. *Management Science, 50*(2), 239–252.

Sharabati, A. A. A., & Jreisat, E. R. (2024). Blockchain technology implementation in supply chain management: A literature review. *Sustainability, 16*(7), 2823.

Sun, Q. Q., Chen, H., Long, R. Y., Li, Q. W., & Huang, H. (2022). Comparative evaluation for recycling waste power batteries with different collection modes based on Stackelberg game. *Journal of Environmental Management, 312*, 114892.

Van Engeland, J., Beliën, J., De Boeck, L., & De Jaeger, S. (2020). Literature review: Strategic network optimization models in waste reverse supply chains. *Omega, 91*, 102012.

van Wyk, B. (July 11, 2023). China's official electric vehicle battery recyclers fight wildcat workshops. https://thechinaproject.com/2023/07/11/chinas-official-electric-vehicle-battery-recyclers-fight-wildcat-workshops/

Wang, Y., Dai, S., Jiang, Z., & Gong, Q. S. (2023). A recycling model based on blockchain and reverse logistics for remanufacturing. *International Journal of Computer Integrated Manufacturing*. https://doi.org/10.1080/0951192X.2023.2294459

Wang, Y., Gao, J., Cheng, T. C. E., Gong, Y., & Wu, T.-H. (2024). Does blockchain technology need to be introduced in the closed-loop supply chain based on product lifecycle management? *Computers & Industrial Engineering, 188*, 109881.

Xie, S., Gong, Y., Kunc, M., Wen, Z., & Brown, S. (2023). The application of blockchain technology in the recycling chain: a state-of-the-art literature review and conceptual framework. *International Journal of Production Research, 61*(24), 8692–8718.

Yue, J., Austin, J., Wang, M.-C., & Huang, Z. (2006). Coordination of cooperative advertising in a two-level supply chain when manufacturer offers discount. *European Journal of Operational Research, 168*, 65–85.

Zhang, Z., Ren, D., Lan, Y., & Yang, S. (2022). Price competition and blockchain adoption in retailing markets. *European Journal of Operational Research, 300*(2), 647–660.

Zheng, B., Yang, C., Yang, J., & Zhang, M. (2017). Pricing, collecting and contract design in a reverse supply chain with incomplete information. *Computers & Industrial Engineering, 111*, 109–122.

Part III
Data-Driven Analytics for Resilient and Adaptive Supply Chains

Chapter 9
Harnessing AI and Big Data to Build a Resilient Supply Chain: An Overview

Azadeh Dindarian

Abstract This chapter explores the transformative role of digital technologies, including Artificial Intelligence (AI), Big Data and Internet of Things (IoT) in shaping supply chain management. As global supply chains grow more complex and vulnerable to disruptions such as geopolitical tensions and global crises, the demand for resilient, adaptive, and efficient systems continues to rise. AI-powered tools, predictive analytics, and real-time data insights enable organizations to shift from reactive problem-solving to proactive risk management, enhancing visibility, agility, and decision-making across the supply chain. Drawing from both academic literature and industry case studies, the chapter offers a well-rounded view of current AI applications in building supply chain resilience. It examines real-world examples from companies like Amazon and Tesco, showcasing how technologies such as demand forecasting, inventory optimization, route planning, and predictive maintenance are being used for competitive advantage. While the benefits are clear, digital transformation also brings challenges. Legacy systems, data quality issues, workforce upskilling, and ethical concerns must be addressed with thoughtful strategies, investment, and robust data governance to ensure sustainable, responsible adoption of these technologies.

Keywords Artificial intelligence · Data analytics · Big data · Digital transformation · Supply chain resilience

A. Dindarian (✉)
SRH Berlin University of Applied Sciences, Berlin, Germany
e-mail: azadeh.dindarian@srh.de

Y. Khojasteh, S. M. M. Jahan Abadi (eds.), *Emerging Technologies in Supply Chains*, International Series in Operations Research & Management Science 373, https://doi.org/10.1007/978-3-032-01218-0_9

9.1 Introduction

In today's rapidly evolving world, supply chain management has become increasingly complex and vulnerable to disruptions. The rise of globalization and advancements in technology have led to vast, interconnected supply networks spanning multiple countries and involving numerous stakeholders. While these developments have improved efficiency and market reach, they have also exposed businesses to greater risks, disruptions, and inefficiencies.

The COVID-19 pandemic provided a stark reminder of how fragile global supply chains can be. Production stoppages, transportation delays, and unpredictable shifts in consumer demand left many businesses struggling to adapt. A significant majority of Fortune 1000 companies reported substantial supply chain disruptions, underscoring the urgent need for enhanced resilience and adaptability (Sherman, 2020). These challenges have highlighted the importance of agile, data-driven, and technology-enabled supply chains that can respond swiftly and effectively to unexpected crises. Beyond pandemic-related disruptions, businesses continue to face mounting risks from geopolitical tensions, economic fluctuations, and climate-related disruptions, making resilience a long-term strategic priority.

Global supply chains are more interconnected than ever, making them susceptible to disruptions across multiple touchpoints. Geopolitical conflicts, labor shortages, regulatory changes, and logistical bottlenecks have increased volatility in supply chain operations (Ivanov & Dolgui, 2021). Research by Hendricks and Singhal (2005) found that companies experiencing supply chain disruptions suffered an average stock market decline of nearly 40%, illustrating the severe financial impact of supply chain failures. This underscores the high stakes of supply chain disruptions and the necessity for proactive resilience strategies.

To effectively address these challenges, businesses are increasingly turning to digital transformation strategies that leverage Artificial Intelligence (AI), Big Data, and advanced analytics. These technologies provide greater visibility, enhanced decision-making, and predictive capabilities, enabling supply chains to shift from a reactive to a proactive approach (Ivanov & Dolgui, 2021).

Advancements in AI, IoT, and real-time data analytics have revolutionized how businesses manage their supply chains. These technologies enhance supply chain visibility, agility and responsiveness and operational efficiency. This shift marks a paradigm change in supply chain management; companies that invest in digital resilience will not only survive disruptions but also gain a competitive advantage in their industries. Beyond immediate challenges, external factors like political instability, climate change, and global health crises continue to reshape supply chain strategies. Businesses must not only prepare for short-term disruptions but also build long-term resilience and sustainability (Ivanov, 2022). To achieve this, supply chains must incorporate flexibility, redundancy and real time data insights (Christopher & Peck, 2004; Tang, 2006).

This chapter delves into the role of digital technologies, particularly AI and Big Data, in strengthening supply chain resilience. Digital transformation is no longer an option but a necessity for businesses aiming to thrive in today's fast-paced environment. AI's ability to automate complex tasks, analyze vast datasets, and generate precise forecasts makes supply chains more efficient and adaptive. Meanwhile, Big Data provides real-time insights that drive strategic decision-making and operational optimization.

To illustrate these concepts, this chapter presents case studies from leading companies such as Amazon and Tesco, which have successfully integrated AI and Big Data into their supply chains. Their experiences highlight both opportunities and challenges associated with digital transformation. Additionally, this chapter explores ethical concerns such as data privacy, cybersecurity, and responsible AI-driven decision-making. As businesses increasingly rely on automation, it is crucial to balance AI's benefits with the risks associated with its adoption.

The remainder of this chapter is structured as follows: The next section examines the importance of digitalization in supply chain management. Section 9.3 discusses the significance of resilience in modern supply chains. In Sect. 9.4, we explore how digital technologies, particularly AI and Big Data, are driving improvements in supply chain resilience. Section 9.5 addresses the challenges and limitations businesses face when implementing digital transformation. Finally, the conclusion summarizes key takeaways and future directions for research and industry practice.

9.2 Why Digitalization in Supply Chain Management Is Important?

Traditional supply chain management, which heavily relies on manual processes and rigid, linear workflows, is struggling to keep pace with today's fast-moving and increasingly complex global economy (Guntuka et al., 2024). These conventional approaches lack the speed, flexibility, and real-time insights needed to effectively manage modern supply chains. A prime example of this vulnerability is the just-in-time (JIT) inventory model, which, while efficient in reducing storage costs, is highly susceptible to unexpected disruptions such as natural disasters and geopolitical conflicts (Carvalho et al., 2011). The COVID-19 pandemic exposed these weaknesses, as industries like automotive and healthcare faced severe shortages due to supply chain bottlenecks (Ivanov & Dolgui, 2021). Without buffer inventory or contingency plans, businesses struggled to keep up, leading to widespread inefficiencies and production delays (Albayrak Ünal et al., 2023).

One of the most significant limitations of traditional supply chain models is their reliance on historical data and static forecasts, which often fail to account for rapidly evolving market conditions (Rahoui et al., 2023). Currently, supply chains are increasingly affected by disruptive events such as extreme weather conditions, raw material shortages, political instability, and fluctuating consumer demand. Without access to real-time insights on supplier performance, shifting demand patterns, or transportation bottlenecks, companies are left reacting to disruptions instead of proactively managing them (McKinsey, 2016). For instance, the Suez Canal blockage in 2021 caused massive delays and financial losses, as traditional supply chain models were unable to swiftly adjust shipping routes (Notteboom et al., 2021). More recently, automotive manufacturers such as Tesla and Volvo were forced to halt production due to geopolitical tensions in the Red Sea region, further emphasizing the vulnerability of global supply chains to political instability (Waldersee et al., 2024). Another major weakness of traditional supply chains is their fragmented structure, where poor communication and lack of integration between suppliers, manufacturers, and logistics providers hinder rapid response to disruptions (Colin et al., 2015).

A critical example is the 2011 Japan earthquake, which had a cascading effect on global supply chains due to the lack of real-time collaboration and supply chain visibility (Park et al., 2013). The slow exchange of information led to delays in production, material shortages, and financial losses, highlighting the urgent need for more integrated and transparent supply chain systems. The lesson from such disasters is clear that supply chains must be digitally connected, enabling instant data sharing and proactive decision-making.

To address these challenges, companies must embrace digital transformation by integrating technologies such as AI, the IoT, and analytics. These technologies enhance supply chain visibility, enable real-time tracking, and support data-driven decision-making, allowing businesses to anticipate and mitigate potential disruptions before they escalate (Colin et al., 2015). Digitalization improves agility and responsiveness, equipping companies with the tools to quickly adapt to changing conditions and minimize vulnerabilities.

More than just an efficiency booster, digitalization is becoming a strategic necessity for supply chain resilience. By leveraging AI, IoT, and advanced analytics, organizations can enhance flexibility, optimize operations, and develop proactive risk management strategies (Colin et al., 2015; Sivakumar & Premrajkumar, 2024). However, transitioning to digital supply chains presents new challenges. Companies must address data privacy concerns, cybersecurity threats, and the ethical implications of automated decision-making (Bag et al., 2023; Möller, 2023). Ensuring responsible and secure technology adoption is critical to maximizing the benefits of digital transformation.

In a world where disruptions are becoming more frequent and unpredictable, digital supply chains offer a smarter, more resilient approach to managing risk and maintaining operational continuity. Businesses that fail to adopt digital solutions risk falling behind, while those that embrace transformation will be better positioned to navigate uncertainty, drive innovation, and sustain long-term growth.

9.3 The Importance of Supply Chain Resilience

As businesses become increasingly aware of the vulnerabilities in their supply chain networks, the importance of supply chain resilience (SCRES) has grown significantly (Munir et al., 2024). Tukamuhabwa et al. (2015) define SCRES as "*the adaptive capability of a supply chain to prepare for and/or respond to disruptions, to make a timely and cost-effective recovery, and therefore progress to a post-disruption state of operations—ideally, a better state than prior to the disruption*". This definition underscores resilience as a proactive strategy, rather than a reactive measure, emphasizing the need for businesses to anticipate, mitigate, and recover from disruptions while continuously strengthening their operations. Instead of merely bouncing back, resilient supply chains evolve to become stronger and more adaptable in the face of future uncertainties.

To build a truly resilient supply chain, businesses must integrate flexibility, redundancy, and real-time visibility into their operations. Digitalization plays a key role in achieving this by enabling real-time data collection and analysis, which allows companies to quickly adapt to disruptions and make informed decisions. According to Ivanov (2024), access to vast amounts of real-time data, combined with insights from past disruptions, provides businesses with new opportunities to enhance supply chain resilience. By leveraging technologies such as predictive analytics, AI-driven automation, and real-time tracking, companies can detect potential risks early and respond effectively.

This capability is more critical than ever as global supply chains become increasingly interconnected and vulnerable to external shocks such as geopolitical conflicts, natural disasters, and global health crises. The COVID-19 pandemic (Sherman, 2020) was a stark reminder of how fragile supply chains can be, with businesses worldwide experiencing severe disruptions due to factory shutdowns, supply shortages, and unpredictable demand fluctuations. Similarly, disruptions in the Electric Vehicle lithium-ion battery (EV LIB) supply chain caused by geopolitical tensions (Ren et al., 2024) and natural disasters such as the 2011 Japan earthquake (Park et al., 2013) further highlight the urgent need for robust, adaptable supply chains.

Resilient supply chains not only withstand crises but also foster long-term business growth and competitiveness. Companies that invest in resilience strategies, such as supplier diversification, greater transparency, and digital transformation are better equipped to navigate uncertainty, maintain operational continuity, and consistently deliver value to customers (Christopher & Peck, 2004). This adaptability is essential for sustainable growth in a volatile business environment (Ivanov, 2022).

Digitalization has emerged as a game-changer in strengthening supply chain resilience, particularly during times of crisis (Zhao et al., 2023). Technologies such as AI, the IoT, digital twins, and advanced data analytics allow businesses to optimize resource allocation, minimize inefficiencies, and reduce costs (Ivanov et al., 2019). Moreover, digital solutions facilitate real-time information sharing and seamless communication among supply chain stakeholders, enabling companies to coordinate responses and make quick, well-informed decisions. Predictive analytics and real-time tracking play a crucial role in early risk detection, helping businesses prepare for disruptions before they escalate (Waller & Fawcett, 2013). The ability to predict and mitigate risks in advance is becoming a defining factor in maintaining supply chain resilience.

Ultimately, investing in digital resilience strategies is no longer optional and it is a competitive necessity. Businesses that embrace smart technologies and data-driven decision-making will not only safeguard themselves against future disruptions but also strengthen their market position. In the face of increasing uncertainty, supply chain resilience is essential and it requires adapting, evolving, and securing long-term success (Spieske & Birkel, 2021).

9.4 Digital Technologies as a Driver in Supply Chain Resilience

In today's highly interconnected and volatile global economy, supply chains face constant disruptions, including natural disasters, geopolitical tensions, supply-demand imbalances, and global health crises. To navigate these uncertainties effectively, integrating digital technologies has become essential for building resilient supply chains. Technologies such as AI, the IoT, and Data Analytics are transforming supply chain management by providing real-time visibility, predictive insights, and enhanced decision-making capabilities (Ivanov, 2022; Zamani et al., 2023). The ability to transition from reactive crisis management to proactive risk mitigation is what makes digitalization a game-changer in modern supply chains.

The adoption of AI-driven predictive analytics enables supply chains to anticipate disruptions and take proactive measures (Waller & Fawcett, 2013). IoT devices and sensors offer real-time tracking, allowing businesses to monitor goods, identify risks, and react swiftly (Saad et al., 2024; Spieske & Birkel, 2021). Digitalization fosters collaboration among supply chain partners, enhancing responsiveness and agility to manage disruptions effectively.

Table 9.1 categorizes the key digital technologies transforming supply chains, their contributions, and references that support their impact:

By leveraging advanced digital solutions, companies can optimize resource allocation, minimize inefficiencies, and implement robust contingency plans. The following sections explore the role of AI and Data Science in strengthening supply chain resilience, with a focus on their practical applications across industries.

Table 9.1 Key digital technologies transforming supply chains

Digital technologies	Contribution to supply chain resilience	References
Artificial Intelligence	Predictive analytics for demand forecasting, automation of processes, enhanced supply chain visibility	e.g., Aslam and Calghan (2023); Ivanov (2022); Waller and Fawcett (2013)
Internet of Things	Real-time monitoring of shipments, predictive maintenance, enhanced coordination among stakeholders	e.g., Saad et al. (2024; Spieske and Birkel (2021)
Data Analytics	Actionable insights, predictive and prescriptive analytics for proactive risk management	e.g., Arunachalam et al. (2018); Tiwari et al. (2018); Zamani et al. (2023)

9.4.1 *Artificial Intelligence in Supply Chains*

Artificial Intelligence is rapidly transforming supply chain management (SCM) by improving efficiency, resilience, and competitiveness. As supply chains become increasingly complex and unpredictable, businesses are leveraging AI to enhance demand forecasting, optimize operations, and manage risks effectively (Ivanov, 2022). AI applications range from predictive analytics to intelligent automation and real-time decision-making, revolutionizing traditional supply chain processes (Aslam & Calghan, 2023; Khedr & Sheeja Rani, 2024). This shift marks a transition from reactive to proactive supply chain management, allowing businesses to anticipate disruptions before they occur, rather than responding to the fact. This section examines the role of AI in supply chains, exploring its applications, decision-making enhancements, and the challenges businesses face in adopting AI-driven solutions.

AI is now an integral part of supply chain management, facilitating optimization (Sahu, 2020), automation (Bhanushali et al., 2024), predictive analytics (Nzeako et al., 2024), and real-time decision-making (Xu et al., 2024). AI-driven solutions improve supply chain responsiveness and efficiency by leveraging real-time insights for faster, data-driven decision-making (van Hoek, 2020). This results in greater agility and reduced human error, particularly in transportation and logistics, where precision and efficiency are critical (Iyer, 2021).

Fosso Wamba et al. (2020) highlight the combined potential of AI, IoT, and Blockchain in improving traceability, coordination, and transparency. AI-powered supply chains also utilize dynamic capabilities, reconfiguring resources rapidly in response to shifting market conditions (Ivanov et al., 2019). As a result, businesses are increasingly using AI for inventory optimization, cost reduction, and lead-time improvement (Albayrak Ünal et al., 2023). Despite AI's capabilities, businesses must carefully evaluate their implementation strategies, as overreliance on AI without adequate oversight can lead to unexpected supply chain vulnerabilities.

9.4.1.1 AI in Inventory Management

Efficient inventory management is a critical component of supply chain success, preventing stockouts, overstocking, and excess holding costs (Ahmed Esmail, 2024). AI applications in inventory management, such as demand forecasting, real-time tracking, and automated replenishment, help eliminate inefficiencies and enhance supply chain resilience (Albayrak Ünal et al., 2023). Below are some the applications of AI to optimize inventory management at Unilever and Walmart. These real-world applications showcase AI's tangible impact on supply chain efficiency and resilience:

- Unilever's AI-powered collaboration with Walmart integrated real-time sales data and forecasts, achieving a 98% product availability rate and increasing sales by 12% within a year (Law, 2024).

- Walmart's AI-driven demand forecasting system predicts customer purchasing patterns, reducing excess inventory and optimizing stock levels (Silverstein, 2020).

9.4.1.2 AI in Demand Forecasting

Demand forecasting, driven by machine learning (ML) models, is one of the most significant applications of AI in inventory management. Traditional inventory management relies heavily on historical data and statistical methods for forecasting demand, which can be inaccurate in the face of fluctuating consumer trends and external disruptions (Albayrak Ünal et al., 2023; Fosso Wamba et al., 2020). AI-driven demand forecasting uses machine learning algorithms to analyze vast datasets, identify patterns, and predict future demand with greater accuracy (Albayrak Ünal et al., 2023; Muthukalyani, 2023). Further, it allows companies to better predict customer needs and manage inventory levels accordingly. However, the effectiveness of AI-driven forecasting depends heavily on data quality and availability, which remains a significant challenge for many supply chains (Muthukalyani, 2023).

9.4.1.3 Real-Time Monitoring with AI-Powered IoT Devices

Real-time monitoring, enabled by AI-powered IoT devices, provides continuous visibility into inventory levels across various locations. This ensures accurate decision-making, especially in smart factory settings (Soori et al., 2023; Ullah et al., 2024). IoT-enabled tracking devices allow suppliers to monitor shipments and provide real-time updates on delivery times and locations, enhancing supply chain analysis and efficiency. These devices contribute significantly to optimizing inventory management, reducing costs, and improving overall operational efficiency in manufacturing environments.

Furthermore, AI, when integrated with IoT, facilitates timely inventory replenishment and helps companies respond more effectively to sudden changes in demand (see Lin, 2023; Peladarinos et al., 2023). Examples of AI in real-time monitoring are:

- UPS equipped 110,000 delivery vehicles with IoT sensors, collecting data on 200+ operational parameters to improve routing and cut fuel costs (Ullah et al., 2024).
- Villegas-Ch et al. (2024) demonstrated a machine learning-based warehouse inventory system, reducing inventory counting time by 45% and increasing accuracy by 9% compared to traditional methods.

These examples showcase AI's potential to revolutionize inventory and logistics operations by improving accuracy, efficiency, and cost savings. However, companies must ensure that AI adoption aligns with their overall supply chain strategies to prevent fragmentation and inefficiencies.

9.4.1.4 AI and the Bullwhip Effect Mitigation

One of the most significant challenges in supply chain management is the bullwhip effect, where small fluctuations in consumer demand cause amplified inventory fluctuations across the supply chain. AI-based inventory optimization tools, combined with machine learning forecasting models, help reduce these distortions and improve demand visibility (Feizabadi, 2022; Muthukalyani, 2023; Weisz et al., 2023). Therefore, AI's Role in "Bullwhip" effect reduction based on the literature can be summarized as:

- Automated data analysis minimizes forecasting errors.
- Machine learning models detect irregular demand patterns and adjust stock levels dynamically.
- AI-driven supplier communication tools improve collaboration across the supply chain.

Despite AI's capabilities, legacy system integration and high implementation costs remain key challenges for businesses (Destino et al., 2022; Irani et al., 2023). Companies must balance AI adoption with practical operational needs to avoid over-reliance on automation.

9.4.1.5 AI in Route Optimization

AI-powered route optimization is transforming supply chain logistics by analyzing real-time data on traffic, weather, and transport capacity to determine the most efficient and cost-effective delivery routes (Gangwani & Gangwani, 2021; Ren et al., 2024). This capability is particularly important for logistics providers aiming to reduce fuel consumption and delivery times which help with sustainability topic in supply chain. Research by Ping et al. (2024) found that AI-based route optimization reduced transportation time by 15% and transportation costs by 10%. Table 9.2 provides examples of application of AI powered route optimization in DHL and UPS.

AI's ability to adapt to real-time disruptions makes it a crucial tool in mitigating last-mile delivery challenges, particularly in urban environments with unpredictable traffic conditions.

Table 9.3 summarizes some of the AI technology use cases in Supply chain management that we have discussed above.

Table 9.2 Examples of application of AI powered route optimization in DHL and UPS

Company	Use case	Impact	Reference
DHL	AI-driven logistics software (Wise Systems)	Improved delivery time accuracy & customer satisfaction	DHL (2023)
UPS	IoT tracking for vehicle fleet optimization	Saved $400 million annually by reducing fuel costs & improving delivery efficiency	Ullah et al. (2024)

Table 9.3 AI technology use cases in Supply chain management

Technology	Use case	Example	Reference
AI in Demand Forecasting	Predicts purchasing behavior and optimizes stock levels	Walmart's AI system improves in-store & online availability	Silverstein (2020)
AI in Route Optimization	Determines best delivery routes using real-time data	UPS saved $400 million annually via IoT tracking	Ullah et al. (2024)
AI in Inventory Monitoring	Enhances warehouse efficiency and tracking accuracy	AI-based warehouse system reduced inventory count time by 45%	Villegas-Ch et al. (2024)
AI in Bullwhip Effect Reduction	Reduces demand distortions and improves stock alignment	Machine learning tools predict fluctuations & adjust inventory	Feizabadi (2022)
Real-Time Monitoring with AI-powered IoT Devices	Enables continuous tracking, enhances decision-making, and optimizes supply chain operations	UPS equipped 110,000 delivery vehicles with IoT sensors, reducing fuel costs and improving efficiency	Ullah et al. (2024)

9.4.2 Data Science as a Driver for Supply Chain Resilience

In today's complex and fast-moving global environment, data science has become a cornerstone of supply chain resilience, offering critical insights, optimizing decision-making, and mitigating risks proactively. Supply chains are increasingly exposed to disruptions, ranging from natural disasters to supply-demand imbalances and geopolitical tensions. To navigate these challenges, businesses must leverage data analytics to gain real-time visibility, predict risks, and develop contingency strategies (Ivanov & Dolgui, 2021). Despite its transformative potential, the effectiveness of data science depends on data quality, accessibility, and integration across supply chain networks. Fragmented data sources, outdated systems, and cybersecurity concerns present significant hurdles that must be addressed for businesses to unlock the full potential of data-driven resilience strategies.

This section explores the role of data science in enhancing supply chain resilience, focusing on three key areas:

- Actionable insights: how data analytics help businesses monitor and improve supply chain performance.
- Predictive and prescriptive analytics: the role of AI-driven analytics in risk management.
- Real-world case studies: practical examples of how data science strengthens supply chain resilience.

9.4.2.1 Role of Data Analytics in Providing Actionable Insights

Data analytics plays a fundamental role in modern supply chain management, allowing businesses to monitor operations in real-time, identify inefficiencies, and implement corrective actions (Tiwari et al., 2018). Companies leveraging big data

Table 9.4 Overview of key benefits of data analytics in supply chain with implemented real case examples

Key benefits of data analytics in supply chains	Examples	References
Real-Time Monitoring	Tesco uses real-time analytics for inventory and logistics management	Marr (2016)
Risk Identification and Mitigation	Home Depot analyzed demand and logistics data to respond effectively to Hurricane Harvey	Güneş (2023)
Efficiency and Cost Reduction	Amazon optimized supply chain efficiency, increasing delivery speed by 75% using AI and big data analytics	Sifted (2024)

analytics have seen significant improvements in customer satisfaction, operational efficiency, and innovation in products and services (Zamani et al., 2023). Table 9.4 gives an overview of key benefits of data analytics in the supply chain with implemented real case examples.

Additionally, data analytics helps organizations adapt to disruptions by leveraging these capabilities to respond effectively to dynamic supply chain conditions (Arunachalam et al., 2018). However, the success of data analytics largely relies on the quality, accessibility, and integration of the data used (Hazen et al., 2014). Furthermore, sustainability-driven supply chains benefit from data analytics. A study by Singh and El-Kassar (2019) found that integrating big data technologies, green supply chain management, and Green human resource management practices leads to enhanced sustainability capabilities and improved firm performance. These findings highlight the increasing role of data-driven decision-making in sustainable supply chain strategies.

9.4.2.2 Predictive and Prescriptive Analytics for Proactive Risk Management

Predictive and prescriptive analytics have become indispensable for managing supply chain risks (See Table 9.5). AI-powered machine learning models analyze historical and real-time data to detect patterns, correlations, and anomalies, helping companies anticipate and prevent disruptions before they escalate (Aljohani, 2023).

- Predictive analytics: Uses historical data to forecast potential disruptions and take preventive action.
- Prescriptive analytics: Provides recommendations based on predictive insights to help companies optimize responses to disruptions (Lepenioti et al., 2020).

While predictive and prescriptive analytics offer tremendous value, they are not without limitations. The accuracy of AI-driven forecasts depends heavily on data quality and reliability. Poor-quality data or biased historical records can lead to inaccurate predictions, underscoring the need for continuous model refinement and human oversight.

The integration of AI-driven predictive models into real-time monitoring systems allows businesses to detect emerging risks and respond swiftly, ensuring greater resilience and agility in volatile markets (Aljohani, 2023).

Table 9.5 The predictive and prescriptive analytics used in supply chain management with examples

Type of analytics	Function	Examples	References
Predictive analytics	Identifies risks, forecasts demand, and optimizes inventory	Walmart predicts shopping trends to adjust stock levels	Silverstein (2020)
Prescriptive analytics	Recommends actions for crisis management, route optimization, and risk mitigation	AI-powered systems suggest alternative suppliers during disruptions	Smyth et al. (2024)

Table 9.6 Examples of the application of big data and analytics and their impact on efficiency and reduction of costs

Company	Data science application	Impact	References
Tesco	Real-time analytics for inventory and demand forecasting	Improved logistics efficiency and reduced costs	Marr (2016)
Amazon	AI-driven supply chain optimization	75% increase in delivery speed	Sifted (2024)
Home Depot	Crisis response planning using big data analytics	Rapid disaster recovery and supply chain adjustments	Güneş (2023)

9.4.2.3 Case Studies on Leveraging Data Science for Improved Supply Chain Resilience

Real-world applications of big data and analytics showcase the effectiveness of data science in building resilient supply chains. One of the most notable examples is Amazon's use of AI-driven demand forecasting during the 2023 Cyber Monday sales. By analyzing real-time consumer behavior and forecasting demand for over 400 million products, Amazon optimized its delivery routes and fulfillment strategies, ensuring smooth operations amid peak shopping periods (Sifted, 2024).

Another remarkable case study is Home Depot's response to Hurricane Harvey in 2017. Using big data analytics, the company assessed demand, logistics, and operational data, allowing them to prioritize essential supplies, reroute shipments, and restore operations quickly (Güneş, 2023). These case studies underscore the growing importance of data science in ensuring supply chain continuity and resilience. However, the rapid evolution of digital tools also raises concerns about data privacy, ethical AI usage, and cybersecurity risks, factors that must be carefully managed as businesses integrate advanced analytics into supply chain operations. Table 9.6 gives examples of the application of big data and analytics and their impact on efficiency and reduction of costs.

9.5 Challenges and Limitations in Digital Transformation

While digital transformation offers substantial benefits for supply chain management, it also presents complex challenges that organizations must navigate. These challenges can be broadly categorized into organizational, technological, and ethical issues amongst others, all of which play a crucial role in determining the success or failure of digital initiatives. Failure to address these issues effectively can result in poor adoption rates, increased operational costs, and even reputational risks

9.5.1 *Organizational Challenges*

One of the biggest hurdles in digital transformation is organizational resistance to change. Employees may hesitate to adopt AI-driven processes due to a lack of familiarity, fear of job displacement, or skepticism about the reliability of automated systems (Hangl et al., 2023; Irani et al., 2023; Ivanov, 2022; Waller & Fawcett, 2013). Organizations must focus on change management strategies, which include transparent communication, upskilling programs, and gradual technology adoption to ease the transition (Waller & Fawcett, 2013).

Another key challenge is skills shortages, as many companies struggle to find employees proficient in AI, IoT, and data analytics (Hangl et al., 2023). Without adequate training and recruitment strategies, businesses risk underutilizing the full potential of digital transformation. Furthermore, a shift toward data-driven decision-making requires a significant cultural transformation, where companies move away from intuition-based decisions to AI-driven analytics (Ameen et al., 2022).

9.5.2 *Technological Challenges*

A major barrier to digital transformation is the complexity of integrating AI and digital technologies into existing legacy systems. Many supply chains still operate on outdated Enterprise Resources Planning (ERP) systems that are not designed to handle the high volume and complexity of AI-driven insights (Hangl et al., 2023; Schwaeke et al., 2024). Data silos, where information is stored in isolated systems across departments, further complicate integration efforts, leading to inconsistent data formats, duplication, and inaccuracies (Ullah et al., 2024). To address these challenges, companies must invest in data standardization protocols and cloud-based solutions that enable seamless information flow across supply chain networks (Ameen et al., 2022). However, these upgrades can be expensive, making it difficult for small and medium enterprises (SMEs) to keep pace with large corporations (Colin et al., 2015). The lack of sufficient, high-quality data is another significant limitation for SMEs, restricting the ability to develop effective AI models

9.5.3 *Ethical and Cybersecurity Concerns*

Digital transformation introduces critical ethical and cybersecurity risks, which companies must proactively manage. One major issue is AI explainability, meaning that many AI-driven decisions lack transparency, making it difficult for businesses to justify actions or detect biases (Dash et al., 2024). If supply chain AI models are trained on biased or incomplete data, they can reinforce systemic inequalities, leading to unfair outcomes in supplier selection, pricing, and inventory distribution.

Additionally, as supply chains generate vast amounts of sensitive operational and customer data, concerns about data privacy and cybersecurity threats are rising (Bag et al., 2023). Cyberattacks targeting supply chains, (such as ransomware attacks on logistics networks) can disrupt entire operations, resulting in financial losses and reputational damage (Dash et al., 2024).

Automated decision-making also presents moral dilemmas, particularly in cases where AI makes choices that impact employment, supplier selection, or environmental sustainability (Manning et al., 2023). Companies must ensure that AI applications align with ethical guidelines and prioritize transparency, fairness, and accountability. Table 9.7 summarizes the major challenges in digital transformation and strategies to overcome them.

Table 9.7 Summary of the major challenges in digital transformation and mitigation strategies

Challenge category	Specific issues	Mitigation strategies	References
Organizational Challenges	Resistance to change, skills shortages, cultural shift	– Comprehensive training programs to develop AI and data analytics skills – Stakeholder involvement to ensure all levels of the organization understand and support digital transformation – Phased implementation strategies to reduce disruption and increase acceptance	Hangl et al. (2023); Irani et al., (2023); Ivanov, (2022); Waller and Fawcett (2013)
Technological Challenges	Legacy system integration, data silos, scalability limitations	– Investing in flexible, scalable IT infrastructure that allows for AI integration – Adopting cloud-based solutions to improve data accessibility and reduce silos – Ensuring real-time data synchronization to enhance supply chain visibility	Ameen et al. (2022); Ivanov et al. (2019); Ullah et al., (2024)
Ethical and Cybersecurity Risks	AI biases, data privacy concerns, automated decision-making ethics	– Developing AI models with explainability and fairness mechanisms to ensure transparency – Implementing robust cybersecurity frameworks to protect against data breaches – Ensuring compliance with ethical AI guidelines and regulatory standards	Bag et al., (2023); Dash et al., (2024); Möller, (2023)

9.6 Conclusions

In this chapter, we examined the impact of AI, Data Science, and Big Data on supply chain management, emphasizing their role in enhancing resilience, efficiency, and adaptability. The increasing complexity of modern supply chains, driven by globalization, geopolitical tensions, and global crises, has underscored the need for digitalization. Advanced technologies enable businesses to transition from reactive to proactive strategies, improving visibility, operational efficiency, and risk mitigation through AI-driven predictive analytics and Big Data insights.

The review highlighted how digital technologies optimize demand forecasting, inventory management, and route planning to enhance supply chain resilience. Data Science further strengthens adaptability by providing actionable insights and enabling predictive and prescriptive analytics for risk management. Case studies from companies such as Tesco and Amazon illustrated how AI and Big Data improve inventory management, streamline logistics, and sustain operational efficiency. These examples demonstrated the tangible benefits of digital transformation in fostering agility and long-term sustainability.

Despite these advancements, several challenges persist. Legacy system integration, data quality issues, organizational resistance, and ethical concerns present significant barriers to widespread adoption. Successful AI deployment requires strategic investment, cultural shifts, and workforce upskilling. Ethical considerations such as data privacy, algorithmic biases, and AI transparency must also be addressed to ensure responsible implementation.

In conclusion, AI, Data Science, and Big Data play a vital role in enhancing supply chain resilience but require careful implementation to maximize their benefits. Future research should focus on improving AI explainability, exploring use cases for Generative AI and Agentic AI, validating digital transformation strategies in real-world settings, and addressing the unique challenges faced by SMEs. Responsible adoption of these technologies will enable agile, future-ready supply chains that drive growth and maintain competitiveness amid global uncertainties.

References

Ahmed Esmail, M. (2024). Inventory management. In T. Bányai (Ed.), *Operations management—Recent advances and new perspectives*. IntechOpen. https://www.intechopen.com/chapters/88430

Albayrak Ünal, Ö., Erkayman, B., & Usanmaz, B. (2023). Applications of artificial intelligence in inventory management: A systematic review of the literature. *Archives of Computational Methods in Engineering, 30*(4), 2605–2625. https://doi.org/10.1007/s11831-022-09879-5

Aljohani, A. (2023). Predictive analytics and machine learning for real-time supply chain risk mitigation and agility. *Sustainability, 15*(20), Article 20. https://doi.org/10.3390/su152015088

Ameen, N., Choudrie, J., Jones, P., & Anand, A. (2022). Innovative technologies and small-medium sized enterprises in times of crisis. *Information Systems Frontiers, 24*(4), 1055–1060. https://doi.org/10.1007/s10796-022-10353-7

Arunachalam, D., Kumar, N., & Kawalek, J. P. (2018). Understanding big data analytics capabilities in supply chain management: Unravelling the issues, challenges and implications for practice. *Transportation Research Part E: Logistics and Transportation Review, 114*, 416–436. https://doi.org/10.1016/j.tre.2017.04.001

Aslam, F., & Calghan, J. (2023). Using NLP to enhance supply chain management systems. *Journal of Engineering Research and Reports, 25*(9), Article 9. https://doi.org/10.9734/jerr/2023/v25i9994

Bag, S., Rahman, M. S., Srivastava, G., Shore, A., & Ram, P. (2023). Examining the role of virtue ethics and big data in enhancing viable, sustainable, and digital supply chain performance. *Technological Forecasting and Social Change, 186*, 122154. https://doi.org/10.1016/j.techfore.2022.122154

Bhanushali, M. M., Bhardwaj, S., Singh, N. K., Vijayalakshmi, P., Mazumdar, N., & Acharjee, P. B. (2024). From automation to optimization: Exploring the effects of AI on supply chain management. In *Utilization of AI technology in supply chain management* (pp. 77–94). IGI Global. https://doi.org/10.4018/979-8-3693-3593-2.ch006

Carvalho, H., Cruz-Machado, V., Carvalho, H., & Cruz-Machado, V. (2011). Integrating lean, agile, resilience and green paradigms in supply chain management (LARG_SCM). In *Supply chain management*. IntechOpen. https://doi.org/10.5772/14592

Christopher, M., & Peck, H. (2004). Building the resilient supply chain. *The International Journal of Logistics Management, 15*(2), 1–14. https://doi.org/10.1108/09574090410700275

Colin, M., Galindo, R., & Hernández, O. (2015). Information and communication technology as a key strategy for efficient supply chain management in manufacturing SMEs. *Procedia Computer Science, 55*, 833–842. https://doi.org/10.1016/j.procs.2015.07.152

Dash, A., Sarmah, S. P., Tiwari, M. K., Jena, S. K., & Glock, C. H. (2024). Cybersecurity investments in supply chains with two-stage risk propagation. *Computers & Industrial Engineering, 197*, 110519. https://doi.org/10.1016/j.cie.2024.110519

Destino, M., Fischer, J., & Trautwein, V. (2022). Demand forecasting in supply chain technology. *McKinsey*. https://www.mckinsey.com/capabilities/operations/our-insights/to-improve-your-supply-chain-modernize-your-supply-chain-it

DHL. (2023). AI last mile delivery: Prospects for logistics. *DHL Freight*. https://dhl-freight-connections.com/en/solutions/ai-last-mile-delivery-prospects-for-logistics/

Feizabadi, J. (2022). Machine learning demand forecasting and supply chain performance. *International Journal of Logistics Research and Applications, 25*(2), 119–142. https://doi.org/10.1080/13675567.2020.1803246

Fosso Wamba, S., Queiroz, M. M., & Trinchera, L. (2020). Dynamics between blockchain adoption determinants and supply chain performance: An empirical investigation. *International Journal of Production Economics, 229*, 107791. https://doi.org/10.1016/j.ijpe.2020.107791

Gangwani, D., & Gangwani, P. (2021). Applications of machine learning and artificial intelligence in intelligent transportation system: A review. In A. Choudhary, A. P. Agrawal, R. Logeswaran, & B. Unhelkar (Eds.), *Applications of artificial intelligence and machine learning* (pp. 203–216). Springer. https://doi.org/10.1007/978-981-16-3067-5_16

Güneş, Ö. A. (2023). *The role of Big Data in logistics and supply chain management*. https://www.threadinmotion.com/en/blog/the-role-of-big-data-in-logistics-and-supply-chain-management

Guntuka, L., Carnovale, S., & Falcone, E. (2024). Supply chain plasticity: A responsive network capability to ensure resilience. *Journal of Business Logistics, 45*(4), e12398. https://doi.org/10.1111/jbl.12398

Hangl, J., Krause, S., & Behrens, V. J. (2023). Drivers, barriers and social considerations for AI adoption in SCM. *Technology in Society, 74*, 102299. https://doi.org/10.1016/j.techsoc.2023.102299

Hazen, B. T., Boone, C. A., Ezell, J. D., & Jones-Farmer, L. A. (2014). Data quality for data science, predictive analytics, and big data in supply chain management: An introduction to the problem and suggestions for research and applications. *International Journal of Production Economics, 154*, 72–80. https://doi.org/10.1016/j.ijpe.2014.04.018

Hendricks, K. B., & Singhal, V. R. (2005). An empirical analysis of the effect of supply chain disruptions on long-run stock price performance and equity risk of the firm. *Production and Operations Management, 14*(1), 35–52. https://doi.org/10.1111/j.1937-5956.2005.tb00008.x

Irani, Z., Abril, R. M., Weerakkody, V., Omar, A., & Sivarajah, U. (2023). The impact of legacy systems on digital transformation in European public administration: Lesson learned from a multi case analysis. *Government Information Quarterly, 40*(1), 101784. https://doi.org/10.1016/j.giq.2022.101784

Ivanov, D. (2022). Viable supply chain model: Integrating agility, resilience and sustainability perspectives—Lessons from and thinking beyond the COVID-19 pandemic. *Annals of Operations Research, 319*(1), 1411–1431. https://doi.org/10.1007/s10479-020-03640-6

Ivanov, D. (2024). Supply chain resilience: Conceptual and formal models drawing from immune system analogy. *Omega, 127*, 103081. https://doi.org/10.1016/j.omega.2024.103081

Ivanov, D., & Dolgui, A. (2021). OR-methods for coping with the ripple effect in supply chains during COVID-19 pandemic: Managerial insights and research implications. *International Journal of Production Economics, 232*, 107921. https://doi.org/10.1016/j.ijpe.2020.107921

Ivanov, D., Dolgui, A., Das, A., & Sokolov, B. (2019). Digital supply chain twins: Managing the ripple effect, resilience, and disruption risks by data-driven optimization, simulation, and visibility. In D. Ivanov, A. Dolgui, & B. Sokolov (Eds.), *Handbook of ripple effects in the supply chain* (pp. 309–332). Springer. https://doi.org/10.1007/978-3-030-14302-2_15

Iyer, L. S. (2021). AI enabled applications towards intelligent transportation. *Transportation Engineering, 5*, 100083. https://doi.org/10.1016/j.treng.2021.100083

Khedr, A. M., & Sheeja Rani, S. (2024). Enhancing supply chain management with deep learning and machine learning techniques: A review. *Journal of Open Innovation: Technology, Market, and Complexity, 10*(4), 100379. https://doi.org/10.1016/j.joitmc.2024.100379

Law, M. (2024, August 2). *AI the new secret ingredient in Unilever's customer recipe*. https://technologymagazine.com/articles/ai-the-new-secret-ingredient-in-unilevers-customer-recipe

Lepenioti, K., Bousdekis, A., Apostolou, D., & Mentzas, G. (2020). Prescriptive analytics: Literature review and research challenges. *International Journal of Information Management, 50*, 57–70. https://doi.org/10.1016/j.ijinfomgt.2019.04.003

Lin, Z. (2023). Digital twins' technology for smart agriculture. In Q. Zhang (Ed.), *Encyclopedia of digital agricultural technologies* (pp. 348–355). Springer. https://doi.org/10.1007/978-3-031-24861-0_191

Manning, L., Brewer, S., Craigon, P. J., Frey, J., Gutierrez, A., Jacobs, N., Kanza, S., Munday, S., Sacks, J., & Pearson, S. (2023). Reflexive governance architectures: Considering the ethical implications of autonomous technology adoption in food supply chains. *Trends in Food Science & Technology, 133*, 114–126. https://doi.org/10.1016/j.tifs.2023.01.015

Marr, B. (2016). Big Data at Tesco: Real time analytics at the UK grocery retail giant. *Forbes*. https://www.forbes.com/sites/bernardmarr/2016/11/17/big-data-at-tesco-real-time-analytics-at-the-uk-grocery-retail-giant/

McKinsey. (2016). *Big Data and the supply chain: The big-supply-chain analytics landscape (Part 1)*. https://www.mckinsey.com/capabilities/operations/our-insights/big-data-and-the-supply-chain-the-big-supply-chain-analytics-landscape-part-1

Möller, D. P. F. (2023). Cybersecurity in digital transformation. In D. P. F. Möller (Ed.), *Guide to cybersecurity in digital transformation: Trends, methods, technologies, applications and best practices* (pp. 1–70). Springer. https://doi.org/10.1007/978-3-031-26845-8_1

Muthukalyani, A. R. (2023). Unlocking accurate demand forecasting in retail supply chains with AI-driven predictive analytics. *International Journal of Information Technology and Management Information Systems, 14*, 48–57.

Munir, M. A., Hussain, A., Farooq, M., Rehman, A. U., & Masood, T. (2024). Building resilient supply chains: Empirical evidence on the contributions of ambidexterity, risk management, and analytics capability. *Technological Forecasting and Social Change, 200*, 123146. https://doi.org/10.1016/j.techfore.2023.123146

Notteboom, T., Pallis, T., & Rodrigue, J.-P. (2021). Disruptions and resilience in global container shipping and ports: The COVID-19 pandemic versus the 2008–2009 financial crisis. *Maritime Economics & Logistics, 23*(2), 179–210. https://doi.org/10.1057/s41278-020-00180-5

Nzeako, G., Akinsanya, M. O., Popoola, O. A., Chukwurah, E. G., & Okeke, C. D. (2024). The role of AI-driven predictive analytics in optimizing IT industry supply chains. *International Journal of Management & Entrepreneurship Research, 6*(5), Article 5. https://doi.org/10.51594/ijmer.v6i5.1096

Park, Y., Hong, P., & Roh, J. J. (2013). Supply chain lessons from the catastrophic natural disaster in Japan. *Business Horizons, 56*(1), 75–85. https://doi.org/10.1016/j.bushor.2012.09.008

Peladarinos, N., Piromalis, D., Cheimaras, V., Tserepas, E., Munteanu, R. A., & Papageorgas, P. (2023). Enhancing smart agriculture by implementing digital twins: A comprehensive review. *Sensors, 23*(16), Article 16. https://doi.org/10.3390/s23167128

Ping, G., Zhu, M., Ling, Z., & Niu, K. (2024). Research on optimizing logistics transportation routes using AI large models. *Applied Science and Engineering Journal for Advanced Research, 3*(4), Article 4. https://doi.org/10.5281/zenodo.12787012

Rahoui, S., Crowe, J., & Mahfouz, A. (2023). A simulation-based TDABC model to manage supply chain costing: A case study. *Winter Simulation Conference (WSC), 2023*, 1653–1664. https://doi.org/10.1109/WSC60868.2023.10408336

Ren, H., Mu, D., Wang, C., Yue, X., Li, Z., Du, J., Zhao, L., & Lim, M. K. (2024). Vulnerability to geopolitical disruptions of the global electric vehicle lithium-ion battery supply chain network. *Computers & Industrial Engineering, 188*, 109919. https://doi.org/10.1016/j.cie.2024.109919

Saad, S. M., Ubeywarna, D., & Zhang, H. (2024). Supply network risk mitigation-Industry 4.0 approach. *Procedia Computer Science, 232*, 3206–3215. https://doi.org/10.1016/j.procs.2024.02.136

Sahu, M. K. (2020). AI-based supply chain optimization in manufacturing: Enhancing demand forecasting and inventory management. *Journal of Science & Technology, 1*(1), 424–426. https://www.thesciencebrigade.com/jst/article/view/349

Schwaeke, J., Peters, A., Kanbach, D. K., Kraus, S., & Jones, P. (2024). The new normal: The status quo of AI adoption in SMEs. *Journal of Small Business Management, 63*(3), 1297–1331. https://doi.org/10.1080/00472778.2024.2379999

Sherman, E. (2020). 94% of the Fortune 1000 are seeing coronavirus supply chain disruptions. *Fortune*. https://fortune.com/2020/02/21/fortune-1000-coronavirus-china-supply-chain-impact/

Sifted. (2024, July). How Amazon is using AI to become the fastest supply chain in the world. *Sifted*. https://sifted.com/resources/how-amazon-is-using-ai-to-become-the-fastest-supply-chain-in-the-world/

Silverstein, S. (2020). *Walmart uses AI to predict demand.* https://www.supplychaindive.com/news/walmart-grocery-AI-demand-operations/585424/

Singh, S. K., & El-Kassar, A.-N. (2019). Role of big data analytics in developing sustainable capabilities. *Journal of Cleaner Production, 213*, 1264–1273. https://doi.org/10.1016/j.jclepro.2018.12.199

Sivakumar, G., & Premrajkumar, E. (2024). Strategic challenges and emerging perspectives in supply chain management. *ShodhKosh: Journal of Visual and Performing Arts, 5*(5), Article 5. https://doi.org/10.29121/shodhkosh.v5.i5.2024.2357

Smyth, C., Dennehy, D., Fosso Wamba, S., Scott, M., & Harfouche, A. (2024). Artificial intelligence and prescriptive analytics for supply chain resilience: A systematic literature review and research agenda. *International Journal of Production Research, 62*(23), 8537–8561. https://doi.org/10.1080/00207543.2024.2341415

Soori, M., Arezoo, B., & Dastres, R. (2023). Internet of things for smart factories in industry 4.0, a review. *Internet of Things and Cyber-Physical Systems, 3*, 192–204. https://doi.org/10.1016/j.iotcps.2023.04.006

Spieske, A., & Birkel, H. (2021). Improving supply chain resilience through industry 4.0: A systematic literature review under the impressions of the COVID-19 pandemic. *Computers & Industrial Engineering, 158*, 107452. https://doi.org/10.1016/j.cie.2021.107452

Tang, C. (2006). Robust strategies for mitigating supply chain disruptions. *International Journal of Logistics: Research and Applications, 9*, 33–45. https://doi.org/10.1080/13675560500405584

Tiwari, S., Wee, H. M., & Daryanto, Y. (2018). Big data analytics in supply chain management between 2010 and 2016: Insights to industries. *Computers & Industrial Engineering, 115*, 319–330. https://doi.org/10.1016/j.cie.2017.11.017

Tukamuhabwa, B. R., Stevenson, M., Busby, J., & Zorzini, M. (2015). Supply chain resilience: Definition, review and theoretical foundations for further study. *International Journal of Production Research, 53*(18), 5592–5623. https://doi.org/10.1080/00207543.2015.1037934

Ullah, I., Adhikari, D., Su, X., Palmieri, F., Wu, C., & Choi, C. (2024). Integration of data science with the intelligent IoT (IIoT): Current challenges and future perspectives. *Digital Communications and Networks*. https://doi.org/10.1016/j.dcan.2024.02.007

van Hoek, R. (2020). Research opportunities for a more resilient post-COVID-19 supply chain—Closing the gap between research findings and industry practice. *International Journal of Operations & Production Management, 40*(4), 341–355. https://doi.org/10.1108/IJOPM-03-2020-0165

Villegas-Ch, W., Navarro, A. M., & Sanchez-Viteri, S. (2024). Optimization of inventory management through computer vision and machine learning technologies. *Intelligent Systems with Applications, 24*, 200438. https://doi.org/10.1016/j.iswa.2024.200438

Waldersee, V., Ringstrom, A., Mannes, M., Waldersee, V., & Mannes, M. (2024). Tesla, Volvo Car pause output as Red Sea shipping crisis deepens. *Reuters*. https://www.reuters.com/business/autos-transportation/tesla-berlin-suspend-most-production-two-weeks-over-red-sea-supply-gap-2024-01-11/

Waller, M. A., & Fawcett, S. E. (2013). Data science, predictive analytics, and big data: A revolution that will transform supply chain design and management. *Journal of Business Logistics, 34*(2), 77–84. https://doi.org/10.1111/jbl.12010

Weisz, E., Herold, D. M., & Kummer, S. (2023). Revisiting the bullwhip effect: How can AI smoothen the bullwhip phenomenon? *The International Journal of Logistics Management, 34*(7), 98–120. https://doi.org/10.1108/IJLM-02-2022-0078

Xu, Z., Guo, L., Zhou, S., Song, R., & Niu, K. (2024). Enterprise supply chain risk management and decision support driven by large language models. *Applied Science and Engineering Journal for Advanced Research, 3*(4), Article 4. https://doi.org/10.5281/zenodo.12670581

Zamani, E. D., Smyth, C., Gupta, S., & Dennehy, D. (2023). Artificial intelligence and big data analytics for supply chain resilience: A systematic literature review. *Annals of Operations Research, 327*(2), 605–632. https://doi.org/10.1007/s10479-022-04983-y

Zhao, N., Hong, J., & Lau, K. H. (2023). Impact of supply chain digitalization on supply chain resilience and performance: A multi-mediation model. *International Journal of Production Economics, 259*, 108817. https://doi.org/10.1016/j.ijpe.2023.108817

Chapter 10
Data-Driven Proactive Inventory Policies Under Demand and Supply Uncertainty in Global Supply Chains

Luca Urciuoli and Mustafa Çagri Gürbüz

Abstract Managing global supply chains presents significant challenges due to the inherent complexities in large networks of suppliers, manufacturers, transportation providers, distributors, retailers, and end consumers. To make the situation worse, supply chain managers face additional challenges due to demand and/or supply-related disruptions, and the ensuing delays and their ripple effects. To address these issues, previous research has proposed both reactive and proactive strategies, such as redundant sourcing, logistics flexibility, and shipment rerouting. Among these, inventory policies play a central role by allowing adjustments to lot sizes and safety stock levels in response to delays. By leveraging real-time visibility, these policies enable the efficient detection of delays and the optimization of inventory allocation. This chapter explores the value of such data-driven inventory policies in improving the performance of global supply chains, specifically when unexpected adverse events like port closures or transportation disruptions occur. In doing so, we develop a simulation model assuming that there is sufficient supply chain visibility that allows the firm to monitor inventory levels and to get real-time updates regarding shipment locations and order progress. The findings demonstrate that data-driven policies lead to some improvements in inventory performance, resulting in higher service levels.

Keywords Supply chain visibility · Inventory policies · Transport risks · Ripple effects · Digitalized information sharing · Data-driven inventory policies

---

L. Urciuoli (✉)
KTH Royal Institute of Technology, Stockholm, Sweden

MIT International Logistics Programme, Zaragoza Logistics Center, Zaragoza, Spain
e-mail: urciuoli@kth.se

M. Ç. Gürbüz
MIT International Logistics Programme, Zaragoza Logistics Center, Zaragoza, Spain

Y. Khojasteh, S. M. M. Jahan Abadi (eds.), *Emerging Technologies in Supply Chains*, International Series in Operations Research & Management Science 373, https://doi.org/10.1007/978-3-032-01218-0_10

10.1 Introduction

Reduced labor costs, access to novel markets and technologies not available domestically, and the expansion of e-commerce have catalyzed the globalization of supply chains. The global economy is heavily reliant on maritime shipping for the transportation of goods. This transport mode is a mandatory choice for intercontinental shipments due to its lower environmental impact and more advantageous economies of scale compared to air freight. Empirical data indicate that ocean shipments are integral to global supply chains, with approximately 90% of goods transported across the ocean to reach consumers (Ilgar & Zagorodnya, 2024). However, the same data indicate the schedule unreliability of this transport mode (i.e., too late or too early deliveries escalating into port congestions and missed transshipments) due to unexpected risks (Jain, 2018; Stamos et al., 2015; Urciuoli & Hintsa, 2020; Sun et al., 2018). The global schedule reliability of this transport mode reached a minimum of 35.6% in 2022 and stabilized to only 50–55% in 2024 (Yue & Mangan, 2024; Safety4Sea, 2025).

Disruptive events in maritime transport and the knock-on effects along the global supply chains cause production halts, stock-outs, and overall increased costs (Singh et al., 2021; Chung et al., 2018; Jain, 2018). Tardiness in the final delivery of the merchandise, even in the case of expedited shipments, clearly leads to customer dissatisfaction, possibly causing potential lost sales (Kouvelis & Li, 2012). For instance, the costs of delayed shipments to retailers due to a slowdown of shipments on West Coast ports in the USA in early 2015 were estimated to be around $3.8–7 billion in 2015 (Jain, 2018). Other studies claim that the value of a shipment decreases by 0.8%, and the likelihood of it being traded decreases by 1%, up to 6% for time-sensitive goods (Yue & Mangan, 2024).

In response to long and unreliable lead times, diverse mitigation strategies have been developed. For instance, companies can request their 3PL providers to add flexibility to their logistics services, e.g., re-routing or expediting shipments using air transport (Jain, 2018). Likewise, inventory managers can implement advanced replenishment strategies, e.g., (1) dual sourcing, i.e., a small proportion of orders is sent to local suppliers, or (2) order splitting, i.e., when delivery takes longer than the constant lead time, a policy consisting of splitting the order between backorders or lost sales (Jain, 2018; Padhi & Mukherjee, 2022; Paul & Rahman, 2018).

Digital solutions, such as end-to-end demand sensing, data pipelines, Estimated Time of Arrival (ETA) for in-transit shipments, and dispatching/tracking technologies, are increasingly being implemented to enhance access to real-time information. These tools enable the efficient detection of delays, facilitating the adoption of data-driven inventory policies. Such policies dynamically adjust lot sizes and safety stock levels based on estimated replenishment delays. Consequently, these advancements improve supply chain responsiveness to unforeseen disruptions. Existing literature acknowledges the positive impacts of these technologies (Zuidwijk & Veenstra, 2015; Urciuoli & Hintsa, 2020); however, they lack empirical prescriptive models demonstrating the potential benefits on inventory performance.

We hypothesize that data-driven inventory policies can use real-time data to adjust reorder points and quantities based on different factors like market conditions, supplier performance, transport delays, etc. Access to these data allows firms to employ predictive analytics to detect changes in demand or replenishment times early on, before small deviations accumulate and lead to a substantial deviation (Ravulakollu et al., 2018; Urciuoli & Hintsa, 2020). Such real-time visibility into inventory (in-transit and at downstream supply chain partners such as warehouses and retailers) help develop data-driven reactions using a combination of prescriptive/predictive analytics (e.g., expediting shipments, reallocation of inventories across several locations) in response to unexpected delays in receiving the in-transit inventory or higher than usual demand at retailers.

Hence, the scope of this chapter is to show the value of visibility enabled by digital transformation in deploying data-driven inventory policies for a global supply chain subject to major/minor disruptions (e.g., port closures and delays in transport). We propose two novel data-driven inventory policies, (1) a Proactive Replenishment Strategy (PRS), that triggers proactive shipments based on historical shipment data and real time monitoring of inventory levels, and (2) the same strategy, PRS, combined with Order Progress Information (OPI), providing location based Estimated Time of Arrivals (ETAs) of in-transit shipments.

The model is tested in a global supply case study consisting of a three-echelon inventory system (i.e., supplier in China, warehouse in Zaragoza/Spain, and three retailers in Bilbao, Valencia, and Sevilla in Spain), where both retailers and the warehouse use continuous review replenishment policies under stochastic demand. Orders are shipped from a supplier located in China to (1) the port of Yantian by road, (2) Barcelona port via sea, (3) the distribution warehouse by rail, and finally, (4) retailers in Spain by road. The goal is to maintain customer satisfaction (i.e., high service levels and timely delivery of orders) without increasing supply chain costs.

10.2 Literature Review

10.2.1 Global Supply Chain Risks and Impact on Performance

Freight transportation in global supply chains is subject to a multitude of risks, which can extend lead times, escalate uncertainty, and occasionally result in the loss or damage of cargo (Urciuoli & Hintsa, 2020; Stamos et al., 2015; Xu et al., 2012). Maritime transport is indicated as one of the less reliable transport modes, and incidents such as traffic congestion or accidents can cause delays or cargo damage (Zuidwijk & Veenstra, 2015; Wagenaar et al., 2021). Adverse weather conditions, such as strong winds and high waves, can impede sea transport, decreasing speed and culminating in an inability to adhere to planned docking schedules during intermediate stops of oceanic shipments (Huang et al., 2025; Xu et al., 2024). Severe meteorological phenomena have the potential to induce unanticipated, temporary closure of seaport operations. An example of this occurred when Hurricane Katrina hit the Port of New Orleans, compelling a cease of operations for more than 2 weeks (Lewis et al., 2013).

10.2.1.1 Risk Impact Amplification and Propagation

The aftermath of such events often manifests in terms of two main occurrences: impact amplification and propagation (Kravchenko et al., 2024; Yin & Tang, 2025). Impacts of transport delays amplify because even marginal delays can precipitate significant disruptions in a meticulously planned transportation process. This is due to the potential for cargo to miss its scheduled stops, thereby necessitating comprehensive reconfiguration and rescheduling of the berths at intermodal ocean stops (Xu et al., 2012). The necessity to reschedule these stops will proportionally augment the delay duration, instigating a chain of rescheduling events along the entire route (Kim et al., 2021). Similarly, delays amplify if the disruption hits a seaport or intermodal hub. The closure of a terminal typically generates protracted queues of cargo vessels awaiting berth access, a situation that may persist for several months and thereby lengthen delays (Lewis et al., 2013; Yin et al., 2024).

Shipment delays can get even worse as a result of the overreaction of buyers, who, in the absence of shipment visibility or updated arrival time estimates, place emergency orders or expedite shipments in transit when shipments are delayed (Hu et al., 2024). The cumulative effect of this behavior results in a surge of orders instigated by buyers in response to delayed shipments. However, the heightened demand for transport exacerbates dwell times at intermodal terminals. This, in turn, escalates congestion at intermodal terminals, reduces capacity, creates queues, and subsequently further impacts the variability of shipment lead times (Fransoo & Lee, 2013; Yin et al., 2024).

10.2.1.2 Absorbing Delays and Avoiding Prolonged Fluctuations

The literature suggests that shipment delays within global supply chains can be absorbed using time buffers in the planned schedule (i.e., schedule padding). Transport planners, aware of delay risks, may opt to incorporate buffer times at intermodal stops to mitigate the uncertainty associated with arrival or departure delays (Eufrásio et al., 2021). This strategy is not a new concept; it is a well-established practice in a variety of scheduling problems, including machine scheduling, project scheduling, flight gate assignment, and flight scheduling on-time performance (Xu et al., 2012; Eufrásio et al., 2021). However, it is well-known that these "*safety lead times or time buffers*" lead to higher safety stocks and, therefore, increase inventory holding costs.

In conclusion, trans-ocean shipments can swiftly experience prolonged fluctuations in lead times, which in turn lead to additional costs that propagate through the supply chain. For example, such situations can lead to administrative expenses associated with the necessary rescheduling of seaport berths and intermodal transshipments (Kim et al., 2021). Similarly, arrival or departure delays at seaports may result in detention or demurrage charges (Fransoo & Lee, 2013; Lewis et al., 2013; Jeong et al., 2025). Cargo can also be lost or damaged; e.g., perishable food items may spoil, or fashion products may become outdated (Urciuoli & Hintsa, 2020).

10.2.2 Modelling Lead Time Uncertainty in Inventory Policies

Lead time uncertainties trigger fluctuations in inventory and cost escalations (Chung et al., 2018). These fluctuations are caused by the combined effect of shipment delays and the inventory policies adopted by the buyers. Basic inventory models undertake static policies, e.g., EOQ (Economic Order Quantity), assuming reliable and instantaneous replenishments (Lewis et al., 2013). Although most models, like the EOQ model, assume that lead time is constant, this assumption is valid when the variance in lead time is negligible compared to the mean or when lead times are generally short.

10.2.2.1 Incorporating Uncertainty

Diverse approaches have been developed incorporating lead time uncertainty. Previous research suggests using lead time distributions based on historical data. Thereby, stochastic inventory models are built assuming random lead time (Song et al., 2000; Chopra & Sodhi, 2004). Tomlin (2006) uses a single-product setting where a firm uses dual-sourcing and shows that suppliers' uptime and disruption intensity are critical factors to decide whether to diversify supply or increase inventory safety stocks. In addition, volume flexibility and contingent rerouting to suppliers, which can ramp up processing capacity, may significantly reduce costs (Tomlin, 2006). Topan and van der Heijden (2020) consider a supply chain with constant lead time but with periodic/planned opportunities to do stock rebalancing to avoid shortages. Paul and Rahman (2018) consider a model where the lead time is normally constant; however, when a 'sudden' delay happens, the decision makers know the duration of the delay (i.e., they do not need to estimate the remaining lead time). Hence, the authors analyze how the lot sizes and safety stocks need to be adjusted during the recovery time window and show that safety stock is critical in recovery from sudden supply delays.

10.2.2.2 Adopting Advanced Data-Driven Inventory Policies

In response to changes in the business environment, detected by real-time information systems, supply chains need to show adaptability for maintaining optimal inventory levels over time and minimizing costs. There exists extensive research considering the use of expedited shipments to minimize the impact of highly uncertain lead times in long supply chains. Kouvelis and Tang (2012) derive the optimal expediting policy with uncertain lead times and analyze the effect on decisions regarding time and size of order placement. Mamani and Moinzadeh (2014) consider the option of expediting shipments in a system with constant lead time but using lead time visibility (i.e., the supply chain is divided into several stages, and decision makers know at what stages the goods are). They analyse the impact of bypassing one or more of the shipping nodes in the supply chain. Kim et al. (2015) present the optimal policy for expediting shipment in a serial supply chain under stochastic lead-times. Hu et al. (2024) emphasize the importance of planning optimal emergency ordering strategies in the presence of supply disruptions in the semiconductor industry to satisfy retailers' demand but also reduce costs due to shortages and lost sales.

Other research investigates how data-driven inventory policies, can be exploited to optimize inventory replenishment. These policies are more responsive to the updated information on the remaining lead time of outstanding orders. Early research by Song and Zipkin (1996) investigated how the inventory decisions can be adapted to the advanced information on lead-time realizations. Pathy and Rahimian (2023) discuss the importance of using adaptive inventory policies with emergency orders and optimized network design, considering a pharmaceutical supply chain operating under lead time uncertainty. Kouvelis and Li (2012) investigate the impact of safety lead-time combined with two additional strategies: (1) Disruption Safety Stock (DSS) and (2) Dynamic Emergency Response (DER) Order, under uncertain lead-times. They find out that the contingency responses (i.e., DSS and DER) reduce the reliance on the safety lead-time. They also show that the safety lead-times must be re-optimized if these two strategies are employed.

Finally, Jain (2018) investigates the value of operational flexibility (expediting shipments and ordering from an alternative nearby supplier when necessary) and lead-time visibility. They test their result on a case study where goods are transported from China to the USA, and lead time can be either 2 months or 4 months (due to potential delays at the Long Beach Port), where full and partial visibility cases are compared to the case with no lead time visibility. They show that the combined effect (operational flexibility with lead time visibility) yields the best result against long lead times and situations in which delays happen frequently.

10.3 Problem Formulation and Model Description

Our work is closely related to the literature on global multi-echelon supply chains with long and uncertain lead times, where decision makers can employ inventory policies to respond to the most up-to-date information on the remaining lead times of the outstanding orders. In this respect, we propose a mitigation strategy that is both proactive (sending proactive replenishment shipments to retailers even before they place their regular orders) and reactive (taking an action when both the delay in delivery is already large enough or inventory level at the retailers are low, to trigger additional shipments from the warehouse to the retailers), using continuous review inventory policies under uncertain demand. We evaluate the effectiveness and value of such data-driven policies in terms of reducing safety stocks or similarly safety lead times, while maintaining high service levels in the presence of unexpected supply disruptions (e.g., port closures) by simulating the global supply chain of a home appliances manufacturer ordering from a supplier in China and delivering goods in Spain. To the best of our knowledge, the model we propose is the first one that considers a multi-echelon network with random demand, stochastic lead times, multiple outstanding orders and order crossing (*i.e.,* when orders placed at the supplier do not necessarily arrive at the warehouse in the order they were placed), and continuous review inventory policies.

We consider a three-echelon supply chain comprising three retailers placing orders for a single product (a household appliance) at a warehouse in Zaragoza, which in turn is served by a supplier in China (Fig. 10.1). The retailers are located in Bilbao, Sevilla, and Valencia, and use continuous review *(Q, R)*—type inventory control policies, ordering Q_j units whenever their inventory position hits the reorder point, R_j. The warehouse, upon receiving an order from any retailer *j*, sends Q_j units, via road transport, to the retailer *j* if there is sufficient inventory.

The warehouse also uses a continuous review *(Q, R)*—type inventory policy to order from the supplier in China. We assume the supplier in China has ample production and transportation capacity and therefore can always fulfill the orders from the warehouse in Zaragoza in full. When the supplier in China receives the order, production starts, and when the products are ready, they are stored in the outbound area of the factory's warehouse (Fig. 10.1). A truck collects the shipment and consigns it to the Port of Yantian, where it waits for the scheduled vessel to the Port of Barcelona. The supplier in China synchronizes the truck shipments with the schedule of the sea vessels to avoid demurrage penalties. Free time for containers in the seaport of Yantian is 1–7 days, hence, the factory ships the order when there are 24 h left before the departure of the next sea vessel (sea vessels depart with a frequency of one vessel every 2 weeks). There is no transportation capacity restriction considered, and they can always load the cargo delivered by the supplier to the Port of Yantian.

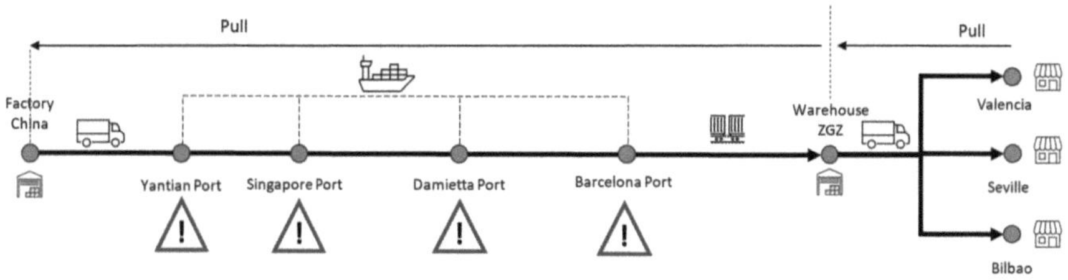

Fig. 10.1 System simulated, orders are shipped from a factory in China by road, then by sea to Barcelona, train to Zaragoza and then by road to the retailers

The sea vessel follows a pre-established route from Yantian to Barcelona, with stops in Singapore and Damietta ports (Fig. 10.1). Along the maritime route, the seaports may unexpectedly close due to factors such as extreme weather events, congestion, or strikes. In such cases, we expect the sea vessel to be delayed by 6 days. Upon arrival at Barcelona port, the container is unloaded and sent to Customs for clearance. Customs operations are modelled using a 1% likelihood for screening/inspection selection. After customs inspection, the containers are moved to the rail terminal, where they wait to be loaded on a rail convoy heading to Zaragoza. The train service operates on a fixed schedule, with departures occurring once per week. It is assumed that trains always have enough capacity to move shipments from the port of Barcelona to Zaragoza. When shipments arrive in Zaragoza, they are moved to the warehouse directly, which is located near the rail track.

10.3.1 Model Development

We first remind the reader that in all the inventory policies that we consider, the retailers and the warehouse in Zaragoza use continuous review (Q, R)—type policies to order from the warehouse in Zaragoza and the factory in China, respectively. However, we model different ways to calculate the reorder point (R) at the warehouse in Zaragoza, and how real-time information regarding inventory levels at the warehouse and the retailers, as well as the status of outstanding orders in the pipeline from China are used to send proactive shipments from the warehouse in Zaragoza to the retailers. A 'proactive shipment' is different to the "regular shipments" from the warehouse to the retailers when retailers order Q_j units. Proactive shipments happen when the warehouse in Zaragoza sends some additional units to the retailers (unsolicited by the retailers) if the real-time inventory level at the warehouse is lower than a certain threshold and/or the potential delay in the delivery of the closest order in the pipeline is larger than a critical value. The rationale of sending proactive shipments is to artificially delay regular orders from the retailers and therefore improve service levels both at retailers and the warehouse. Hence, in our analysis, we consider the following four data-driven inventory policies (Table 10.1):

As explained above, these inventory policies differ in terms of how the reorder point at the warehouse in Zaragoza is calculated (i.e., *conservative* v *lean*), how proactive shipments from the warehouse in Zaragoza to retailers happen based on the real-time inventory levels at the retailers (PRS inventory policy), and how proactive shipments from the warehouse in Zaragoza to retailers happen based on the real-time inventory levels at the retailers as well as the expected delay in the delivery of the closest outstanding order in the pipeline from the factory in China (PRS-OPI).

Our experimental design consists of six cases in total, varying how the reorder point at the warehouse in Zaragoza is calculated and what kind of real-time information is used to send proactive shipments to retailers. An overview of the six cases

Table 10.1 Summary of inventory policies

Inventory Policy	Description
Conservative	The warehouse uses a *continuous review (Q,R)—type policy*, and there are no proactive shipments. The reorder point at the warehouse is calculated assuming a "very large constant lead time" (specifically, the lead time is assumed to be equal to the historical average lead time plus two standard deviations). It is conservative in the sense that the warehouse places orders at the factory in China earlier than normal (i.e., safety lead time).
Lean	The warehouse uses a *continuous review (Q,R)—type policy*, and there are no proactive shipments. The reorder point at the warehouse is calculated assuming that lead time from China is "random" and the safety stocks are computed using the historical average and standard deviation of the lead time. It is lean in the sense that stock levels are calculated using the average (optimistic view of delivery times) historical lead time and the standard deviation of lead time. Consequently, inventory levels are lower with the *lean* policy (compared to the *conservative* policy).
Proactive Replenishment Strategies (PRS)	The warehouse uses a *continuous review (Q,R)—type policy* for regular shipments (to fulfill demand from retailers when they order Q_j units). However, in addition to the regular shipments, the warehouse continuously monitors the inventory level at the warehouse as well as at the retailers and sends proactive shipments to the retailers when the inventory level at the warehouse is less than a certain threshold. The proactive shipment quantities depend on the real-time inventory levels at the retailers, prioritizing the retailers that need larger shipments.
Proactive Replenishment Strategy with Order Progress Information (PRS-OPI)	The warehouse uses a *continuous review (Q,R)—type policy* for regular shipments (to fulfill demand from retailers when they order Q_j units). However, in addition to the regular shipments, the warehouse continuously monitors the inventory level at the warehouse as well as at the retailers and the order progress information (where the outstanding orders are in the pipeline), and sends proactive shipments to the retailers when the inventory level at the warehouse is less than a certain threshold, considering the expected delay in delivery from China (based on real-time Estimated Times of Arrival (ETA) from China). The proactive shipment quantities depend on the real-time inventory levels at the retailers and the expected delay in delivery, prioritizing the retailers that need larger shipments.

used in our simulation model is given in Table 10.2, and those cases are summarized in Table 10.3.

Following the approach outlined in Simchi-Levi and Kaminsky (2008), we define the reorder points respectively for the *conservative* and *lean* inventory policies. The reorder point (R_Z^c) for the warehouse in Zaragoza, where a conservative policy is used (*case 1a*), is given by:

$$R_Z^c = \mu_D^{ZGZ} \times \mu_{L^e} + z \times \sigma_D^{ZGZ} \sqrt{\mu_{L^e}} = \mu_D^{ZGZ} \times \left(U_L + L_r\right) + z \times \sigma_D^{ZGZ} \times \sqrt{U_L + L_r} \qquad (10.1)$$

Table 10.2 Overview of simulation cases

Simulation cases	Definition
Case1a	The reorder point at the warehouse in Zaragoza is calculated using the *conservative inventory policy* (i.e., assuming a very large constant lead time from the supplier in China), using Eq. (10.1). There are no proactive shipments in this case.
Case1b	The reorder point at the warehouse in Zaragoza is calculated using the *lean inventory policy* (i.e., assuming a random lead time from the supplier in China, using the historical average and standard deviation), using Eq. (10.2). There are no proactive shipments in this case.
Case2a	The *conservative inventory policy* is applied to determine the reorder point for the Zaragoza warehouse, consistent with the approach used in Case 1a. However, the warehouse manager implements a *PRS*, whereby all inventory on hand in Zaragoza is dispatched to retailers ahead of their orders when the real-time inventory level at the warehouse in Zaragoza is less than a certain threshold (threshold calculated using Eq. (10.3)).
Case2b	The *lean inventory policy* is applied to determine the reorder point for the Zaragoza warehouse, in alignment with the approach used in Case 1b, where the *PRS* is implemented.
Case3a	The *conservative inventory policy* is applied to determine the reorder point for the Zaragoza warehouse and the warehouse manager implements a *PRS.* However, different to Case 2a, in addition to the real-time information on the inventory levels at the warehouse/retailers, the *Order Progress Information,* using the Estimated Time of Arrivals (ETA), is also considered to determine 'when' proactive shipments must be triggered and 'what the shipment quantities' would be (determined according to Eqs. (10.5) and (10.6)).
Case3b	The *lean inventory policy* is applied to determine the reorder point for the Zaragoza warehouse and the warehouse manager implements a *PRS* and the *Order Progress Information* as explained in Case3a.

Table 10.3 Summary of the designed experiments (⊗ = adopted, ○ = not adopted)

	Conservative	Lean	PRS	PRS-OPI
CASE 1a	⊗	○	○	○
CASE 1b	○	⊗	○	○
CASE 2a	⊗	○	⊗	○
CASE 2b	○	⊗	⊗	○
CASE 3a	⊗	○	⊗	⊗
CASE 3b	○	⊗	⊗	⊗

where,

- R_Z^c : reorder point for the Zaragoza warehouse using a conservative policy
- $\mu_D^{ZGZ} = \sum_{3}^{j=1} \mu_D^j$: average (aggregate) demand per time unit across all retailers
- μ_{L^e} : echelon leadvfrom the supplier to the warehouse in Zaragoza plus the lead time from Zaragoza to any retailer) = $U_L + L_r$

- $U_L = \mu_L + 2 \times \sigma_L$: constant lead time for the Zaragoza warehouse in *case 1a* (the "actual" lead time in our simulation is still randomly generated).
- L_r: the constant lead time from the warehouse in Zaragoza to any retailer.
- $\sigma_D^{ZGZ} = \sqrt{\sum_{3}^{j=1} \left(\sigma_D^j\right)^2}$: standard deviation of the aggregate demand at Zaragoza Warehouse, assuming demands at different retailers are independent of one another.
- z: standard normal score corresponding to a 90% cycle service level.

When a 'Lean policy' is followed (*case 1b*), the reorder point (R_Z^l) is computed as follows:

$$R_Z^l = \mu_D^{ZGZ} \times \mu_{L^e} + z \times \sqrt{\left(\sigma_D^{ZGZ}\right)^2 \times \mu_{L^e} + \left(\mu_D^{ZGZ}\right)^2 \times \sigma_{L^e}^2} = \mu_D^{ZGZ}\left(\mu_L + L_r\right) + z \times \sqrt{\left(\sigma_D^{ZGZ}\right)^2 \times \left(\mu_L + L_r\right) + \left(\mu_D^{ZGZ}\right)^2 \times \sigma_L^2} \quad (10.2)$$

where,

- R_Z^l : reorder point Zaragoza warehouse using a lean policy.
- μ_L: average lead time (from the factory in China to the warehouse in Zaragoza).
- σ_L: standard deviation of the lead time (from the factory in China to the warehouse in Zaragoza).

In both cases (*case 1a* and *case 1b*), when the inventory position is below the computed reorder point, the warehouse will place an order at the factory in China. A comparison of *cases 1a* and *1b* will help understand the relative impact of safety lead-time (*case 1a* by assuming that the lead time is constant but much larger than the historical average) versus safety stock (*case 1b* by treating the lead time as a random variable, using historical averages).

The PRS needs to have real-time access to inventory data from the retailers as well as orders' historical Estimated Time of Arrival (ETA). PRSs are utilized by the inventory planner when the inventory level in Zaragoza falls below a can-ship threshold, ψ, computed for the conservative (*case 2a*) and lean policy (*case 2b*) scenarios, using Eq. (10.3) below. This particular policy was inspired by the '*allocation policy*' where a warehouse enters a rationing mode and reserves some inventory in anticipation of future demands from retailers, which was studied by Schwarz (1976). Our model does not consider rationing. However, by sending proactive shipments to retailers (note that these shipments are not solicited by the retailers), we artificially delay the future orders from retailers to some extent, reserving the available on-hand inventory for the retailers that need it the most. The can-ship condition is defined as follows:

$$Ioh_{ZGZ} \leq \psi = R_Z^c - \mu_D^{ZGZ} \times \mu_{L^e} \left(\text{for } case\, 2a\right) \text{and}$$
$$Ioh_{ZGZ} \leq \psi = R_Z^l - \mu_D^{ZGZ} \times \mu_{L^e} \left(\text{for } case\, 2b\right). \quad (10.3)$$

The shipment quantity to be shipped to the retailer j when the 'can-ship condition' is satisfied is computed as:

$$Sh_j = \max\left\{\min\left\{w_j \cdot Ioh_{ZGZ}; R_j + Q_j^r - IP_j\right\}, 0\right\}, \forall j = 1 \ldots n\, retailers \quad (10.4)$$

where,

- $R_Z^c, R_Z^l =$ reorder *point Zaragoza warehouse; c conservative, l, lean policy*
- $Sh_j =$ rationed *units sent to retailer j*
- $R_j =$ retailer *j's reorder point*
- $Q_j^r =$ retailer *j's order quantity*
- $IP_j =$ inventory *position of retailer j*
- $T_j = \dfrac{IP_j - R_j}{\mu_D^j}, \forall \text{retailer } j = 1,2,3$
- $V_j = \dfrac{\max(T_1, T_2, \ldots, T_n)}{T_j}, \forall j = 1,2,3$
- $w_j = \dfrac{V_j}{\sum_{j=1}^{n} V_j}, \forall \text{retailer } j = 1,2,3$

Finally, we consider the OPI with the Proactive Replenishment Strategy (PRS-OPI policy). This means that the planner in the warehouse can receive location-based estimates of order arrival time delays (with hourly frequency). The expected delays are computed using an algorithm that considers the possibility of '*order-crossing*' (when multiple orders are outstanding in the pipeline, an order placed earlier can arrive at the warehouse in Zaragoza later than another order placed later in time):

1. *Check the position of the order (this is not necessarily the first order in terms of the time of order placement, but it is the order that is "closest" to the destination).*
2. *Compute the expected remaining time to destination,* $ETA_{dynamic}$
3. *Compute expected delay as* $E(delay) = ETA_{dynamic} - ETA_{static}$

The ETA_{static} consists of a database with average times spent in the seaports before arriving at the warehouse in Zaragoza, as well as when moving from factory to Yantian, between seaports, and finally between Barcelona and Zaragoza (Fig. 10.1). These values were statistically computed using historical data of orders shipped from China to Zaragoza during a period of 3 years, and the ETA_{static} is the expected remaining lead time for the first order (in terms of the "time of order placement").

When this functionality is used, the proactive replenishment shipments to the retailers are triggered by a modified condition:

$$Ioh_{ZGZ} \leq \psi = R_Z^c - \mu_D^{ZGZ} \times \left(\mu_{L^e} - E(delay)\right) \text{for } case\, 3a$$

$$Ioh_{ZGZ} \leq \psi = R_Z^l - \mu_D^{ZGZ} \times \left(\mu_{L^e} - E\left(delay\right)\right) \text{for } case\,3b \tag{10.5}$$

When Eq. (10.5) is satisfied, either when using a conservative, $R_Z^c\left(case\,3a\right)$, or a lean reorder policy, R_Z^l (*case 3b*), as computed with Eq. 10.1 and Eq. 10.2, the shipment quantities are computed using the following set of Eqs. (10.6):

$$Sh_j = \begin{cases} \max\left\{\min\left\{w_j \cdot Ioh_{ZGZ}; R_j + Q_j^r - IP_j\right\}, 0\right\}, if\ \mu_D^{ZGZ} E\left(delay\right) > Ioh_{ZGZ} \\ \max\left\{\min\left\{\mu_D^j \times E\left(delay\right); R_j + Q_j^r - IP_j\right\}, 0\right\}, otherwise \end{cases},$$
$$\forall j = 1 \ldots n\, retailers \tag{10.6}$$

10.3.2 Objectives, Input Parameters, and Assumptions

Our objective is first to understand if the safety lead time alone (i.e., the conservative scenario assuming a constant but large lead time) is good enough to maintain the desired service levels at the retailers as well as in Zaragoza (*case 1a* vs *case 1b*). Second, we intend to discern the impact of proactive shipments to the retailers, i.e., PRS (*cases 2a and 2b* compared to *cases 1a and 1b*, respectively). The purpose is to determine if the warehouse in Zaragoza can improve its service to the retailers by having access to information about their inventory levels. Finally, we evaluate whether proactive shipments, combined with Order Progress Information, PRS-OPI, help improve service levels by better responding to potential delays in deliveries (*cases 3a and 3b* compared to *cases 2a and 2b*, respectively). The numerical experiment design would help decision-makers choose the appropriate model depending on the costs and efforts required to initiate proactive shipments, incentivize retailers to share information about their inventory levels, and obtain the most up-to-date information about the remaining lead times of the outstanding orders in the pipeline for enabling proactive shipments.

The transportation time for the sea shipment section of the whole journey is estimated based on the operator's historical data of shipments on this selected route (i.e., Yantian to Barcelona).[1] Table 10.4 shows the average and standard deviation of the time required to produce and ship the goods from the manufacturing facility in China to Yantian, Barcelona, and finally the central warehouse in Zaragoza. This data is used to compute static ETAs and, consequently, the expected delays *E*(*delay*), in cases 3a and 3b.

The input parameters that are exogenous to our model are presented in Table 10.5. Some of these have been informed by the manufacturing firm but manipulated for confidentiality reasons. The seaport closure likelihood, L_{sc}, has been set at 20% to

[1] https://www.marinetraffic.com/

Table 10.4 Average and standard deviations of shipment time (hours) from the factory in China to the Zaragoza warehouse

	Mean (h)	SD (h)
Factory production and storage	166.72	98.84
Truck transport factory to Yantian	23.67	1.54
Port of Yantian	141.22	163.74
YNT-SNG	106.69	3.28
Port of Singapore	22.19	3.15
SNG-DMT	275.84	149.61
Port of Damietta	20.55	5.72
DMT-BCN	113.91	3.62
Port of Barcelona	163.35	42.36
Train BCN—ZGZ	9.52	0.83
Total	**1043.69**	**246.63**

Table 10.5 Simulation parameter settings

Parameter	Description	Value
I_0^{ZGZ}	ZGZ Initial Inventory on-hand	1550 units
Q_{zgz}	Zaragoza reorder quantity	850 units
R_Z^c	Reorder point conservative scenario	1827
R_Z^l	Reorder point lean scenario	1655
SL	Target Service Level	90%
R_j	Retailers reorder points	$R_{Bilbao} = 6$ $\{R_{Sevilla} = 4\ units$ $R_{Valencia} = 6$
Q_j	Retailers reorder quantities	50 units
I_0^j	Retailers' initial inventory on hand	80 units
L_{sc}	Seaport Closure Likelihood	20%
t_{SC}	Seaport closure time	6 days
μ_L	Average lead time factory in China to Zgz	1043.69 h (43.48 days)
L_r	Lead time warehouse to retailers	4 h
Retailers demand	$N \sim \left(\mu_D^j :, \sigma_D^j\right), \forall j = 1,2,3$	
D_1	Bilbao demand	N~(13, 6) units/day
D_2	Sevilla demand	N~(11, 4) units/day
D_3	Valencia demand	N~(12, 7) units/day

represent a worst-case scenario and stress-test our model. We aim to simulate an extreme scenario characterized by a globally turbulent environment, during which logistics hubs experience congestion, unexpected closures due to strikes, labor shortages, and other disruptions. Given the experience with the COVID-19 pandemic, which affected seaports globally, causing closures or delays due to illness and labor shortages, this scenario could be seen as plausible. In addition, ongoing geopolitical tensions, trade wars causing fluctuations in trade volumes, and extreme weather conditions, which are expected to increase due to climate change, create a context in which logistics hubs will have to increasingly contend with unexpected events, leading to frequent delays and congestion.

Finally, we make the following additional assumptions in the model:

- We allow for partial fulfillment of the retailer orders when there is insufficient inventory at the warehouse in Zaragoza.
- We do not allow lateral transshipments among retailers.
- There could be multiple outstanding orders, and orders may cross.
- The lead time (from the supplier to the warehouse in Zaragoza) is random. We assume that the lead time from Zaragoza to the retailers is constant.
- We assume that the lead time from Zaragoza to each retailer is the same. We also assume that demand at different retailers is independent and can have different means and variances.
- The additional holding cost the retailers incur due to the units sent in a proactive shipment is borne by the warehouse in Zaragoza.
- The company does not use emergency shipments (e.g., air freight).
- All the entities (retailers and the warehouse in Zaragoza) employ continuous review *(Q, R)* type policies for replenishment. The values of Q are given (exogenous).
- The inventory manager at the warehouse in Zaragoza has access to the inventory information at the retailers as well as the location of the shipment in transit.
- The inventory manager in Zaragoza has the authority to send unsolicited units (proactive shipments) to the retailers without the retailers asking for them.
- There is no limit on the storage space, either at the warehouse in Zaragoza or the retailers.
- An additional delay of 144 h (6 days) is added to shipments that have experienced a seaport closure.
- We ignore the possibility of losing some of the goods in transit due to potential inspection by the Customs in Zaragoza and goods being seized and disposed.

An agent-based discrete event model (ABM-DES) is developed in AnyLogic. The agents used are the retailers, the central warehouse, the factory, the sea vessels and seaports, the orders sent from retailers to the warehouse as well as from the warehouse to the factory, and the trucks moving cargo in Spain and China. The discrete event simulation is used to model the processing of orders at the central warehouse and the transport of cargo on sea, rail, and road. The simulation is run over a period of 4 years, and 15 model runs are used to simulate each of the 6 cases/scenarios considered (Table 10.3).

10.4 Results

Table 10.6 expounds the averages and standard deviations of order lead times when $L_{sc} = 20\%$ and compares these values with those measured in the absence of seaport closure events. It can be observed that the average times and standard deviations in the seaports increase significantly, except for the road and rail transport, respectively, to the port of Yantian and the warehouse in Zaragoza. The average time spent in Singapore, Damietta, and Barcelona increased significantly, while it only increased moderately in the Port of Yantian. The simulation also shows an increment of the lead time to move between seaports, mostly because of the waiting time that sea vessels spend for the reopening of the seaport and the necessary berth reallocation (Table 10.6).

Table 10.7 shows the inventory performance of the Zaragoza warehouse and three retailers when we simulate what we consider the *baseline scenario*, i.e., when conservative and lean policies are adopted, cases 1a and 1b, with or without the possibility of seaport closure. We can observe that the conservative policy helps maintain high performance, even when risks increase (possibly by adjusting the inventory levels adequately in response to such risks). In addition, shifting from the conservative to the lean policy, there is a slight deterioration in performance (using the most common customer service measures in inventory management: average cycle service level and fill rate). This reduction in service level becomes more significant when we simulate using a seaport closure likelihood of 20% (Table 10.7). Hence, in line with previous research, adopting a lean policy, which typically aims to reduce inventory and streamline processes, results in a small but measurable decrease in the resilience and the ability to meet customer demands promptly and effectively.

Table 10.6 Mean and standard deviations of shipping times, with $L_{sc} = 20\%$ (compared with Table 10.1, M_Diff: Mean Difference, SD_Diff: Standard Deviation Difference)

	Mean	M_Diff	SD	SD_Diff
Factory production and storage	161.56	−5.16	93.39	−5.45
Truck transport factory to Yantian	23.69	0.02	1.47	−0.07
Port of Yantian	148	6.78	164.69	0.95
YNT-SNG	140.79	34.1	41.91	38.63
Port of Singapore	67.23	45.04	43.59	40.44
SNG-DMT	324.16	48.32	145.24	−4.37
Port of Damietta	52.69	32.14	23.81	18.09
DMT-BCN	163.57	49.66	49.54	45.92
Port of Barcelona	194.34	30.99	76.59	34.23
Train BCN—ZGZ	9.6	0.08	0.75	−0.08
Total	1285,63	241.97	640.98	168.29

Table 10.7 Inventory performance in the central warehouse (Zaragoza) and retailers in Bilbao, Seville, and Valencia (average values), respectively, when the seaport closure likelihood, L_{sc}, is 0% vs 20%

	Conservative (Case 1a)		Lean (Case 1b)	
Echelon	$L_{sc} = 0\%$	$L_{sc} = 20\%$	$L_{sc} = 0\%$	$L_{sc} = 20\%$
Zaragoza				
Service Level	90.6	90.5	83.97	83.67
Fill Rate	90.81	90.8	84.12	83.81
Bilbao				
Service Level	86.9	87.1	77.73	77.31
Fill Rate	87.04	87.2	77.85	77.46
Seville				
Service Level	86.9	87.1	77.73	77.31
Fill Rate	86.9	87.1	77.73	77.31
València				
Service Level	86.9	87.1	77.73	77.31
Fill Rate	87.04	87.2	77.85	77.46

10.4.1 Impacts of PRS

Using a Proactive Replenishment Strategy implies that a company can access real-time information about inventory levels at Zaragoza and retailers, as well as historical ETA data. During the simulated runtime of the conservative policy, we observe that the warehouse initiates 40 proactive shipments, corresponding to a total of 1563 units shipped to the three retailers. The average service level of the central warehouse is 90.73%, and the fill rate is 91.48% (about a 1% increment), showing a performance increment when compared with the figures of the baseline scenarios (Fig. 10.2, top). When a lean planning policy is adopted (Case2b), the number of proactive shipments is 39, corresponding to 1406 units. The average service level is 83.87% and fill rate is 83.91%, in line with the argument that lowered inventory levels are more vulnerable to disruptions. Yet, when compared with the baseline scenarios, we observe a moderate service level improvement (about 0.2%).

Similarly, the service levels of the three retailers benefit from the Proactive Replenishment Strategy. When conservative policy is used, Bilbao has an average service level of 88.24% (average fill rate 88.34%), Sevilla, 88.26% (average fill rate 88.31%), and Valencia, 88.31% (average fill rate 88.34%) (Table 10.8). When using lean policies, we observe a reduction in the performance, but this time, there are significant improvements when compared with the baseline scenario (about 1%) (Table 10.7).

10.4.2 Impacts of PRS and OPI

In this section, we simulate and collect inventory performance figures when using the Proactive Replenishment Strategy in combination with Order Progress Information. When a conservative (Case 3a) policy is followed, the warehouse uses

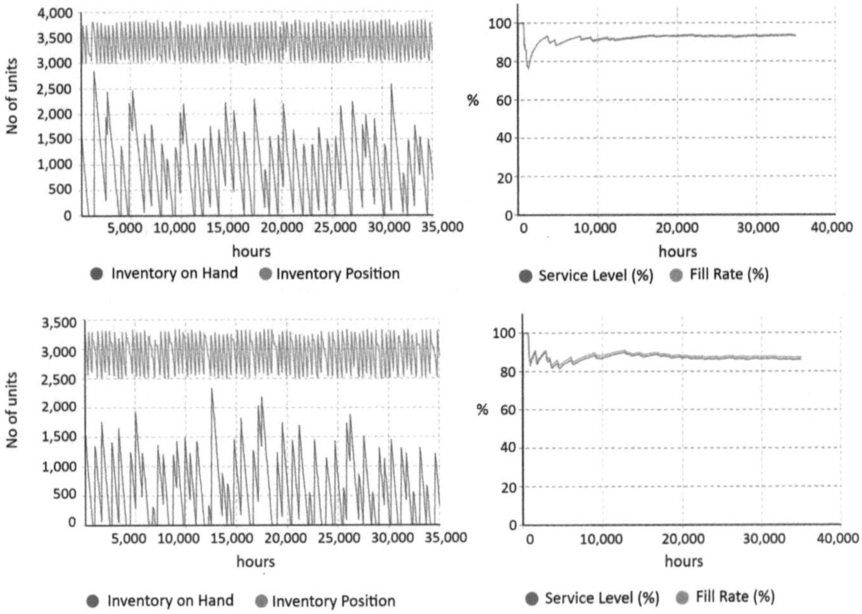

Fig. 10.2 Inventory on hand, inventory position (SKU/hour) (left) and service level (%) (right). Case 2a, conservative policy (top) and Case 2b lean policy (bottom), $L_{sc} = 20\%$

Table 10.8 CASE 2a and 2b, inventory performance analysis at the three retailers, Bilbao, Sevilla, and Valencia (average values reported, $L_{sc} = 20\%$)

	Conservative & PRS (Case 2a)		Lean & PRS (Case 2b)	
	Service Level (%)	Fill Rate (%)	Service Level (%)	Fill Rate (%)
Bilbao	88.24	88.34	78.23	78.33
Sevilla	88.26	88.31	78.23	78.23
Valencia	88.31	88.34	78.24	78.33

the PRS 39 times, shipping 1518 units. The average service level is 91.23%, and the average fill rate is 92.07%, showing improvements (Fig. 10.3, top). In a lean planning scenario, the PRS is used 39 times, during which 1492 units are shipped to the retailers. Also, in this case, the performance of the central warehouse improves when compared with the PRS and baseline cases, i.e., average service level 84.2% and average fill rate 84.6% (Fig. 10.3, bottom).

The PRS-OPI strategy leads to a reduction in the retailers' service levels and fill rates. This means that, in alignment with the improvements in the central warehouse, the overall ability to meet customer demands promptly and to fulfill orders directly from stock without backordering has improved under the PRS-OPI strategy (Table 10.9). This indicates that the usage of the Proactive Replenishment Strategy combined with OPI is more effective when lean policies are adopted, i.e. when the supply chain is more vulnerable.

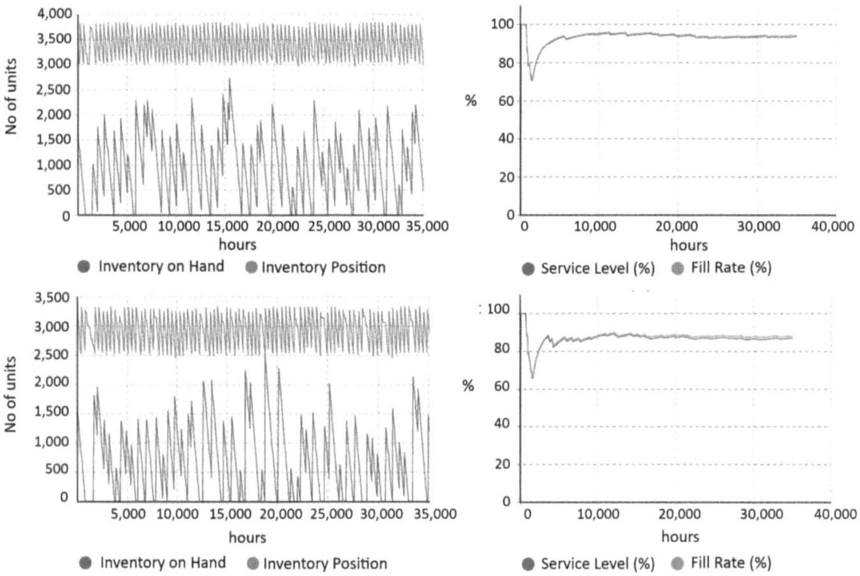

Fig. 10.3 Inventory on hand, inventory position (SKU/hour) (left) and service level (%) (right). Case 3a conservative planning (top) and Case 3b, lean planning (bottom), $L_{sc} = 20\%$

Table 10.9 CASE3a and 3b, performance analysis of Bilbao, Sevilla, and Valencia (average values, $L_{sc} = 20\%$)

	Conservative & PRS-OPI (Case 3a)		Lean & PRS-OPI (Case 3b)	
	Service Level (%)	Fill Rate (%)	Service Level (%)	Fill Rate (%)
Bilbao	89.33	89.39	79.02	79.23
Sevilla	89.33	89.33	79.02	79.23
Valencia	89.33	89.39	79.92	79.23

10.5 Discussion and Conclusion

In this study, we investigate the case of a global supply chain consisting of a three-echelon inventory system under the threat of seaport disruption. A simulation model is developed and tested in 6 cases: the first two consider the adoption of static conservative and lean policies, comparing safety lead time versus safety stock (cases 1a and 1b). Next, we simulate the cases of using proactive replenishments with historical shipment data and real-time data about inventory levels at the retailers (cases 2a and 2b); and, finally, two additional cases, where real-time order progress information is combined with proactive replenishments and real-time data about inventory levels. Our results show the value of implementing data-driven inventory policies, leading to improvements in customer service levels.

Across all simulated scenarios, it is evident that the conservative policy yields better inventory performance compared to the lean policy, especially as disruption risks increase. This observation aligns with previous research, which suggests that lean strategies heighten the vulnerability of supply chains (Ahmed & Rashdi, 2021). Additionally, PRS prove to be instrumental in improving and maintaining desired service levels. These results align with those of Topan and van der Heijden (2020), who claim that proactive shipments (which are cheaper compared to reactive shipments as they are planned shipments at pre-determined periods) are most effective to reduce downtime without increasing shipment costs substantially.

Another key finding is that the combination of PRS and PRS-OPI (Order Progress Information) consistently enhances performance. This implies that firms are enabled to use lead time visibility to "rectify" the situation (in case of potential delivery delays) and maintain desired service levels. Jain (2018) shows that up-to-date information about the remaining lead time does not necessarily lead to more orders being expedited (actually, they observe that fewer orders are expedited). In addition, Jain (2018) find that lead time visibility combined with operational flexibility benefits the firm as the inventory levels are lower (i.e., lead time visibility is a substitute for higher inventory levels). Our simulation model confirms these findings.

Existing research highlights the benefits of digital visibility platforms by discussing their impacts on investment costs. Kouvelis and Li (2012) note that visibility reduces reliance on safety lead-time, thus lowering shipment costs. Jain (2018) suggests that partial visibility suffices to gain most of the operational flexibility benefits, reducing the need for full lead-time visibility and making it feasible for firms lacking financial or technical resources. Based on our findings, the PRS option performs effectively across all scenarios considered. The additional OPI feature further provides added value. One can, therefore, infer that full lead-time visibility, which may be substantially more expensive to achieve than partial lead-time visibility, may not provide sufficient marginal return on investment.

The simulations reveal several key managerial implications. Decision-makers need to carefully weigh the benefits of lean policies against the potential risks of supply chain disruptions. While lean strategies can improve efficiency and reduce inventory costs, they also increase the vulnerability of supply chains to unexpected events, such as seaport closures. Therefore, managers must implement proactive replenishment strategies and maintain adequate safety stock to mitigate these risks.

Some limitations can be identified. Proactive shipment systems may incur higher costs due to increased transportation requirements. However, these costs can potentially be offset by reducing the need for back-order shipments. To minimize expenses, we recommend optimizing delivery routes by consolidating visits to multiple retailers within a single trip.

Another challenge lies in sending shipments to retailers that did not explicitly request them, as this could be difficult to implement. Nevertheless, our findings demonstrate that retailers can maintain high service levels—and ensure customer satisfaction—without holding excessive inventory. This can be achieved by allowing the Zaragoza warehouse access to their inventory levels and enabling proactive shipments when inventory runs critically low or when significant delivery delays are anticipated.

Future research should undertake a more comprehensive analysis of both major and minor risks, quantifying these based on available data regarding their frequency and impacts. Additionally, this study does not address the effects of inspections and potential seizures of cargo imported into Spain. Given the significant impact these factors may have on performance, future research should focus on analyzing their potential impacts and evaluating additional mitigation measures.

References

Ahmed, W., & Rashdi, M. Z. (2021). Understanding the influence of lean and agile strategies on creating firms' supply chain risk management capabilities. *Competitiveness Review: An International Business Journal, 31*, 810–831.

Chopra, S., & Sodhi, M. S. (2004). Managing risk to avoid supply-chain breakdown. *MIT Sloan Management Review, 46*, 53.

Chung, W., Talluri, S., & Kovács, G. (2018). Investigating the effects of lead-time uncertainties and safety stocks on logistical performance in a border-crossing JIT supply chain. *Computers & Industrial Engineering, 118*, 440–450.

Eufrásio, A. B. R., Eller, R. A., & Oliveira, A. V. (2021). Are on-time performance statistics worthless? An empirical study of the flight scheduling strategies of Brazilian airlines. *Transportation Research Part E: Logistics and Transportation Review, 145*, 102186.

Fransoo, J. C., & Lee, C. Y. (2013). The critical role of ocean container transport in global supply chain performance. *Production and Operations Management, 22*, 253–268.

Hu, M., Liu, X.-X., & Jia, F. (2024). Optimal emergency order policy for supply disruptions in the semiconductor industry. *International Journal of Production Economics, 272*.

Huang, G., He, Z., Zhao, P., & Zhang, C. (2025). A review of climate-related disasters impact on global shipping carbon emissions. *Transportation Research Part D: Transport and Environment, 139*, 104553.

Ilgar, O. & Zagorodnya, Z. 2024. *Deep dive: The importance of oceans in our global supply chains* [Online]. Forbes. Accessed October, 2024, from https://www.forbes.com/sites/sap/2024/07/08/deep-dive-the-importance-of-oceans-in-our-global-supply-chains/#:~:text=Deep%20Dive%3A%20The%20Importance%20Of%20Oceans%20In%20Our%20Global%20Supply%20Chains,-SAPBRANDVOICE&text=While%20most%20people%20think%20of,the%20ocean%20to%20reach%20you

Jain, A. (2018). Responding to shipment delays: The roles of operational flexibility & lead-time visibility. *Decision Sciences, 49*, 306–334.

Jeong, Y., Kim, G., & Shin, Y. (2025). Uncertain demurrage and detention decisions in hinterland container management. *International Transactions in Operational Research*.

Kim, A., Park, H.-J., Park, J.-H., & Cho, S.-W. (2021). Rescheduling strategy for berth planning in container terminals: An empirical study from Korea. *Journal of Marine Science and Engineering, 9*, 527.

Kim, C., Klabjan, D., & Simchi-Levi, D. (2015). Optimal expediting policies for a serial inventory system with stochastic lead time. *Production and Operations Management, 24*, 1524–1536.

Kouvelis, P., & Li, J. (2012). Contingency strategies in managing supply systems with uncertain lead times. *Production and Operations Management, 21*, 161–176.

Kouvelis, P., & Tang, S. Y. (2012). On optimal expediting policy for supply systems with uncertain lead-times. *Production and Operations Management, 21*, 309–330.

Kravchenko, K., Gruchmann, T., Ivanova, M., & Ivanov, D. (2024). Responding to the ripple effect from systemic disruptions: Empirical evidence from the semiconductor shortage during COVID-19. *Modern Supply Chain Research and Applications, 6*, 354–375.

Lewis, B. M., Erera, A. L., Nowak, M. A., & Chelsea, C. W. (2013). Managing inventory in global supply chains facing port-of-entry disruption risks. *Transportation Science, 47*, 162–180.

Mamani, H., & Moinzadeh, K. (2014). Lead time management through expediting in a continuous review inventory system. *Production and Operations Management, 23*, 95–109.

Padhi, S. S., & Mukherjee, S. (2022). Optimal portfolio choices to split orders during supply disruptions: An application of sport's principle for routine sourcing. *Decision Sciences, 53*, 1068–1087.

Pathy, S. R., & Rahimian, H. (2023). A resilient inventory management of pharmaceutical supply chains under demand disruption. *Computers & Industrial Engineering, 180*.

Paul, S. K., & Rahman, S. (2018). A quantitative and simulation model for managing sudden supply delay with fuzzy demand and safety stock. *International Journal of Production Research, 56*, 4377–4395.

Ravulakollu, A. K., Urciuoli, L., Rukanova, B., Tan, Y.-H., & Hakvoort, R. A. (2018). Risk based framework for assessing resilience in a complex multi-actor supply chain domain. *Supply Chain Forum: An International Journal, 2018*. Taylor & Francis, 266–281.

Safety4sea. (2025). *Sea-Intelligence: Liner schedule reliability improves in November 2024* [Online]. Accessed February, 2025, from https://safety4sea.com/sea-intelligence-liner-schedule-reliability-improves-in-november-2024/

Schwarz, L. B. (1976). An examination of the United States Air Force (Q, R) policies for managing depot-base inventories.

Simchi-Levi, D., & Kaminsky, P. (2008). *Designing and managing the supply chain: concepts, strategies and case studies*. McGraw-Hill.

Singh, S., Kumar, R., Panchal, R., & Tiwari, M. K. (2021). Impact of COVID-19 on logistics systems and disruptions in food supply chain. *International Journal of Production Research, 59*, 1993–2008.

Song, J.-S., Yano, C. A., & Lerssrisuriya, P. (2000). Contract assembly: Dealing with combined supply lead time and demand quantity uncertainty. *Manufacturing & Service Operations Management, 2*, 287–296.

Song, J.-S., & Zipkin, P. H. (1996). Inventory control with information about supply conditions. *Management Science, 42*, 1409–1419.

Stamos, I., Mitsakis, E., Salanova, J. M., & Aifadopoulou, G. (2015). Impact assessment of extreme weather events on transport networks: A data-driven approach. *Transportation Research Part D: Transport and Environment, 34*, 168–178.

Sun, X., Chung, S., Chan, F. T., & Wang, Z. (2018). The impact of liner shipping unreliability on the production–distribution scheduling of a decentralized manufacturing system. *Transportation Research Part E: Logistics and Transportation Review, 114*, 242–269.

Tomlin, B. (2006). On the value of mitigation and contingency strategies for managing supply chain disruption risks. *Management Science, 52*, 639–657.

Topan, E., & Van Der Heijden, M. C. (2020). Operational level planning of a multi-item two-echelon spare parts inventory system with reactive and proactive interventions. *European Journal of Operational Research, 284*, 164–175.

Urciuoli, L., & Hintsa, J. (2020). *Can digital ecosystems mitigate risks in sea transport operations?—estimating benefits for supply chain stakeholders*. Maritime Economics & Logistics.

Wagenaar, J., Fragkos, I., & Zuidwijk, R. (2021). Integrated planning for multimodal networks with disruptions and customer service requirements. *Transportation Science, 55*, 196–221.

Xu, X., Li, Z., Jiang, X., Huang, X., Lu, Y., & Li, L. (2024). Port berth allocation and microgrid cluster joint optimization scheduling based on master-slave game. *Journal of Cleaner Production, 452*, 142220.

Xu, Y., Chen, Q., & Quan, X. (2012). Robust berth scheduling with uncertain vessel delay and handling time. *Annals of Operations Research, 192*, 123–140.

Yin, J., Khan, R. U., Wang, X., & Asad, M. (2024). A data-centered multi-factor seaport disruption risk assessment using Bayesian networks. *Ocean Engineering, 308*, 118338.

Yin, X., & Tang, W. (2025). The bullwhip effect, market competition and standard deviation ratio in two parallel supply chains. *Computers & Chemical Engineering, 192*, 108916.

Yue, Z., & Mangan, J. (2024). A framework for understanding reliability in container shipping networks. *Maritime Economics & Logistics, 26*, 523–544.

Zuidwijk, R. A., & Veenstra, A. W. (2015). The value of information in container transport. *Transportation Science, 49*, 675–685.

Chapter 11
Enhancing Supply Chain Resilience Using Digital Twins

Sirish Kumar Gouda and Prashanth Chandrasekharan

Abstract Supply Chains across the world are facing an increasing number of disruptions due to man-made as well as natural causes. Adding to this complexity is the global nature of supply chains with ever-changing consumer demands. Such complex supply chains require digital solutions with the ability to track, visualize, forecast, and act in an agile, efficient, and effective manner. Supply chain digital twins provide a comprehensive solution for managing these risks and developing resilient supply chains. In this chapter, we use a combination of primary and secondary data to understand how these digital twins build supply chain resilience with the help of other technologies. We also understand the capabilities that digital twins extend to supply chains to make them reliable, responsive, agile, and, in turn, resilient. We provide a suitable architecture for a supply chain digital twin combining different technology assets of an organization. Our data also highlights the importance of people, processes, and technology in making Supply Chain Digital Twins successful.

Keywords Supply chain resilience · Digital twin · Control tower · Empirical study

11.1 Introduction

With an increasing demand for products produced in one place to be made available around the world, supply chains have become global (Christopher & Peck, 2004). Given the myriad stakeholders such as suppliers, freight forwarders, logistic service providers, customs, distributors, wholesalers, and retailers from different parts of the world involved in most supply chains, it is not just global macro-economic and

S. K. Gouda (✉)
Indian Institute of Management Tiruchirappalli, Tiruchirappalli, Tamil Nadu, India
e-mail: sirish@iimtrichy.ac.in

P. Chandrasekharan
Tata Consultancy Services, Chennai, Tamil Nadu, India

Y. Khojasteh, S. M. M. Jahan Abadi (eds.), *Emerging Technologies in Supply Chains*, International Series in Operations Research & Management Science 373, https://doi.org/10.1007/978-3-032-01218-0_11

geopolitical shocks, but even local events have serious ripple effects in the supply chain of firms (DesJardine et al., 2019). For instance, a report by Statista indicates that the semiconductor shortage alone led to a potential revenue loss of 210 million dollars for the automotive industry (Statista, n.d.). Similarly, UNCTAD reports that the recent disruption in shipping routes in the Red Sea, Black Sea, and Panama Canal due to geopolitical wars and droughts has led to a reduction of 49% of monthly transits (UNCTAD, 2024).

Owing to these challenges, global supply chains are investing heavily in solutions related to the visibility of what's happening across various nodes and networks of the organization and beyond. One of the most used technologies to visualize supply chains virtually is the Supply Chain Digital Twin (SCDT).

There are several definitions of supply digital twins as per academia and practice. Gartner defines Digital Twin as "a digital representation of a real-world entity or system" (Gartner, 2018). Given the costs involved in handling a challenging situation in real time, a digital twin provides a risk-free, non-destructive, effective, and efficient way to simulate probable real-time situations to train any system to perform better. As the supply chain of any organization is a complex system that can be represented digitally, Anylogistix defines an SCDT as "A Supply Chain Digital Twin is a detailed simulation model of an actual supply chain which uses real-time data/snapshots to forecast supply chain dynamics" (anyLogistix, 2025). Ivanov et al. (2023) define SCDT as a virtual system that is a combination of the digital visualization of the system, technology to capture data of the system and a data analytics engine for better decision-making. As can be seen from the definition, a digital twin needs to have a dynamic model that absorbs real-time data to develop insights useful for planning future events. A sample SCDT model covering supplier and customer integration is provided below in Fig. 11.1. As can be seen from the

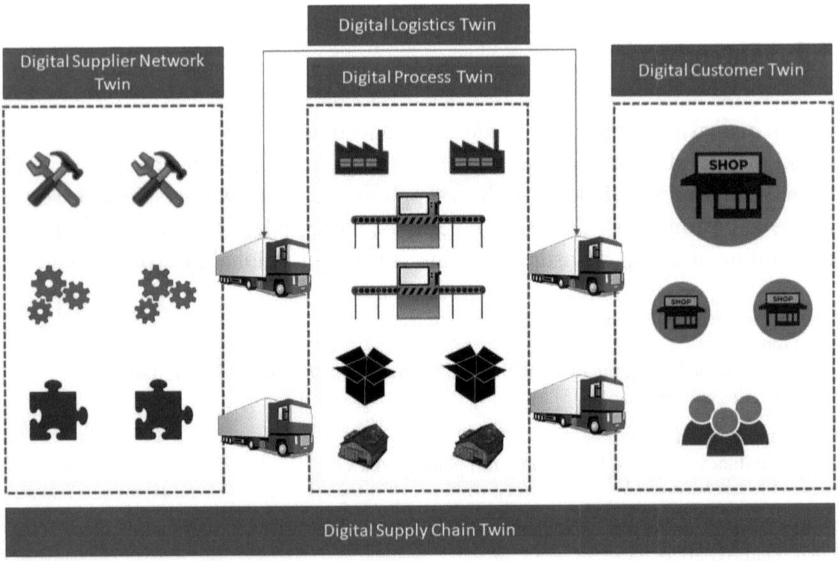

Fig. 11.1 High-level model of Supply Chain Digital Twin

figure, an SCDT model involves a supplier network twin, a process twin, a logistics twin, and a customer twin.

Given the strengths of an integrated and intelligent digital twin, it is our strong conviction that they can be deployed as a powerful tool to counter disruptions in supply chains with varied likelihood of occurrence and impact. In this chapter, we study the role of digital twins in enhancing supply chain resilience by analyzing primary responses from eight in-depth interviews of digital twin experts and secondary data from publicly available case studies. We investigate the following research questions as part of our study

(a) What is the role of digital twins in enhancing supply chain resilience (consisting of Reliability, Responsiveness, and Agility)?
(b) What capabilities of digital twins help supply chain resilience (consisting of Reliability, Responsiveness, and Agility)?
(c) What is a suitable architecture and data model for a typical supply chain digital twin?

In the next section, we look at the extant literature on supply chain resilience and supply chain digital twins to motivate our study. We then analyze the primary and secondary data collected to arrive at recurring themes supporting our thesis of the impact of digital twins on supply chain resilience. We finally provide a conclusion and future research directions.

11.2 Literature Review

11.2.1 Different Technologies in Supply Chain

Technology plays a vital role in supply chains. By optimizing the use of resources, it increases the efficiency of the firms' operations (Patterson et al., 2003). Supply chain processes involve several iterative actions that can be automated using technology (Shah, 2009). Technology also plays an important role in capturing the data generated by different stakeholders and machines at various touchpoints in the supply chain. It also helps organizations to coordinate seamlessly with different upstream and downstream partners in the supply chain (Yang et al., 2021).

Traditionally, enterprise technologies have played a role in digitizing several processes of supply chain. As per APICS Supply Chain Operational Reference (SCOR) Model,[1] we have mapped the different technologies adopted by firms in each process in Table 11.1.

[1] The SCOR Model used in this Table is an earlier version referred to as SCOR 12 which is one of the most popular SCOR models used in practice. The current version SCOR 14.0 contains the Processes—Plan, Source, Order, Fulfill, Return and Orchestrate.

Table 11.1 Different Technologies in SCM in each process. Adapted from Bowersox et al. (2020) and Shah (2009)

Process	Technologies Adopted
Plan	Materials requirement planning (MRP), collaborative planning Forecasting and replenishment (CPFR)
Source	Purchase order automation (POA), e reverse auctions (e-RA), Electronic data interchange (EDI), supplier relationship Management (SRM)
Make	Manufacturing resource planning (MRP II), capacity requirement Planning (CRP), quality management (QM), inventory management (IM)
Deliver and return	Warehouse management systems (WMS), transportation Management (TM)
Enable	Order tracking, point of Sale (POS), customer relationship Management (CRM)

Table 11.2 Different capabilities of technologies helpful in Supply Chains

Technology	Capabilities
Blockchain	Visibility, transparency
Artificial intelligence	Accuracy, automation
Machine learning	Prediction, optimization
Co-bots	Automation, productivity, efficiency
Additive manufacturing	Automation, customization, flexibility

While these technologies remain in use, several new technologies have revolutionized supply chains over the last decade. Additive Manufacturing, Sensors, Industrial Internet-of-Things (IIOT), co-bots, Artificial Intelligence and Machine Learning, Blockchain, etc., have disruptively transformed the way supply chains are managed by organizations (Frazzon et al., 2021). These technologies provide different capabilities to supply chains, making them more efficient and effective as discussed in Table 11.2.

Industry 5.0 brings back the human element into these technologies, termed the cyber-physical systems focusing on sustainability and resilience. For these cyber-physical systems to operate at scale and solve the challenging problems that supply chain disruptions pose to organizations, a 'living and learning' virtual system such as digital twin is needed.

11.2.2 Supply Chain Digital Twin

While digital twins are not a recent phenomenon and have been around for the last three decades, it is the proliferation of the Internet of Things (IoT) that has made them popular (Freese & Ludwig, 2024). One of the first applications of digital twins was in the aviation industry to create a virtual system of turbines in product lifecycle management (Grieves, 2019). Given their capabilities, digital twins have several applications in the fields of manufacturing (Leng et al., 2021), transportation (Feng et al., 2023), and supply chain management (Le & Fan, 2023).

SCDT can be defined as a virtual model of the end-to-end supply network, internal processes and distribution network that represents its state at any moment in time (Burgos & Ivanov, 2021; Ivanov & Dolgui, 2021). Research and practice have been using simulation models that seem similar to Digital twins. However, they have some important differences (Burgos & Ivanov, 2021):

(a) Digital twins work with real-time or near-real-time data from various sources of the physical supply chains while simulations use archival data.
(b) Simulations run in waves over discrete intervals of time while digital twins run and update continuously as data flows through various sources.
(c) Digital twin can seamlessly integrate scenario analysis, prediction, and optimization while simulations are restricted usually for what-if analysis.

While Digital Twin is a virtual model, it is also an integrating platform of several technologies that support the key purpose of the application. Sensors for data collection, cloud servers for data backup, AI/ML and advanced analytics for data analysis, augmented and virtual reality for visualization and simulation for prescriptive analytics are integrated into a platform that is also represented as a digital twin (Tozanli & Saénz, 2022; Freese & Ludwig, 2024). Some of the research that highlights the application of digital twin technologies for various supply chain uses is provided in Table 11.3.

11.2.3 Supply Chain Resilience

Over the last decade, several internal and external events such as accidents, natural disasters, terrorist attacks, and pandemics have made it increasingly difficult for supply chains to be free from disruptions (Dolgui & Ivanov, 2021; Katsaliaki et al., 2022). These disruptions have a deleterious impact on firms' market performance as well as operational performance (Hendricks & Singhal, 2005). Given the increasing frequency and omnipresence of these disruptions, practitioners as well as researchers, have begun to explore the various facets of supply chain risk management (SCRM) (Christopher & Peck, 2004; Dolgui & Ivanov, 2021; Ivanov, 2024b).

Kumar et al. (2010) define supply chain risks as "the potential deviations from the initial overall objective that, consequently, trigger the decrease of value-added activities at different levels" (p 3717). Jüttner et al. (2003) categorize risks faced by firms into three categories—(a) *internal risks*—arising within the organization, (b) *supply chain risks* internal to the firm's supply chain but outside its boundaries, and (c) *external supply chain risks* impacting the firm from the events in the external

Table 11.3 Summary of select literature on use of digital twins in supply chains

Research Study	Method	Tools used	Context	Summary of Research
Marmolejo-Saucedo (2020)	Case study / simulation	ALX software	COVID-19 epidemic simulation in pharma industry	Design and development mechanism of digital twin for pharma supply chain proposed using a combination of simulation, optimization and prediction tools
Burgos and Ivanov (2021)	Simulation	AnyLogistix software supply chain simulation	Multi-stage supply chain for a German retailer	Discrete event simulation model used to study the impact of COVID-like pandemics on food supply chains' resilience
Freese and Ludwig (2024)	Conceptual/ empirical	Semi-structured interviews with five experts	Logistics service providers, IT and consulting	Conceptual framework developed using design science approach to outline layers, dimensions, stakeholders and technology dependence
Ivanov and Dolgui (2021)	Conceptual	–	–	A conceptual and technological framework proposed to manage disruption risk and enhance resilience using SCDT and Industry 4.0 tools. An integrated decision-support system (DSS) for developing a proactive resilient SC using SCDT.
Badakhshan and Ball (2023)	Simulation	Simulation with machine learning (decision tree)	Three-echelon typical FMCG supply chain	Discrete event simulation model used to study the use of digital twins in managing inventory and cash under physical and financial disruption
Longo et al. (2023)	Simulation	Simulation using Anylogistix	–	Simulation-based decision-making framework proposed to manage disruption and sustainability risks in Agri-food supply chains
Park et al. (2021)	Conceptual/ empirical	Distributed agent-based simulation	Automobile manufacturing supply chain	Designed and evaluated an architectural framework of cyber-physical supply chain digital twin system in a make-to-order manufacturing supply chain

(continued)

Table 11.3 (continued)

Research Study	Method	Tools used	Context	Summary of Research
Singh et al. (2023)	Empirical	Structural equation modelling (SEM)	Manufacturing supply chains from India	Explored mediating role of supply chain resilience on the relationship between use of digital twins and sustainability of supply chains using primary survey data collected from manufacturing professionals

environment. Based on the classification of Gouda and Saranga (2018), we categorize supply chain risks as follows

(i) *Inbound risk or Supply risk* emanating from disruptions upstream to the firm
(ii) *Process risk* resulting from disruptions or failures inside the boundaries of the firm, and
(iii) *Outbound or Customer risk* based on disruptions in distributional logistics while reaching the customer.

Supply chain risk management has three major elements: a) Risk Identification b) Risk Assessment, and c) Risk Mitigation (Kern et al., 2012). Once any firm identifies and categorizes different supply chain risks, the process of risk assessment requires the firm to identify the probability of occurrence of a risk event and its corresponding impact on the firm's operations/value (Kleindorfer & Saad, 2005; Manuj & Mentzer, 2008; Kumar et al., 2010). Based on the assessment of the various risk, firms invest several preventive and reactive risk mitigation efforts (Thun & Hoenig, 2011; Gouda & Saranga, 2018). While both preventive and reactive risk mitigation strategies are planned by the firm before a risk event happens, preventive risk mitigation strategies while focusing on reducing the probability of occurrence of an event provide benefits before any risk event occurs and reactive risk mitigation strategies have an impact after the risk event occurs as they reduce the impact of a risk event (Thun & Hoenig, 2011). These risk mitigation efforts include choosing a reliable supplier, supplier monitoring, having backup supply sources, backup warehouses and transportation mechanisms, having large safety inventory, maintaining extra capacity, internal and external training programmes etc. (Chopra & Sodhi, 2004).

Supply Chain Resilience is the ability of the firm to withstand these internal and external sources of risks and be able to ensure business continuity. Wieland and Durach (2021) define Supply Chain Resilience as "the capacity of a supply chain to persist, adapt, or transform in the face of change." (p 316). Literature has looked at several aspects of supply chain resilience—such as Stability (DesJardine et al., 2019), Agility (Lee, 2004, Lee 201), Robustness (Cohen & Kouvelis, 2021), and Adaptability (Hillmann & Guenther, 2021; Phadnis, 2024). SCOR identifies key supply chain performance metrics that all organizations need to track. Among those metrics, Reliability, Responsiveness, and Agility are found responsible for achieving Supply Chain Resilience (ASCM, 2022).

While the above discussed supply chain risk mitigation efforts and other strategic actions by firms can help make supply chains resilient, technology plays an important integrating role in this. Modern technologies such as Industrial Internet of Things (IIoT), Artificial Intelligence and Machine Learning, along with SCDTs play an important role in helping supply chains become resilient (Lee, 2021; Ivanov & Dolgui, 2021; Longo et al., 2023).

In this study, based on the literature survey, we use the constructs: *reliability*, *responsiveness*, and *agility* to measure supply chain resilience and look at how SCDTs can contribute to the same. As can be seen from Table 11.3, the extant literature has studied the impact of Digital twins on Supply Chain resilience primarily in a conceptual sense or using simulation studies (Burgos & Ivanov, 2021; Ivanov & Dolgui, 2021). We try to fill the gap in the literature by using an empirical study to establish the relationship between these two using primary data (in-depth interviews) and secondary data (publicly available case studies).

11.3 Methodology and Analysis

11.3.1 Data Description

In-depth interviews were conducted with eight experts between July 2024 and February 2025 to gather qualitative aspects associated with supply chain digital twins. We reached out to our respondents based on their expertise, as identified through their LinkedIn profiles. We provided them with the context of the study and explained the purpose of the data collection to them over a Zoom call or via email. We then shared a Google form with questions categorized into different sections based on the topic of inquiry. The Google form had questions related to:

(a) Industry details—Type, Size and Complexity
(b) Digital transformation efforts by the firm
(c) Supply Chain Digital Twins
(d) Organizational enablers and deterrents to use of Supply Chain Digital Twins
(e) Firms' actions towards Supply Chain Resilience

The detailed questionnaire is provided in the Appendix. Based on the responses of the experts, codes were developed to arrive at themes linking elements of supply chain digital twins with resilience.

We also collated secondary data in the form of some prominent case studies from several sources such as websites of companies that provide SCDT solutions (such as Anylogistix, Coupa etc.), expert blogs, and trade and business publications (such as CIO, Wall Street Journal, The Economist etc.).

11.3.2 Data Analysis

Extant literature (e.g., Lee, 2004, 2021; Wieland & Durach, 2021) have studied the different aspects of resilience—reliability, responsiveness, and agility. Reliability of a supply chain system is defined as its ability to perform various logistical tasks as per expectation (ASCM, 2022). Responsiveness is defined as the ability of the supply chain to dynamically meet the changing demands of the environment with the speed at which customer expects (Lee, 2004; Wieland & Durach, 2021; ASCM, 2022) while Agility is defined as the supply chain's flexibility to cater to variabilities and uncertainties in the system in an efficient manner (Lee, 2021). On the aspects of supply chain resilience, we asked our respondents questions about—the importance of resilience in supply chains, and ways in which digital twins and other technologies make supply chains resilient.

Our respondents, on average, have more than 10 years of experience in supply chain management, supply chain consulting, and technology applications in supply chains. They represent varied industries, including machine manufacturing, pharma, auto ancillaries, semiconductors, and energy management with their firms' turnover ranging between hundred million US dollars to greater than one billion US dollars.

We follow a qualitative content analysis procedure prescribed by Tangpong (2011). Our sampling unit can be broadly defined as a unit demarcated for selective inclusion in the analysis. In our study, we used one respondent for the primary data and one document for the secondary data as a sampling unit. The coding process involved identifying all relevant words and phrases that provided an understanding of the various aspects of supply chain resilience from our sampling units and dividing them into different aspects of supply chain resilience identified earlier based on our literature survey.

Our respondents indicated that supply chains need to become resilient to cater to the changing needs of the customers, to ensure business continuity, to meet regulatory requirements, to achieve cost and quality advantage, and to achieve sustainability goals. While discussing supply chain digital twins' role in resilience, they used the following keywords—'real-time monitoring/visibility/signaling', 'predictive analytics', 'scenario-planning', 'custom constraints modeling', 'simulation modeling', 'transparency', 'quick decision-making', 'optimized operations', 'enhanced responsiveness', 'operational efficiency', 'business continuity', 'robustness', 'proactive risk management', 'rapid response', 'decision-support', 'scalability', 'flexibility', 'reduced errors', 'real-time data-integration', 'swift response', 'security', 'better decision-making', 'continuous improvement', 'early indications', 'better visibility', 'accuracy', 'forecasting', 'network planning'. Based on the definitions of the different dimensions of resilience as defined earlier, we categorize these keywords into themes of reliability, responsiveness, and agility.

When we looked at the case studies of different firms such as ICP Group, Microsoft, Nestle, Mars, Walmart, Bayer Crop Science, Singapore Land Transport Authority, Schlumberger, dmdrogerie mrkt, from the publicly available sources as mentioned above, we found the usage of the following keywords in conjunction with SCDT –.

Table 11.4 Keywords from Primary and Secondary Data for different themes of Resilience

Themes	Definition (literature)	Role of Digital Twins (keywords from primary and secondary data)	Capabilities of Digital twins (keywords from primary and secondary data)
Reliability	Ability to perform various logistical tasks as per expectation	Business continuity, early indications, robustness of supply chain, greater visibility, early issue discovery	Real-time monitoring, predictive analytics, end-to-end visibility, security, network planning,
Responsiveness	Ability of the supply chain to dynamically meet the changing demands of the environment with the speed at which customer expects	Continuous improvement, scalability, effective Interruption response	Forecasting, real-time data integration, decision support
Agility	Flexibility to cater to variabilities and uncertainties in the system in an efficient manner	Rapid/swift/quick response, optimized operations, optimized processes, reduced errors, reduced waste, accuracy, enhanced responsiveness, accelerated growth	Simulation modeling, custom constraints modeling

'Optimized processes', 'real-time data', 'real-time traffic monitoring', 'controlled congestion', 'effective interruption response', 'early issue discovery', 'predictive analytics', 'demand estimation', 'optimized inventory levels', 'enhanced flow', 'scenario planning', 'greater visibility', 'reduced waste', 'end-to-end visibility', 'accelerated growth'.

Our first research question was about the role of digital twins in enhancing supply chain resilience. From our primary and secondary data, Table 11.4 showcases the role SCDT plays in increasing resilience. Through our second research question, we intended to inquire about the capabilities of digital twins to develop agility, stability, and adaptability. For instance, through the capabilities of *real-time monitoring* and *end-to-end visibility*, supply chains get *early indications* leading to *early issue discovery* thus developing the *reliability* of supply chains. Similar aspects can be observed in the *responsiveness* and *agility* elements of supply chain resilience in Table 11.4.

As can be seen from the above set of keywords from primary and secondary data, the keywords converge to benefits that are accrued to supply chains due to the applications of SCDT and the host of other associated technologies on the platform. While the Digital Twin virtual platform gives greater visibility for the various stakeholders in the supply chain, other technologies such as the cloud servers and IIoT allow capture of the right data, and analytics engines with AI/ML models allow the stakeholders to build simulation models with multiple scenarios, forecast uncertainties and allow them to get the benefits of reduced errors/waste, optimized operations, and accelerated growth.

11.4 Discussion

Organizations' investments in technologies such as digital twins involve significant efforts in digital transformation from various stakeholders. Ivanov (2024a) conceptualized a 7-element framework for the implementation of digital twins for extracting the maximum value for different stakeholders in the supply chain. These elements are—Management, People, Organization, Modeling, Task, Scope, and Technology. Our primary data also provided support for various aspects mentioned above. As can be observed from our data provided in Table 11.5, SCDT replacement requires planned phasing out of legacy systems with a focus not only on immediate benefits but focus on Return on Investment (ROI) and the value to the different stakeholders. From a people perspective, as this task involves keeping in the loop different upstream and downstream supply chain partners, stakeholder engagement is key. As emphasized by the experts in our respondent pool and the several case studies we analyzed, the success of SCDT lies in the availability and collection of high-quality data from various nodes of the supply chain network. Given that this involves a significant overhaul of systems, change management is a key aspect like all digital transformations.

Table 11.5 Key themes from Primary Data on Elements supporting Digital Twin adoption

Elements from Ivanov (2024a)	Definitions	Key themes from our primary Data
Management	Management principles for technology adoption	ROI analysis, cost-benefit analysis, value to stakeholders
People	Stakeholders and their perspectives	Stakeholder engagement, change champions, skilled workforce
Organization	Larger organizational aspects to be considered while undergoing Digital transformation	Alignment with business goals, governance, replacing legacy systems, agile methodology, modular architecture
Modeling	Mathematical and computer models	Descriptive, predictive, prescriptive analytics, simulation, MLOps
Task	Tasks to be focused on for the success of digital tools	Monitoring and evaluation, training and support, upskilling, automation, clear communication, domain knowledge
Scope	Scope of the application of the digital tools	Pilot projects, scalability, interoperability, connectivity
Technology	Technology aspects under consideration	Data quality/accuracy, data security, privacy, IoT, AI, API integration, cloud and hybrid solutions

From a larger organizational perspective, SCDT requires upfront mapping of this transformation with larger business goals/strategies of the firm and the supply chain. SCDT investments also require the identification of a governance framework for data storage and data use, given that critical information across the supply chain from different partners needs to be collected and analyzed for the best output. Given that several technologies are part of the larger SCDT platform, a modular architecture is an ideal choice.

To make the best use of the data collected, a strong analytics engine with capabilities of describing the data for as-is visualization, predicting future events based on the forecasting models, and simulating various probabilistic future events through scenario analysis is a must. While Ivanov (2024a) defines scope based on applications in different business settings, we look at the important tasks from an organizational perspective to be focused on to make SCDT a success. This involves focusing on the scalability of the models across plants, warehouses, retail stores, etc., interoperability of data and models for various business settings, and connectivity across different nodes in the supply chain network. As discussed earlier, SCDT standalone cannot provide all the benefits expected without integration with other key technologies such as cloud technologies, IIoT, AI/ML, and analytics engines, etc.

11.4.1 Supply Chain Digital Twin Architecture

Our third research question is about a suitable architecture and data model for a typical supply chain digital twin. Coupa, a popular supply chain software service provider categorizes digital twins into two types[2]—Product Digital Twin and Control Tower Digital Twin. While a product twin is a virtual replica of a physical entity, such as a machine or a product, a control tower digital twin is a virtual system replicating an entire network such as a supply chain system. As our focus is on a supply chain digital twin, we consider a control tower digital twin. A typical SCDT architecture is provided in Fig. 11.2.

For an SCDT investment, it is necessary to start with a data mapping exercise, identifying data source systems, databases, data pipelines, transformation logics, historical data, APIs, and UX aspect which is represented as a 'Data layer' in Fig. 11.2. Once the required data is established and sources identified, a data repository is created in the cloud, which accesses all the data sources and updates databases in real time. Front end applications need to be developed, that transformed, analyzed and populated the data on the UI for user decision making. Once the dashboard is established and business processes are tuned to the new ways of operating, the analytics and simulation capabilities are developed. Machine learning algorithms are deployed to analyze historical data for developing quantitative forecasts

[2] https://www.coupa.com/blog/the-complete-guide-to-supply-chain-digital-twins/

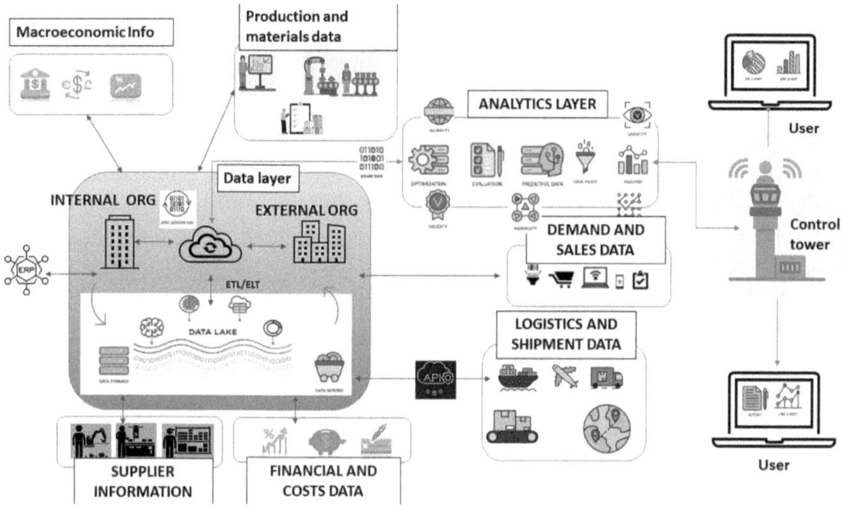

Fig. 11.2 Sample representation of SCDT architecture

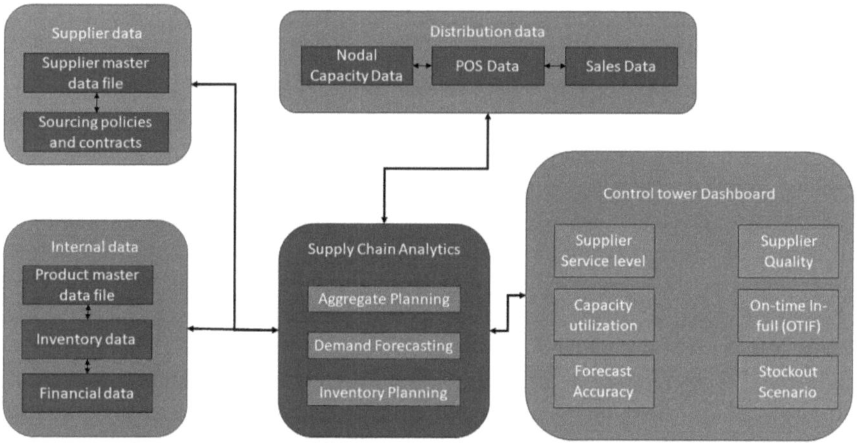

Fig. 11.3 Sample data model for an SCDT

and demand planning. This is referred to as the 'Analytics layer' in the above fig. A control tower represented in the figure above is a centralized panel that provides visibility and control over every supply chain activity after integrating data from different sources and the decision support from the analytics engine. This control tower can provide real-time insights to different user departments.

A sample data model for an SCDT is provided in Fig. 11.3. Simulation capabilities for specific aspects such as component availability, orders in transit, orders delivered in full, backorder quantity, backorder value, inventory quantity and value etc. are developed based on key parameters. For instance, users can also specify an Order delivered in full % and compute the back-order volume estimate based on past data. With automated analytics and AI/ML models running in the background, backorder estimation accuracy improves tremendously, and orders are placed with the suppliers much before a specific backorder is triggered for a particular component. From a supply chain responsiveness angle, by the time a backorder is triggered, the component would already be in production, and thereby, the days of backorder would be significantly reduced. As data accuracy improves, the simulation capabilities also improve along with it.

Real-time visibility of orders, component subassemblies, spare parts and integration with logistics providers for shipment-related information through GPS and smart RFID technologies enables any company to foresee supply chain disruptions before they occur and take spontaneous decisions to reduce supply chain risks and manage operations efficiently. Integrating customers with the system is the next level of collaboration, where the customers can see the real-time status of their orders and plan their operations accordingly. By extending the visibility, risks can be reduced for customers significantly, thereby improving long term strategic partnerships.

11.5 Conclusions

In this chapter, we discuss about the application of Digital Twins in achieving Supply Chain Resilience and provide evidence for the same using evidence from primary and secondary data sources. Our primary data analysis reveals the importance of several factors, such as people, process, and technological aspects in the success of SCDT implementation. Our literature review revealed a dearth of empirical literature in the space of SCDT which our study tried to fulfil. We interviewed experts in the field of SCDT to understand the capabilities it renders to supply chain teams and the role it plays in aiding supply chains to become reliable, responsive, and agile. We also looked at various external sources such as exemplary case studies, blogs, and white papers from secondary data sources, to understand how different aspects of SCDT enabled resilience. We also provide a sample control architecture and data model for SCDT implementation.

Our research's contributions are limited by our access to knowledge experts in the field and the availability of case studies in the public domain. Future research can study the real-world implementation of digital twins in supply chains and understand the true benefits, enablers, and barriers to the adoption of these technologies. Future research can also collect subjective (survey-based) as well as objective data on the successes and failures of digital twin implementations in supply chains. There are several advancements in the field of cloud computing, data architecture, and AI/ML modeling, which might also reshape the way digital twins are adopted and used in supply chains which future research can focus on.

Appendix

Questionnaire used for Primary Data Collection

Industry details

(a) Which industry do you represent?
(b) What is your organization's annual turnover?
(c) How complex is your supply chain in terms of global presence?
(d) Briefly explain the dynamics of the industry that you are currently building the product for

Digital Transformation Efforts by Firms

(e) What, in your view, is contributing to the growing interest in Supply Chain Digital transformation?
(f) How did the idea of digital twin as a solution for the business problem that you are working on, come about?
(g) Briefly explain how you have been able to link the solution design with the organization and supply chain strategy of the business?

Supply Chain Digital Twins

(h) Briefly explain the architectural considerations (such as applications, networks, data, infrastructure etc.) and how you have been able to unify supply chain data availability across diverse platforms and functions?
(i) How have you been able to adapt modern and emerging technologies into the digital twin solution?
(j) How can companies reduce the technology debt when they have a lot of legacy systems, while building a digital twin system which will source data from these applications?
(k) What are some of the key technical challenges that companies encounter when embarking on implementation of Digital twins for supply chains?
(l) What are some of the key strategic considerations for investing in digital twin technology? How do companies avoid " shiny toy" syndrome while going for cutting edge technologies?
(m) How do organizations manage business process change management when implementing cutting edge solutions such as digital twins?
(n) How does Digital Twin improve resilience of a supply chain? How do companies adapt to changing business realities through cutting edge technology?

On Supply Chain Resilience

(o) Why is it important for supply chains to be adaptable and resilient?
(p) What are the levers that businesses are using, to build resiliency into their supply chains?
(q) What are the key elements of digital twins that can make supply chains resilient?

(r) How do digital twins leverage other technologies to make firms resilient?
(s) Apart from digital twins, are there any other technologies that can improve supply chain resiliency and adaptability?

References

anyLogistix, (2025). *Supply chain digital twins*. [online] Anylogistix.com. Accessed Mar 27, 2025, from https://www.anylogistix.com/resources/white-papers/supply-chain-digital-twins/

ASCM. (2022). *Supply chain operations reference model SCOR digital standard*. [online] Accessed Mar 25, 2025, from https://www.ascm.org/globalassets/ascm_website_assets/docs/intro-and-front-matter-scor-digital-standard2.pdf

Badakhshan, E., & Ball, P. (2023). Applying digital twins for inventory and cash management in supply chains under physical and financial disruptions. *International Journal of Production Research, 61*(15), 5094–5116.

Bowersox, D. J., Closs, D. J., Cooper, M. B., & Bowersox, J. C. (2020). *Supply chain logistics management*. Mcgraw-hill.

Burgos, D., & Ivanov, D. (2021). Food retail supply chain resilience and the COVID-19 pandemic: A digital twin-based impact analysis and improvement directions. *Transportation Research Part E: Logistics and Transportation Review, 152*, 102412.

Chopra, S., & Sodhi, M. S. (2004). Managing risk to avoid supply-chain breakdown. *MIT Sloan management review*.

Christopher, M., & Peck, H. (2004). Building the resilient supply chain. *The International Journal of Logistics Management, 15*(2), 1–14.

Cohen, M. A., & Kouvelis, P. (2021). Revisit of AAA excellence of global value chains: Robustness, resilience, and realignment. *Production and Operations Management, 30*(3), 633–643.

DesJardine, M., Bansal, P., & Yang, Y. (2019). Bouncing back: Building resilience through social and environmental practices in the context of the 2008 global financial crisis. *Journal of Management, 45*(4), 1434–1460.

Dolgui, A., & Ivanov, D. (2021). Ripple effect and supply chain disruption management: New trends and research directions. *International Journal of Production Research, 59*(1), 102109.

Feng, H., Lv, H., & Lv, Z. (2023). Resilience towarded digital twins to improve the adaptability of transportation systems. *Transportation Research Part A: Policy and Practice, 173*, 103686.

Frazzon, E. M., Freitag, M., & Ivanov, D. (2021). Intelligent methods and systems for decision-making support: Toward digital supply chain twins. *International Journal of Information Management, 57*, 102281.

Freese, F., & Ludwig, A. (2024). A conceptual framework for supply chain digital twins– Development and evaluation. *International Journal of Logistics Research and Applications*, 1–23.

Gartner. (2018). *Digital Twin*. [online] Accessed Mar 24, 2025, from https://www.gartner.com/en/information-technology/glossary/digital-twin

Gouda, S. K., & Saranga, H. (2018). Sustainable supply chains for supply chain sustainability: Impact of sustainability efforts on supply chain risk. *International Journal of Production Research, 56*(17), 5820–5835.

Grieves, M. W. (2019). Virtually intelligent product systems: Digital and physical twins. *Complex systems engineering: Theory and practice.*

Hendricks, K. B., & Singhal, V. R. (2005). Association between supply chain glitches and operating performance. *Management Science, 51*(5), 695–711.

Hillmann, J., & Guenther, E. (2021). Organizational resilience: A valuable construct for management research? *International Journal of Management Reviews, 23*(1), 7–44.

Ivanov, D. (2024a). Conceptualisation of a 7-element digital twin framework in supply chain and operations management. *International Journal of Production Research, 62*(6), 2220–2232.
Ivanov, D. (2024b). Two views of supply chain resilience. *International Journal of Production Research, 62*(11), 4031–4045.
Ivanov, D., & Dolgui, A. (2021). A digital supply chain twin for managing the disruption risks and resilience in the era of industry 4.0. *Production Planning & Control, 32*(9), 775–788.
Ivanov, D., Dolgui, A., Blackhurst, J. V., & Choi, T. M. (2023). Toward supply chain viability theory: From lessons learned through COVID-19 pandemic to viable ecosystems. *International Journal of Production Research, 61*(8), 2402–2415.
Jüttner, U., Peck, H., & Christopher, M. (2003). Supply chain risk management: Outlining an agenda for future research. *International Journal of Logistics: research and applications, 6*(4), 197–210.
Katsaliaki, K., Galetsi, P., & Kumar, S. (2022). Supply chain disruptions and resilience: A major review and future research agenda. *Annals of Operations Research*, 1–38.
Kern, D., Moser, R., Hartmann, E., & Moder, M. (2012). Supply risk management: Model development and empirical analysis. *International Journal of Physical Distribution & Logistics Management, 42*(1), 60–82.
Kleindorfer, P. R., & Saad, G. H. (2005). Managing disruption risks in supply chains. *Production and Operations Management, 14*(1), 53–68.
Kumar, S. K., Tiwari, M. K., & Babiceanu, R. F. (2010). Minimisation of supply chain cost with embedded risk using computational intelligence approaches. *International Journal of Production Research, 48*(13), 3717–3739.
Le, T. V., & Fan, R. (2023). Digital twins for logistics and supply chain systems: Literature review, conceptual framework, research potential, and practical challenges. *Computers & Industrial Engineering*, 109768.
Lee, H. L. (2004). The triple-a supply chain. *Harvard Business Review, 82*(10), 102–113.
Lee, H. L. (2021). The new AAA supply chain. *Management and Business Review, 1*(1), 173176.
Leng, J., Wang, D., Shen, W., Li, X., Liu, Q., & Chen, X. (2021). Digital twins-based smart manufacturing system design in industry 4.0: A review. *Journal of Manufacturing Systems, 60*, 119–137.
Longo, F., Mirabelli, G., Padovano, A., & Solina, V. (2023). The digital supply chain twin paradigm for enhancing resilience and sustainability against COVID-like crises. *Procedia Computer Science, 217*, 1940–1947.
Manuj, I., & Mentzer, J. T. (2008). Global supply chain risk management. *Journal of Business Logistics, 29*(1), 133–155.
Marmolejo-Saucedo, J. A. (2020). Design and development of digital twins: A case study in supply chains. *Mobile Networks and Applications, 25*(6), 2141–2160.
Patterson, K. A., Grimm, C. M., & Corsi, T. M. (2003). Adopting new technologies for supply chain management. *Transportation Research Part E: Logistics and Transportation Review, 39*(2), 95–121.
Park, K. T., Son, Y. H., & Noh, S. D. (2021). The architectural framework of a cyber physical logistics system for digital-twin-based supply chain control. *International Journal of Production Research, 59*(19), 5721–5742.
Phadnis, S. (2024). A review of research on supply chain adaptability: Opening the black box. *Journal of Business Logistics, 45*(1), e12370.
Shah, J. (2009). *Supply chain management, texts and cases*. Pearson Education India.
Singh, G., Singh, S., Daultani, Y., & Chouhan, M. (2023). Measuring the influence of digital twins on the sustainability of manufacturing supply chain: A mediating role of supply chain resilience and performance. *Computers & Industrial Engineering, 186*, 109711.
Statista. (n.d.). *Automotive industry revenue and production loss forecast*. [online] Accessed Mar 27, 2025, from https://www.statista.com/statistics/1280639/automotive-industry-revenue-and-production-loss-forecast/

Tangpong, C. (2011). Content analytic approach to measuring constructs in operations and supply chain management. *Journal of Operations Management, 29*(6), 627–638.

Thun, J. H., & Hoenig, D. (2011). An empirical analysis of supply chain risk management in the German automotive industry. *International Journal of Production Economics, 131*(1), 242–249.

Tozanli, Ö., & Saénz, M. J. (2022). Unlocking the potential of digital twins in supply chains. *MIT Sloan Management Review, 63*(4), 1–4.

UNCTAD. (2024). *Navigating troubled waters: Impact to global trade of disruption of shipping routes in the Red Sea, Black Sea and Panama Canal*, UNCTAD. [online] Accessed Mar 26, 2025, from https://unctad.org/publication/navigating-troubled-waters-impact-global-trade-disruption-shipping-routes-red-sea-black

Wieland, A., & Durach, C. F. (2021). Two perspectives on supply chain resilience. *Journal of Business Logistics, 42*(3), 315–322.

Yang, M., Fu, M., & Zhang, Z. (2021). The adoption of digital technologies in supply chains: Drivers, process and impact. *Technological Forecasting and Social Change, 169*, 120795.

Chapter 12
A Framework for Enhancing Supply Chain Resilience Through Metaverse and Blockchain Technologies

Yu Cui and Prakash J. Singh

Abstract This chapter explores the potential of Metaverse environments, underpinned by Web 3.0 ecosystems and Decentralized Autonomous Organization (DAO) mechanisms, to enhance supply chain resilience. The study examines the evolution of the metaverse concept, the emergence of Web 3.0, and the characteristic features of DAO mechanisms. It proposes a novel conceptualization of a decentralized Metaverse predicated on supply chain resilience within the Web 3.0 ecosystem, with DAO mechanisms at its core. The chapter presents case studies from Japan to illustrate the practical applications and potential benefits of this approach. These examples demonstrate how Metaverse initiatives, blockchain technology, and Web 3.0 principles can be leveraged to improve supply chain management, enhance data security, and foster innovation.

Keywords Metaverse · Web 3.0 · DAO · Japanese case studies

12.1 Introduction

Since the 2010s, humanity has entered a period of robust development within the Fourth Industrial Revolution (Okwu et al., 2022). In the contemporary landscape characterized by the profound integration and evolution of digital technologies—including 5G network, cloud computing, the Internet of Things (IoT), big data analytics, artificial intelligence (AI), blockchain, and extended reality (XR) encompassing virtual reality (VR), augmented reality (AR), and mixed reality

Y. Cui (✉)
Faculty of Sociology, Kansai University, Suita, Osaka, Japan
e-mail: yucui@kansai-u.ac.jp

P. J. Singh
Department of Management and Marketing, University of Melbourne, Parkville, VIC, Australia

Y. Khojasteh, S. M. M. Jahan Abadi (eds.), *Emerging Technologies in Supply Chains*, International Series in Operations Research & Management Science 373, https://doi.org/10.1007/978-3-032-01218-0_12

(MR)—the digital transformation of industries and the reconfiguration of digital supply chains have emerged as focal points of attention across both industrial sectors and broader societal spheres (French et al., 2021; Hahn, 2020).

The conceptual genesis of Metaverse, an amalgamation of the aforementioned frontier technologies, has precipitated the advent of a novel synchronous supply chain management paradigm (Dolgui & Ivanov, 2023). This innovative model, characterized by its simultaneous operation in both corporeal and virtual domains, permeates the entire supply chain continuum—encompassing design, production, logistics, and services—through comprehensive digitalization, networkization, and intelligentization processes (Queiroz et al., 2023).

In contrast to the conventional digital transformation of traditional industries, Metaverse paradigm places greater emphasis on mapping and expanding the operational capabilities of physical industries within virtual spaces (Alpala et al., 2022). Through collaborative work and simulated operations in virtual environments, this approach guides the efficient functioning of tangible business processes, thereby empowering various stages and scenarios within the supply chain (Ramanathan, 2014). This empowerment enables supply chain enterprises to achieve their objectives of cost reduction and enhanced production efficiency (Ali et al., 2021). Furthermore, it facilitates highly efficient coordination both within and among supply chain enterprises, contributing to high-quality industrial development and enabling a further upgrade in intelligent supply chain management.

In tandem with the synergistic advancements in next-generation information technologies, the conceptual framework and extensional boundaries of the metaverse undergo continuous refinement and expansion (Bhattacharya et al., 2023). The metaverse transcends the confines of a unitary product or technology, instead embodying a nascent paradigm of future industrial ecosystem evolution, predicated on the convergence of multifarious digital technologies (Asif & Hassan, 2023).

This emergent concept is predicated upon a substratum of next-generation intelligent networks, IoT architectures, and distributed ledger technologies (Xevgenis et al., 2023). It leverages natural human-computer interfaces and AI systems as interactive modalities, engendering a new industrial morphology characterized by the seamless integration of virtual and physical domains (Siddiqui et al., 2020).

This emergent framework not only enables virtual interactions among users that approach verisimilitude, but also catalyzes cross-sectoral convergence through multifaceted scenarios. It engenders transformative capabilities across a spectrum of industries, including healthcare, education, entertainment, and manufacturing, potentially permeating virtually all dimensions of human societal production and quotidian existence (see Table 12.1).

Conversely, as delineated in the aforementioned table, concurrent with the robust proliferation of Metaverse, the dominion over numerous domains of Metaverse development and application—with the exception of specific sectors such as gaming and social networking—remains predominantly within the purview of

Table 12.1 Representative examples of Metaverse in various fields

Domain	Development Summary	Representative Examples
Gaming	Pioneering immersive experiences; integrating blockchain for in-game economies	Fortnite (epic games): Virtual concerts and events; Decentraland: Blockchain-based virtual world; Roblox: User-generated content platform
Social media	Creating virtual spaces for social interaction; implementing avatar systems	Facebook horizon (meta): VR social platform; VRChat: Social VR application; AltspaceVR (Microsoft): VR events and meetups
Education	Developing virtual classrooms and interactive learning environments	Engage: VR education and training platform; Minecraft: Education edition; ClassVR: VR headsets for classroom use
Healthcare	Implementing VR for training, therapy, and patient care	OssoVR: Surgical training platform; Psious: VR therapy for mental health; AccuVein: AR vein visualization
Retail	Creating virtual shopping experiences; implementing digital try-on technology	IKEA place: AR furniture visualization; Walmart's virtual store (Roblox); Gucci garden (Roblox): Virtual fashion experience
Real estate	Offering virtual property tours; developing digital land markets	Matterport: 3D virtual property tours; the sandbox: Virtual real estate platform; NextHome: VR home tours
Workplace	Developing virtual offices and collaboration tools	Spatial: AR/VR workspace; Microsoft mesh: Mixed reality collaboration; glue: VR collaboration platform
Entertainment	Creating immersive experiences for concerts, movies, and events	Wave: Virtual concerts platform; MelodyVR: VR music experiences; Bigscreen VR: Social movie watching
Finance	Implementing blockchain and cryptocurrencies; creating virtual banks	Somnium space: VR world with its own economy; JPMorgan's onyx lounge in Decentraland; CryptoVoxels: Ethereum-based virtual world
Tourism	Offering virtual travel experiences and cultural explorations	Ascape VR: Virtual travel experiences; Google earth VR; Timelooper: Historical site reconstructions
Manufacturing	Utilizing AR/VR for design, training, and operations;implementing digital twins	Siemens NX: VR for product design and visualization; Boeing: AR-assisted manufacturing and training; NVIDIA Omniverse: Collaborative 3D design and simulation platform
Logistics	Enhancing supply chain visibility; optimizing warehouse operations; improving last-mile delivery	DHL: AR-powered vision picking in warehouses; UPS: VR driver training program; Amazon: AR for efficient package scanning and sorting

large-scale digital platforms, spearheaded by the Magnificent Seven,[1] and global corporate behemoths (Cheng et al., 2022). This concentration of control presents a significant impediment to the most fundamental and crucial aspects of information sharing and real-time collaboration in the processes of enterprise digital transformation and digital supply chain reconfiguration (Tana & Chai, 2023).

In the context of the metaverse environments predicated on third-party platforms, participating entities are unable to establish comprehensive trust and collaborative sharing mechanisms, thereby impeding the effective realization and expansion of the metaverse's original objectives. Moreover, during exigent circumstances such as pandemics or seismic events, the supply chain system demonstrates a critical inability to rapidly and systematically reallocate resources and reinitiate a fully automated, intelligent, decentralized, and self-regulating supply chain system.

This chapter addresses the challenges and impediments encountered in the industrial digital transformation and reconfiguration of digital supply chains within the context of the metaverse concept. It proposes a paradigm shift within the Web 3.0 ecosystem, advocating for the establishment of Decentralized Autonomous Organization (DAO) mechanisms. This approach aims to redirect the focus of Metaverse environment construction from an overemphasis on immersive, multi-scenario modalities towards a system architecture predicated on the core value proposition of blockchain technology—namely, decentralized, self-governing, and intelligent systems.

This approach aims to ensure and facilitate proactive information sharing and real-time collaboration among all participating entities. In the event of exigent circumstances, it enables the full realization of the digital supply chain's potential within the Web 3.0 ecosystem. By activating DAO mechanisms, supported by a decentralized, self-governing, and intelligent Metaverse model, this framework can expedite the mitigation of disruptions and facilitate the rapid restoration of normal operations within real-world supply chain systems.

This chapter presents a systematic synthesis of the development and current status of Metaverse-related technologies. Furthermore, it provides an explication of the role played by Metaverse in the advancement and evolution of digital supply chains at the present stage, as well as an analysis of the challenges encountered in this process. The chapter concludes with two examinations of pioneering Japanese exemplars, substantiating and investigating the viability of this conceptualization while identifying future research imperatives.

[1] The term "Magnificent Seven" in the context of digital platforms refers to a group of prominent technology companies that have come to dominate the global digital landscape. These companies, namely Apple, Amazon, Alphabet (Google's parent company), Meta (formerly Facebook), Microsoft, Nvidia, and Tesla represent the vanguard of technological innovation and digital market capitalization in the contemporary era.

12.2 Evolution and Development of the METAVERSE

12.2.1 Conceptual Foundations of the Metaverse

The conceptual genesis of Metaverse can be traced to Neal Stephenson's 1992 speculative fiction opus, "Snow Crash". Within this seminal text, Stephenson conceptualized the Metaverse as a virtual tridimensional realm existing in parallel with the corporeal world, wherein human entities traverse the digital landscape via virtualized proxies denominated as Avatars (Abbate et al., 2022). Howard Rheingold, a prominent network sociologist, published his seminal work "The Virtual Community" in 1993. This text elucidated the emergence of virtual communities and collectives, constituted by social networks of individuals engaging in digital communication, knowledge dissemination, information exchange, and reciprocal empathy through computer-mediated platforms (Turner, 2005).

The dawn of the twenty-first century witnessed the proliferation of "Second Life", a pioneering online virtual world (Zack, 2007). This digital ecosystem afforded users the opportunity to construct and manipulate interactive digital proxies, while simultaneously enabling the procurement of virtual artifacts through the utilization of tangible financial instruments (Gent, 2022). The successful initial public offering of Roblox in March 2021, predicated on the metaverse concept, heralded a pivotal moment in the inauguration of the metaverse epoch (Kim, 2021). This was followed by Facebook's strategic rebranding to Meta in October 2021, signaling a paradigm shift towards Metaverse-centric business development. These seminal events catalyzed the metaverse's formal integration into the industrial landscape and its ascendance as a locus of multidisciplinary attention.

In the context of Metaverse, underpinned by 5G high-speed communication infrastructure, the voluminous data generated through digital economic transactions is instantaneously transmitted to robust, stable, secure, and accessible cloud repositories (Lee et al., 2021). Cloud computing and artificial intelligence paradigms facilitate the dynamic allocation and management of data resources, while edge computing, exploiting its user-proximate and low-latency attributes, guarantees an optimal user experience (Huynh et al., 2022). This technological synergy engenders an authentic digital ecosystem characterized by the intelligent interconnectivity of all entities and digital twin manifestations (Aloqaily et al., 2022).

Extended Reality (XR) technologies engender tridimensional human-computer interfaces, facilitating augmented immersive experiences (Salvetti et al., 2023). Virtual Reality (VR) constitutes an advanced simulated environment that synchronizes users' perceptual experiences in virtual and physical domains (Fadzli et al., 2020). Augmented Reality (AR) pivots on the superimposition of virtual data onto identified physical entities via intelligent apparatuses, actualizing visual representation and enabling real-time interplay between digital and corporeal information streams (Balakrishnan et al., 2021). Mixed Reality (MR) constitutes an avant-garde visualization paradigm that synthesizes corporeal and virtual realms (Warin & Reinhardt, 2022). This innovative perceptual ecosystem facilitates the gradual dissolution of delineations between VR and AR modalities, engendering simultaneous existence and reciprocal engagement of tangible and digital entities (Alnajim et al., 2023). Holographic technologies further attenuate the dichotomy between physical and virtual domains, thereby establishing a robust substrate for the substantive actualization of the metaverse concept.

The metaverse facilitates the manifestation of corporeal entities and individuals as virtualized digital avatars, engendering a nascent virtual societal paradigm (Hamidouche et al., 2023). The IoT instantiates omnipresent interconnections between physical and virtual constructs, effectuating a bidirectional mapping between tangible reality and Metaverse (Guan et al., 2023). The tactile IoT, an exemplar of IoT technology, substantially augments immersive modalities within this digital ecosystem (Jha et al., 2023).

Furthermore, Metaverse constitutes an artifact of human societal praxis in the digital epoch, intricately interwoven within the fabric of specific economic production and quotidian existence (Bhattacharya et al., 2023). Supply chains manifest as loosely coupled aggregates within the complex social network architecture, each amalgamating multiple disparate industrial processes. The social production and dissemination of supply chains as artifacts accentuate the social relationships and forces inherent in industries. Metaverse, as an emergent artifact, can only ascertain its ecological niche for persistence and evolution through industrialization, which delineates its societal role and function. In the age of ubiquitous interconnectivity, inter-organizational network associations have become increasingly intricate, with the holistic supply chain system exhibiting pronounced network effects. Ergo, value generation from an industrial perspective exhibits systemic characteristics of network effects, with supply chain network structures emerging as pivotal conduits for the genesis and metamorphosis of the metaverse value (Bag et al., 2023).

12.2.2 Metaverse in Supply Chain

In the current phase of supply chain management, the implementation of Metaverse technologies exhibits a disproportionate emphasis on digital immersion. However, at the level of production relations, a significant concern emerges: the data generated through user interactions is predominantly monopolized by leading digital platforms, epitomized by the Magnificent Seven, and industry-dominant enterprises. This concentration of data control presents a substantial barrier to widespread adoption. Consequently, a multitude of potential users are deterred from engaging with and applying Metaverse technologies due to insufficient assurances regarding data privacy protections (Tukur et al., 2023). This dichotomy between immersive capabilities and data governance issues underscores a critical challenge in the integration of Metaverse concepts within supply chain systems, potentially impeding the realization of its full transformative potential (Kouhizadeh et al., 2021).

Moreover, due to factors such as corporate monopolization, a preponderance of Metaverse applications are constructed within centralized network architectures.

This prevalent structural paradigm engenders significant vulnerabilities within supply chain systems, rendering them susceptible to catastrophic failure in the face of sudden, devastating disruptions (Cui et al., 2019). The centralized nature of these networks creates a single point of failure, potentially leading to a state of near-complete paralysis across the entire supply chain system when confronted with extreme exogenous shocks. This inherent fragility underscores the pressing need for more resilient, decentralized architectural approaches in the design and implementation of Metaverse-enabled supply chain systems (Masip-Bruin et al., 2021).

In the wake of the COVID-19 pandemic outbreak, global supply chain systems experienced a protracted period of semi-paralysis, extending over a considerable duration (Golan et al., 2020). This unprecedented disruption not only inflicted catastrophic impacts on supply chain-related enterprises but also severely compromised the operational efficacy and efficiency of information systems across diverse industries intricately linked to these supply chains (Wang, 2023). The pervasive and prolonged nature of this disruption underscores the systemic vulnerabilities inherent in contemporary supply chain architectures and their associated information systems.

Another paradigmatic incident occurred in July 2024, when an operational oversight by CrowdStrike, a prominent U.S.-based cybersecurity firm, precipitated a widespread systemic failure (Krebs on Security, 2024). This event resulted in the protracted inoperability of countless information systems globally, with a particular impact on Microsoft Windows systems, persisting for several dozen hours. The systems affected manifested in a state colloquially known as the "blue screen of death", indicative of critical system failures. The repercussions of this incident were particularly acute across critical services, such as airlines, banks, and healthcare providers, which experienced severe disruptions to their normal functioning. This cascading failure inflicted catastrophic impacts on an innumerable multitude of end-users and stakeholders, underscoring the far-reaching consequences of centralized vulnerabilities in our increasingly interconnected digital ecosystem and highlighting the potential for localized errors to propagate globally in our hyper-connected technological landscape.

This incident compels a critical reassessment of the inherent vulnerabilities and potential detriments associated with the prevalent centralized Metaverse systems. It serves as a catalyst for sober reflection on the systemic risks embedded within current architectures. Concurrently, this event provides compelling empirical evidence substantiating the imperative and urgency of developing Metaverse environments predicated on decentralized, self-regulatory paradigms. The stark ramifications of centralized failures underscore the exigency of transitioning towards more resilient, distributed architectures in the construction of future Metaverse frameworks. This shift not only promises enhanced system robustness but also aligns with principles of distributed governance and user empowerment, potentially mitigating the risks of cascading failures and centralized points of vulnerability.

In the subsequent sections, we will elucidate and analyze the innovative Metaverse model that encompasses the Web 3.0 ecosystem, predicated on blockchain technology as its foundational architecture, and employs Decentralized Autonomous Organizations (DAOs) as the operational mechanism for supply chain systems. This explication will provide a comprehensive examination of the symbiotic relationship between these advanced technological paradigms and their potential to revolutionize supply chain management within Metaverse context.

12.3 Emergence of Web 3.0: Historical and Conceptual Analysis

12.3.1 Historical Milestones of Web 3.0

The concept of Web 3.0 was initially propounded by Jeffrey Zeldman in 2006. At that juncture, the development of Web 2.0 was reaching its zenith on a global scale (Boulos & Wheeler, 2007). Consequently, the Web 3.0 paradigm postulated by Zeldman did not garner widespread attention or critical examination from the academic and technological communities.

The concept of Web 2.0 was first articulated by Tim O'Reilly in 2004. In contrast to the predominantly 'read-only' nature of Web 1.0, Web 2.0 introduced a paradigm shift towards a 'read-write' model of internet interaction (Murugesan, 2007). This evolution catalyzed the proliferation of interactive applications, including but not limited to instant messaging platforms, social networking sites, and e-commerce portals.

Web 2.0, however, is susceptible to information leakage, data manipulation, and the emergence of monopolistic internet platforms (Fuchs, 2011). The business models of internet conglomerates are predominantly predicated on the monopolization of user data and privacy. Moreover, in both Web 1.0 and Web 2.0 paradigms, the vast quantities of data generated on the internet are proprietary to websites or platforms, with users devoid of data ownership rights (Fuchs, 2011). Consequently, even when these data generate value, users are precluded from benefiting from the associated rights and interests.

In 2014, Gavin Wood, co-founder of Ethereum, reintroduced the concept of Web 3.0 in his blog post titled "DApps: What Web 3.0 Looks Like" on his platform "Insights into Modern World". Wood's definition of Web 3.0 primarily emphasizes its nature as a decentralized, distributed network of interconnected entities, fundamentally underpinned by blockchain technology (Tennakoon & Gramoli, 2022). As next-generation internet technologies, epitomized by Ethereum, garner significant global attention, the concept of Web 3.0 is gaining increasing traction within both industrial and academic spheres.

In juxtaposition to the Client/Server architecture characteristic of Web 2.0, which predominantly relies on centralized Internet Data Centers (IDCs) for server support, Web 3.0 introduces a paradigm shift through its distributed node architecture underpinned by blockchain technology (Suryavanshi et al., 2023). This innovative structural approach facilitates the deployment of code in closer proximity to the user endpoints, thereby enabling the provision of personalized security and privacy-preserving computational services (Kim et al., 2023). The decentralized nature of Web 3.0's architecture represents a fundamental departure from the centralized data processing models of its predecessor, offering enhanced data sovereignty and computational autonomy to end-users. This architectural transformation not only mitigates the risks associated with centralized data repositories but also engenders a more resilient and user-centric digital ecosystem. The shift towards edge computing in Web 3.0 further accentuates the potential for reduced latency, improved privacy, and more efficient resource utilization in the delivery of web services.

12.3.2 Core Principles and Technological Underpinnings

The initial conceptualization and implementation of Web 3.0 were predominantly predicated on smart contract platforms such as Ethereum and Solana, which are categorized as Layer 1 (L1) protocols. These foundational blockchains serve as the primary infrastructure for decentralized applications and transactions (Pierro & Tonelli, 2022). However, as the volume of data and transactions processed by L1 protocols began to exceed their inherent capacity limitations, a need for scalability solutions emerged. In response to this challenge, multiple Layer 2 (L2) scaling protocols were developed and deployed atop the Ethereum network. These L2 solutions are designed to alleviate the congestion and capacity constraints of the underlying L1 network by offloading a significant portion of transaction processing and computation (Sguanci et al., 2021). This hierarchical approach to blockchain architecture represents a critical evolution in the Web 3.0 ecosystem, enabling enhanced throughput, reduced latency, and improved overall network efficiency while maintaining the security guarantees of the base layer.

The proliferation of Layer 1 (L1) and Layer 2 (L2) networks has precipitated the necessity for interoperability solutions, colloquially referred to as bridges, to facilitate seamless value transfer across disparate blockchain ecosystems (Hafid et al., 2020). These cross-chain bridges function as high-throughput conduits, enabling users to transfer assets and information from one blockchain to another with enhanced efficiency and reduced friction (Zarick et al., 2021). The emergence of these interoperability protocols represents a critical development in the blockchain landscape, addressing the inherent limitations of isolated blockchain networks and fostering a more interconnected and fluid Web 3.0 ecosystem. By enabling cross-chain communication and value transfer, these bridges not only enhance the overall utility and liquidity of digital assets but also contribute to the realization of a truly interconnected decentralized internet.

The cloud computing platforms that have been largely monopolized by entities such as Amazon Web Services (AWS), Microsoft's Azure, and Google Cloud during the Web 2.0 era are poised to be supplanted by distributed data centers in the Web 3.0 paradigm (Saraswat & Tripathi, 2020). This transition represents a fundamental shift in the architecture of data storage and computation, moving from centralized to decentralized models. In the realm of Web 3.0 storage solutions, several projects have emerged as exemplars of this new approach, notably including Arweave, Filecoin, and Crust (Guidi et al., 2022). These initiatives leverage blockchain technology and distributed networks to create more resilient, censorship-resistant, and user-controlled data storage systems (Anand et al., 2023). This evolution in data infrastructure not only challenges the hegemony of established cloud service providers but also aligns with the core principles of Web 3.0, such as data sovereignty, enhanced privacy, and democratized access to computing resources.

In the domain of identity authentication, Decentralized Identifiers (DIDs) serve as the digital identification and passport for Web 3.0 users. DIDs can be conceptualized as a novel form of globally unique identifier, characterized by their versatility in application not only to individuals but also to a wide array of entities and objects (Alzahrani, 2020). This universality underscores the potential of DIDs to revolutionize identity management across diverse sectors. Within the Web3.0 ecosystem, prominent projects exemplifying the implementation of DIDs include LIT and ICX, among others (Petcu et al., 2023). Concurrently, the identity authentication sector encompasses domain name systems, notably represented by initiatives such as Ethereum Name Service (ENS) and Terra Name Service (TNS). These systems play a crucial role in enhancing user accessibility and interaction within the decentralized web environment (Jain et al., 2022).

The advent of Web 3.0 is transforming the landscape of internet ownership, establishing more egalitarian production dynamics, and enabling a more equilibrated allocation of value, thereby enhancing the cohesion of groups united by shared goals.

DAOs, as the most representative manifestation of Web 3.0, embody decentralized and self-governing characteristics while fully leveraging cutting-edge technologies such as smart contracts and artificial intelligence. Consequently, DAOs have emerged as a novel mechanism for inter-enterprise digital and intelligent collaboration, as well as supply chain management.

12.4 Characteristic Features of Dao Mechanisms

12.4.1 Decentralization, Autonomy, and Openness

Vitalik Buterin, Ethereum's co-founder, propounded the concept of Decentralized Autonomous Organizations (DAOs) in 2014. Buterin conceptualized DAOs as entities with automation and intelligence at their core, while human agency is consigned to the margins (Sims, 2019). The quintessential characteristics of DAOs comprise decentralization, autonomy, and horizontal openness (Dimitri, 2023). The pivotal operational elements of DAOs incorporate smart contracts, tokenized incentivization, and decentralized governance protocols.

In the context of DAOs, organizational governance is executed through autonomous computational processes where code supersedes legal structures as the primary guarantor of organizational norms (Faqir-Rhazoui et al., 2020). The decentralized operation of these processes is facilitated by blockchain technology, while the algorithmic rules impose constraints on the behavior of organizational constituents. The demarcation between participants and owners in DAOs becomes indistinct, as participants typically also hold tokens (Faqir-Rhazoui et al., 2020). Beyond compensation for project contributions, participants share in the economic benefits accruing from organizational growth, a convergence of interests that further reinforces organizational consensus.

The mechanism of DAOs is characterized by informational lucidity, with the preponderance of project codebases being publicly accessible, facilitating universal access to organizational intelligence (Sharma et al., 2023). This model diverges from conventional supply chain governance by dismantling informational asymmetries, thus catalyzing intra-organizational competition (Feichtinger et al., 2023). This mechanism enables adept individuals with community endorsement to expeditiously attain influential roles, stewarding operational initiatives and maximizing the efficacy of their expertise and social capital.

The mechanism of DAOs is predicated on principles of openness and flexibility, enabling participants to simultaneously contribute to multiple DAOs and disengage at their discretion (Chotkan et al., 2022). This architectural feature engenders dual advantages: primarily, it facilitates a more dynamic and frequent resource flux between DAOs compared to traditional corporate entities, engendering particularly profound supply chain information exchange, thereby catalyzing innovation, optimizing resource allocation, and propelling the proliferation of the Web 3.0 ecosystem (Siaw & Sarpong, 2021). Additionally, the frictionless ingress and egress of participants ensures the coalescence of individuals with congruent objectives, while allowing for the facile departure of those misaligned with organizational trajectories, thus accelerating the attainment of organizational consensus.

12.4.2 Tokenized Incentives and Smart Contracts for Supply Chain Collaboration

The architectural feature of DAOs eschews tightly coupled organizational structures in favor of real-time, loosely connected networks. This configuration directly empowers DAOs to optimally deploy their decentralized decision-making and governance protocols during exigencies or in scenarios where systemic failures in traditional oligopolistic entities create leadership lacunae (Sharma et al., 2023). Operationally, constituents holding a singular or predetermined quantum of native DAO tokens are empowered to propound smart contract amendments, initiatives, or investment theses and participate in their ratification (Zaczyk, 2023). The specific mechanisms exhibit heterogeneity across DAOs, thus catalyzing organizational evolution and expansion.

The genesis and dissemination of DAOs are intrinsically contingent upon the substrate of Web 3.0—a blockchain-centric, secure, and trustworthy internet ecology (Rathor et al., 2023). Simultaneously, DAOs constitute quintessential elements and pioneering organizational archetypes within the Web 3.0 ecosystem. The user-centric data governance model intrinsic to DAOs engenders robust protection of data privacy and security, thus facilitating the architecting and operationalization of Web 3.0's trusted network ecosystem (Bambacht & Pouwelse, 2022).

Within the novel economic paradigm and value creation modalities engendered by Web 3.0's blockchain-centric architecture, DAOs leverage smart contract-driven automatic value distribution mechanisms to provide critical instrumentalities for Web 3.0's autonomy, self-operation, and self-evolution (Bambacht & Pouwelse, 2022). Consequently, DAOs play a pivotal role in the construction and maintenance of an open, mutually beneficial Web 3.0 ecosystem. The establishment of the Web 3.0 ecosystem facilitates the convergence of data ownership rights with blockchain architecture and cross-chain integration, thereby providing novel trust mechanisms and solutions for enhancing digital supply chain resilience (Saberi et al., 2018). Moreover, it furnishes a comprehensive theoretical foundation and technological framework for the reconfiguration of the foundational network infrastructure underpinning Metaverse.

12.5 A Novel Conceptualization of the METAVERSE Based on Dao Mechanisms within the Web 3.0 Ecosystem

In light of the foregoing discussion, we hereby propose a novel conceptualization of a decentralized Metaverse predicated on supply chain resilience within the Web 3.0 ecosystem, with the DAO mechanisms at its core. Primarily, the construction of this Metaverse, operating under the Web 3.0 paradigm and DAO mechanisms, necessitates the reconfiguration of four fundamental Metaverse systems (see Fig. 12.1).

12.5.1 Reconfiguration of the Metaverse Systems

Meta-grid

The meta-grid constitutes the substratum upon which the metaverse operates, integrating a compendium of supportive technologies including communication, data storage, computational processing, and networking protocols. This infrastructure forms the fundamental underpinning of the metaverse system. In this context, advanced communication technologies, exemplified by 5G, enable the expeditious transmission of substantial data quantities, facilitating the low-latency transfer of voluminous information streams.

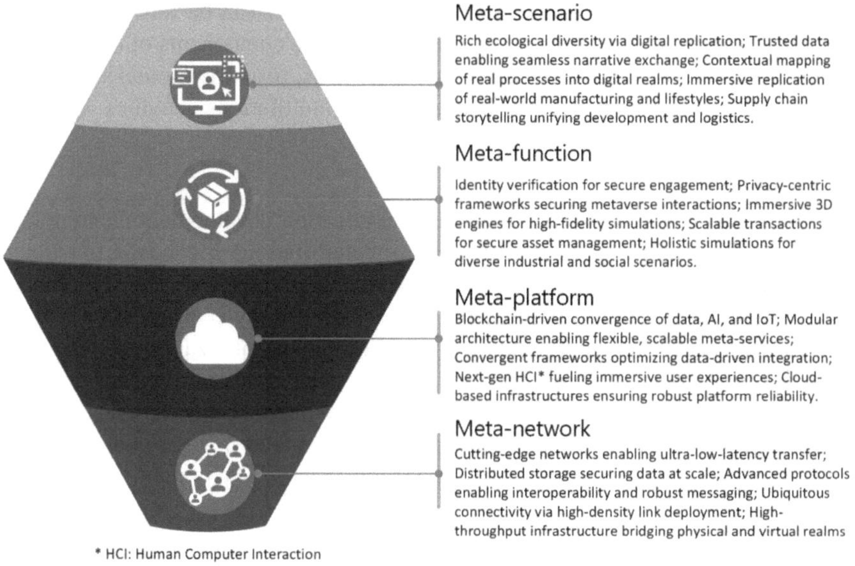

Fig. 12.1 The New Metaverse Systems

The implementation of distributed data storage architectures safeguards the confidentiality and integrity of voluminous data sets. Concomitantly, advanced networking protocols facilitate device integration and interoperability, while simultaneously enabling robust message transmission within the metaverse system. Leveraging cutting-edge communication network technologies, the meta-grid expedites the realization of ubiquitous connectivity characterized by high traffic and link densities, facilitating optimal interplay between physical and virtual realms. In the context of the metaverse, the quality of user experience is intrinsically linked to low-latency performance and immersive rendering capabilities.

Meta-platform

The meta-platform represents the technological substratum underpinning the multifarious functionalities of the metaverse. With blockchain technology as its nucleus, it synthesizes a constellation of advanced technologies, including big data analytics, cloud computing architectures, artificial intelligence paradigms, IoT frameworks, human-computer interaction modalities, and information security protocols. This amalgamation engenders system-level foundational technological competencies for meta-services, constituting a pivotal nexus in the realization of functional heterogeneity within the metaverse system.

The amalgamation of next-generation information technologies within the meta-platform surpasses simplistic aggregation, mandating the genesis of a data-centric convergent architectural framework to optimally harness data's latent value. The ubiquitous components—encompassing networking, storage, and computational elements—derived from these cutting-edge technologies can be subsumed into the meta-grid substrate. Simultaneously, the idiosyncratic components of diverse technologies can be modularized as autonomous services, fostering mutual support and synergistic collaboration, thereby buttressing the multifarious activities within the meta-function stratum.

Meta-function

Meta-services represent an aggregation of application development technological paradigm that facilitates the manifestation of multifarious Metaverse functionalities. This technological paradigm encompasses a spectrum of advanced methodologies, including identity verification protocols, privacy-centric computational frameworks, environmental rendering algorithms, incentivization architectures, content management systems, transaction processing mechanisms, asset administration platforms, 3D rendering engines, and simulation cases. These integrated technologies provide the requisite technological substrate for the construction of meta-narrative and the realization of application-specific functionalities.

The metaverse represents a colossal data corpus, encompassing a heterogeneous array of informational constituents, including but not limited to identity attributes, social network topologies, economic circumstances, industrial processes, and environmental topographies. The functional dynamics of the metaverse necessitate addressing a multitude of complex issues, such as biometric recognition, asset confidentiality, cybersecurity, and environmental visualization. Meta-function proffer comprehensive solutions to this diverse spectrum of challenges.

Meta-narratives

Meta-narrative constitutes the digital virtualization and replication of societal production and lifestyle paradigms, thereby engendering the multifaceted ecological diversity inherent in the metaverse. These narratives predominantly encompass the metaverse applications of manufacturing processes, entertainment modalities, social interaction frameworks, financial systems, and energy sector paradigms.

In the context of meta-narrative, digital fabrication methodologies enable the transposition of multifarious physical world paradigms into the digital-virtual domain. Supply chain-centric narrative, encompassing product development and design processes, manufacturing operations, sales strategies, and logistical frameworks, can be instantiated as meta-narrative within the metaverse system. The implementation of distributed ledger technologies facilitates trusted data interoperability and value transfer mechanisms across these heterogeneous meta-narratives.

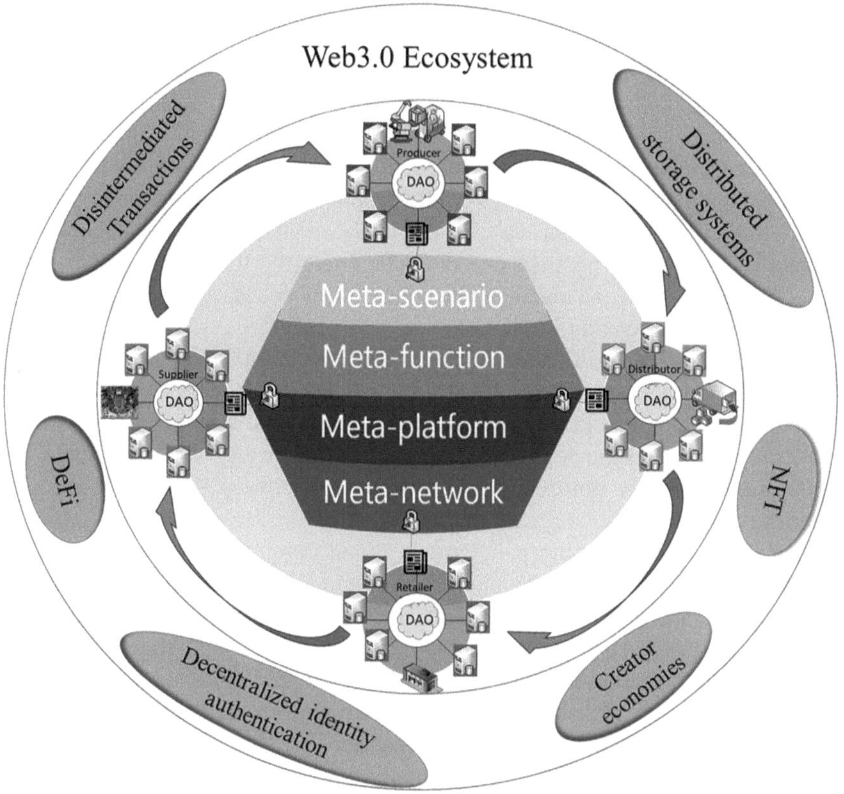

Fig. 12.2 A Framework of Metaverse-based on DAO mechanisms within the Web 3.0 ecosystem

12.5.2 DAO-Led Collaboration in Metaverse Supply Chains

Concurrent with the construction of these four metaverse systems, supply chain constituents—encompassing suppliers, manufacturers, distribution channel operators, and end consumers—collaboratively operate and maintain these metaverse systems under their respective DAO mechanisms (see Fig. 12.2). Furthermore, through the Web 3.0 ecosystem, these entities share characteristics such as digital wallets, decentralized identity authentication, privacy-preserving computation, distributed storage, disintermediated transactions, creator economies, Non-Fungible Tokens (NFTs), and Decentralized Finance (DeFi).

In the context of Web 3.0 and Metaverse ecosystems, digital wallets emerge as the quintessential point of ingress. The salient divergence between Web 3.0 and conventional financial architectures resides in the disintermediation of account-centric proxy paradigms, enabling direct user custody of assets. This transformative reconfiguration is predominantly actualized through the implementation of digital wallet technologies.

The implementation of distributed identity authentication systems engenders the codification and denotation of digital assets. Blockchain-centric distributed storage architectures maximize the efficiency of dormant storage resource utilization, mitigating energy expenditure in cloud-based computational processes for voluminous data. The immutable characteristics of blockchain's on-chain information substantiate digital content provenance. Concomitantly, consensus protocols, interchain operability, and smart contract execution facilitate the transference and transactional processes of digital assets, while concurrently preserving the rights and entitlements of content creators.

The utilization of DeFi features enables the invocation and aggregation of basal protocols underpinning diverse applications. Concurrently, synthetic asset platforms facilitate the blockchain representation of tangible assets. This convergence effectively attenuates the demarcation between on-chain and off-chain frameworks, while simultaneously obfuscating the distinction between virtual and corporeal realities.

In contrast to the leasing paradigm of cloud computing, Web 3.0 ecosystem introduces a business model predicated on execution and access frequency-based remuneration, facilitating extensive privacy-preserving computational processes. Within metaverse systems, voluminous corporate assets and transaction-related confidential information are susceptible to unauthorized disclosure, misappropriation, or exploitation. Privacy-preserving computation technologies enable the execution of calculations and analyses on obfuscated data, thereby ensuring the secure utilization of Metaverse-generated information.

Decentralized Exchange (DEX) protocols facilitate transactional processes for Web 3.0 users via Decentralized Applications (DApps), enabling autonomous management of private keys and digital assets in peer-to-peer transactions. In juxtaposition to Centralized Exchange (CEX) paradigms, DEX architectures demonstrate superior security measures, augmented privacy safeguards, and comprehensive user-centric asset governance.

Non-Fungible Tokens (NFTs), a quintessential expression of the Web 3.0 paradigm, facilitate the individualized representation of digital intellectual property rights, smart contractual agreements, and virtual real estate, thereby providing a robust framework for addressing the multifaceted demands of a diverse user base. NFTs constitute unique, blockchain-recorded data constructs that are inherently non-replicable, non-interchangeable, and indivisible, functioning as authenticators of veracity or exclusivity for specific digital assets.

The creator economy introduces the concept of digital scarcity, reestablishing creators' market agency in terms of pricing and transactional rights. This model reconceptualizes creative endeavors as investment activities rather than purely altruistic pursuits, while simultaneously framing audience engagement as consumptive behavior. Moreover, the creator economy facilitates the implementation of innovative, programmable economic models, engendering wealth redistribution across the creative atmosphere. Through the establishment of novel value-creation pathways, this economic framework empowers creators to not only retain ownership of their intellectual output but also to acquire stakeholder status within their creative platforms.

12.6 Case Study

12.6.1 Metaverse Initiatives in Japanese Automotive Manufacturing

A prominent entity in vehicular manufacturing, designated as Corporation A, is expediting its DX through Metaverse. This strategic reorientation is manifested through the submission of quaternary intellectual property safeguards related to Web 3.0 protocols to the United States Patent and Trademark Office (USPTO). Such actions signify a substantial shift towards digital economic paradigms and virtual consumer engagement modalities. The intellectual property protections, encompassing multiple proprietary designations, indicate a prospective transformation in automotive marketing methodologies. This development represents a comprehensive integration of VR, NFTs, and blockchain systems. Corporation A's strategic vision incorporates digital representations of sartorial items, automotive products, and miscellaneous commodities, with intentions to establish virtual commercial venues and exchange platforms for blockchain-based tokens (see Fig. 12.3).

Corporation A distinguishes itself as one of the pioneering entities to incorporate blockchain technologies, exemplified by NFTs, into the automotive supply chain. This innovative approach seeks to compensate users for their data contributions via data equity mechanisms, thereby effectuating the legitimate attribution of user data value. Vehicle proprietors can amass NFTs through dual methodologies: cultivation-centric mining or mileage-based mining, contingent upon the sanctioned utilization of "vehicular and driving behavioral data". Upon attaining a predetermined threshold of NFTs, proprietors gain the capacity to exchange these tokens for either hardware or software functional enhancements for their vehicles.

Corporation A's venture into Metaverse demonstrates its strategic anticipation of shifting consumer paradigms. A salient exemplification is the "A FIRM HYPE LAB", a digital commercial space inaugurated in March 2023. This platform facilitates a comprehensive vehicular acquisition process, encompassing product selection through contractual finalization, entirely within the confines of Metaverse. The aforementioned digital platform, characterized by perpetual accessibility, enables users to traverse the virtual environment utilizing personalized digital representations. Participants can engage in discourse with simulated customer service entities and interact with 3D renderings of prospective Corporation A automotive products.

The process encompasses initial product evaluation, consultations, virtual test drives, and culminates in the execution of purchase agreements for corporation A vehicles. The A FIRM HYPE LAB, which is now entering its experimental phase,

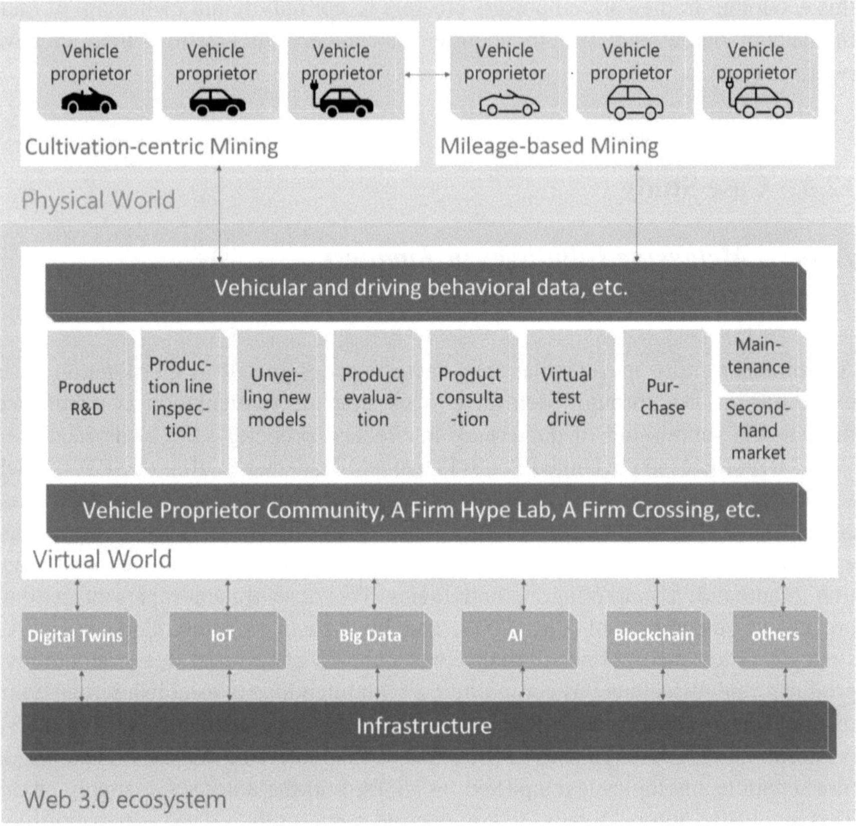

Fig. 12.3 Metaverse applications in corporation A

represents an innovative paradigm wherein consumers can engage in the entire automotive acquisition process, from initial product evaluation to contractual finalization, within Metaverse. This collaborative initiative between corporation A and its affiliated distributors leverages the accumulated expertise in digital technologies and Metaverse to establish a novel platform. This construct aims to synthesize physical and virtual realities, thereby offering consumers an unprecedented experiential modality. Concurrently, this venture serves as a proving ground for potential alternative methodologies in vehicular retail distribution.

Corporation A is pioneering innovative communication and methodological approaches throughout various stages of vehicular development and unveiling processes, leveraging emergent platforms and software technologies. Noteworthy implementations include the utilization of VR in automotive design processes, the application of MR to enhance proficiency in production line inspection protocols, and the deployment of a virtual exhibition space, designated "A FIRM CROSSING", within the metaverse for the purpose of unveiling new vehicle models. These initiatives exemplify the corporation's strategic integration of advanced technological paradigms across its operational spectrum.

The implementation of this immersive experiential paradigm, grounded in automotive motifs, facilitates Corporation A's establishment of a social network infrastructure predicated on the vehicle proprietor community. Through the utilization of NFTs as virtual tokens within this construct, it articulates the embryonic configuration of a Metaverse economy situated within the Web 3.0 ecosystem. This innovation constitutes a pivotal advancement in the digital metamorphosis of the automotive supply chain.

In its prospective operational paradigm, Corporation A will leverage the Web 3.0 ecosystem as its fundamental substrate to instantiate virtual contractual nexuses among its corporate entity, upstream and downstream supply chain conduits, suppliers, distribution sector, and end customers. This strategic initiative is designed to attenuate trust-based impediments among diverse supply chain stakeholders, consequently engendering a more efficacious and resilient autonomous operational framework for the supply chain.

12.6.2 Innovative Approaches in Japan's Printing Supply Chain

Corporation B, a preeminent entity in Japan's printing industry, is dedicating substantial resources to the development of a digital printing supply chain, while concurrently expanding its operations in adjacent domains such as Metaverse and blockchain technologies. The corporation has long-standing expertise in technologies related to Metaverse, with particular emphasis on expressive techniques and security protocols. Notably, expressive technologies constitute a core competency within the printing sector, and Corporation B possesses a significant portfolio of proprietary technologies and methodologies in this domain. For instance, the corporation's advanced rendering techniques, capable of generating high-fidelity 3D representations from computer-aided design (CAD) data, find application in the realm of Metaverse (see Fig. 12.4).

While Corporation B's core operations center on print-based communication modalities, the emergence of Metaverse represents an indispensable paradigm in the evolution of communicative mediums, succeeding traditional print and Internet-based platforms. The corporation's foray into this domain is exemplified by "METAPA", a service analogous to a department store within a Metaverse. This initiative encompasses two primary facets: providing operational support for corporate entities establishing virtual retail presences within Metaverse, and concurrently offering a mobile application that facilitates consumer engagement in Metaverse.

Additionally, the corporation offers "MiraVerse", a platform facilitating the construction of Metaverse for enterprise applications. This system enables organizations to design, implement, and manage metaverse spaces for diverse purposes, including promotional activities, cultural heritage simulations, architectural visualization, and urban planning. By publishing MiraVerse-generated environments on

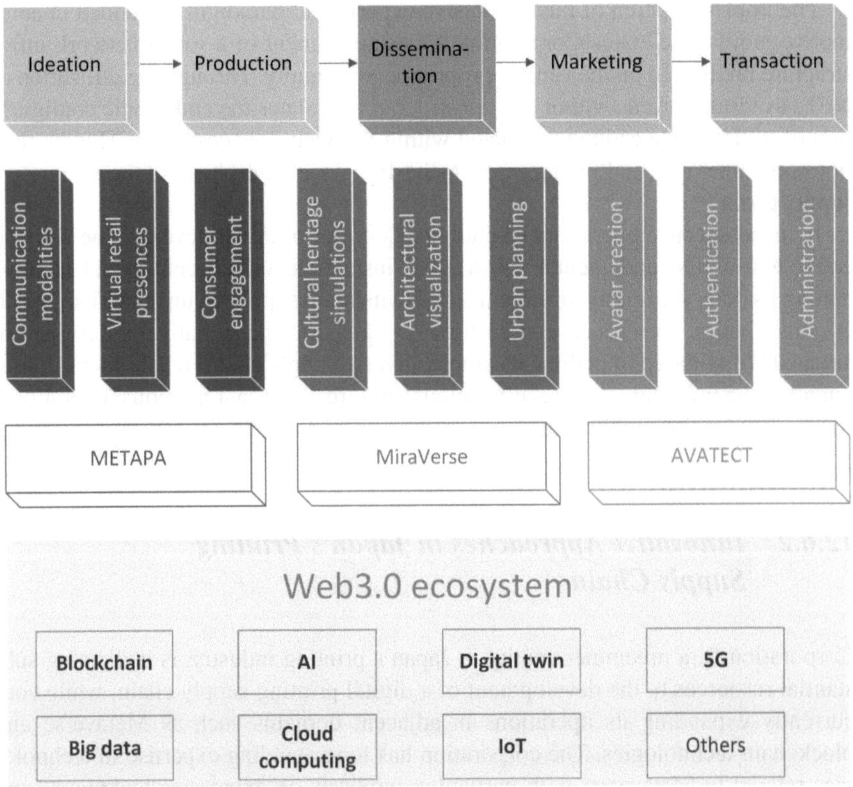

Fig. 12.4 Metaverse and blockchain applications in Corporation B

the World Wide Web, enterprises can efficiently disseminate accurate information to consumers, thereby substantially influencing purchasing decisions. Moreover, the platform's architecture allows for decentralized, autonomous management of 3D product data by individual entities. This feature contributes to a reduction in update-related expenses and diminished lead times during product modifications. Furthermore, it enhances privacy protection measures and facilitates rapid response and recovery across the entire supply chain in exigent circumstances.

Furthermore, the corporation has developed two additional services: "Metaclone Avatar", an automated avatar generation system, and "AVATECT", a comprehensive avatar management platform. Metaclone Avatar employs advanced AI algorithms to facilitate the creation of highly accurate digital representations of individuals from a single photographic input. AVATECT provides an integrated platform for avatar creation, authentication, and administration. This platform not only enables the storage of avatar data in conjunction with associated metadata but also incorporates features such as digital watermarking to ensure the authenticity and integrity of these digital entities.

Within the context of the digital business epoch, data has ascended to become the paramount productive element for corporate entities. As a corollary, the evolution of supply chain management systems in the digital publishing sector will inevitably gravitate towards architectures that prioritize the protection of data privacy, the preservation of data sovereignty, and the fortification of data security.

The burgeoning development of digital technologies, including AI, IoT, Cloud Computing, Digital Twins, and 5G, has precipitated the interconnection of a burgeoning array of sensors, localization systems, social media platforms, and Metaverse-related apparatuses. This interconnectedness is engendering a substantial conflation of the corporeal and virtual domains in the digital economic epoch. Of particular significance is the metamorphosis in the modality of productive resources, with virtual assets ascending to pre-eminence. User profiles, credit scores, intellectual property rights, user engagement, critiques, experiential data, and even digital identities now constitute valuable virtual data assets.

These assets, which defied quantification or valuation in traditional and contemporary economic frameworks, can now be concretized as wealth through the tokenization process within the Web 3.0 ecosystem. Blockchain-centric technologies facilitate the authentication and validation of ownership claims. Specifically, blockchain infrastructures, through consensus protocols and smart contracts, can dispense token-based incentives to various stakeholders in the digital publishing industry's supply chain, proportionate to their contributions. This incentivization framework ensures synergistic success among supply chain participants and augments product generation, propagation, and monetization.

Corporation B has effectively implemented a distributed ledger technology solution accelerator, developed by a blockchain enterprise, to enhance the security of healthcare-related data across IoT and cloud computing infrastructures. This initiative represents a significant advancement in the practical application of blockchain technology to enterprise-level use cases, with particular emphasis on the protection of individual privacy. Corporation B, in collaboration with the blockchain technology firm, has validated the efficacy of an experimental Proof of Concept (PoC). This PoC demonstrates the utilization of blockchain protocols for device authentication processes and the secure transmission of medical information.

Corporation B and the blockchain enterprise implemented a distributed ledger technology solution accelerator to deploy an application facilitating data exchange between IoT devices and cloud infrastructure. The corporation conducted PoC in conjunction with its manufacturing facility, focusing on the transmission of biometric data indicative of emotional stress from wearable devices through an IoT gateway to a cloud-based system. The blockchain enterprise's proprietary data security protocols enabled corporation B to achieve enhanced levels of security in healthcare-related applications.

Company B, as a constituent entity within the digital publishing industry's supply chains, is formulating a supply chain strategy that incorporates Web 3.0 as the underlying architecture for the next-generation digital publishing network ecosystem. This strategic initiative entails the establishment of a decentralized database system, which will serve as the foundation for the development of internet infrastructure. Conventional databases and ancillary software applications will be deployed on this operational framework. The operating system will assume responsibility for transaction data monitoring, regulatory adherence verification, and information integrity assurance. Transactional data will be immutably recorded on the blockchain, with kernel-level execution limited to account and transaction-specific processes.

Within the context of the Web 3.0 paradigm, the publishing industry's supply chain—encompassing ideation, production, dissemination, marketing, and transactional processes—integrates token-centric reward systems. These systems, founded upon predefined consensus mechanisms and smart contracts, are engineered to catalyze the evolution of the digital publishing sector. This framework has the potential to galvanize the proactive engagement and passion of all supply chain stakeholders in the advancement of the digital publishing industry, consequently engendering an innovative digital publishing supply chain archetype rooted in the tenets of mutual prosperity, transparency, synergy, and ecologically sound progression.

12.6.3 Achieving Supply Chain Resilience

Corporations A and B offer compelling illustrations of how immersive digital platforms, blockchain-based data governance, and decentralized decision-making structures can collectively strengthen supply chain resilience. Their recent initiatives integrate Web 3.0 technologies, virtual reality interfaces, and automated coordination mechanisms in ways that align with established digital transformation trends, suggesting that key performance metrics, such as lead times, inventory coordination, and adaptability to disruptions, can experience significant gains compared to conventional processes (METI, 2022; Dwivedi et al., 2022).

In the automotive context, Corporation A has directed substantial resources toward developing an immersive product-design ecosystem, enabling concurrent engineering and smoother interactions among geographically distributed suppliers (Balance, 2023). VR-driven collaborative platforms can significantly shorten design review cycles, translating to a substantial reduction in overall time-to-market (Lawson et al., 2016). This advantage stems from real-time data exchange and automated contract execution, which together mitigate communication lags and eliminate redundant validations. These observations align with the broader theoretical premise that transparency, immutability, and decentralized oversight foster higher trust among parties and expedite decision-making, particularly in extended global supply chains (Ivanov, 2021).

Corporation B's Metaverse-driven distribution strategy likewise underscores the benefits of linking immersive collaboration with blockchain-centric data management. Cross-functional teams overseeing production, logistics, and inventory have reported tighter synchronization and reduced delays, demonstrating noticeable improvements improvement in process efficiency based on comparative assessments from the digital publishing. The integration of decentralized data nodes further supports dynamic re-routing of orders during unexpected fluctuations in demand, consistent with prior findings that real-time visibility and distributed governance can alleviate the bullwhip effect (Choi et al., 2019). These advancements not only optimize routine operations but also enhance the robustness of crisis-response protocols by eliminating single points of failure.

Corporation A and B have explored automated governance through DAO-like frameworks, which allow consensus-based rule enforcement and resource allocation. Such mechanisms notably reduce dependence on top-down directives and foster resilience in scenarios where human intervention becomes infeasible or delayed. In line with earlier research, participants in these pilot implementations have highlighted the seamless coordination and transparency afforded by on-chain logs and smart contracts (Lehner et al., 2023; Sharma et al., 2024). This decentralized model effectively streamlines the negotiation and adjustment of supply commitments, expediting critical decisions in response to supply chain shocks.

In summary, the initiatives at Corporation A and B exemplify the broader potential of Web 3.0–enabled Metaverse environments to enhance visibility, trust, and agility in supply chain management. Quantitative and qualitative insights from related pilot studies in manufacturing and publishing suggest that immersive collaboration, real-time data synchronization, and automated governance collectively contribute to meaningful efficiency gains and improved operational continuity, particularly under conditions of uncertainty and disruption. This holistic perspective supports the assertion that Metaverse- and DAO-centric architectures are well poised to advance supply chain resilience across diverse industries in the evolving digital landscape.

12.7 Discussion

Examination of the preceding case studies elucidates that, transcending the domains of social interaction and gaming, the automotive industry, printing and myriad other sectors possess the potential to evolve towards a Metaverse economy fortified by the Web 3.0 ecosystem. Exemplifying this, Corporation A's after-sales service utilizes AR/VR technologies to furnish virtual remote assistance for vehicle proprietors precluded from physical attendance, thereby amplifying the efficacy of incident analysis (Aslam et al., 2023). Alternatively, through the deployment of MR and virtual scenario development, immersive virtual reality retail environments can be engineered. These can proffer consumers simulated in-store consultations, test drives, or interior modification experiences via AI and human digital projection technologies. This paradigm not only attenuates the financial burden associated with establishing and operating physical retail spaces but also obviates consumer commute time, thus further enhancing transactional efficiency.

Regarding the enhancement of supply chain resilience, the augmentation provided by the Web 3.0 ecosystem enables the automotive Metaverse to liberate itself from the spatial, temporal, and distance-related constraints of the physical realm.

Furthermore, in the face of unforeseen events or exceptional circumstances, it facilitates the expeditious transition of the entire supply chain to a fully Metaverse-centric modality. Additionally, within a decentralized framework, it ensures the security and uninterrupted operation of the comprehensive supply chain system.

At present, Metaverse's evolution resides in its incipient phase, requiring extensive refinement and advancement in domains including computational capacity, rights attestation, and XR technologies (Bhattacharya et al., 2023). Within the automotive sector, while manifold technical impediments in the Web 3.0 sphere may decelerate integration into Metaverse economy, the burgeoning ingress of nascent enterprises into the automotive industry through cross-sectoral synthesis is progressively metamorphosing automotive products into next-generation interactive terminals (Aslam et al., 2023). Predicated on intelligent evolution, these products persistently innovate, engendering novel business paradigms that synergize Metaverse concepts with automotive attributes. The metaverse concept's significant referential value for the automotive industry's future trajectory is manifest, both in terms of qualitative enhancement and efficiency augmentation in supply chain management, and in inspiring unprecedented business models.

On the other hand, the advent of blockchain technology has engendered a proliferation of distributed public ledgers and databases. The ubiquitous dissemination and redundant storage of pertinent data across the network ensures data parity among all decentralized nodes, thereby disrupting the erstwhile data monopolization by dominant platforms and mitigating impediments to data sharing (Haque & Rahman, 2020). Simultaneously, nodes within the blockchain ecosystem employ asymmetric cryptography, rigorous protocols, and consensus algorithms to maintain data integrity. The immutability of validated information confers attributes of authenticity, transparency, security, and provenance (Roy et al., 2019). This blockchain-mediated trust paradigm supplants individual, organizational, and institutional trust with machine-derived credibility. The resultant disintermediated transactional ecosystem facilitates abbreviated transactional pathways, more immediate exchanges, diminished transaction costs, and augmented operational efficiency within the digital publishing sector.

With specific reference to augmenting supply chain resilience, within the Web 3.0 ecosystem epitomized by blockchain technology, stakeholders in the publishing industry supply chain can attain reciprocal trust while facilitating real-time information dissemination via Metaverse, thereby precluding the circulation of erroneous data within the supply chain. This engenders a perfect synchronicity between virtual and tangible supply chains, enhancing the capability for pre-emptive risk identification. Furthermore, it empowers the entire supply chain to swiftly and efficaciously reallocate resources in the face of unforeseen circumstances, thus ensuring operational continuity.

12.8 Conclusions

Metaverse environments, transcending traditional physical constraints, enable interactions in a new dimension unfettered by space, time, or distance. When integrated with the Web 3.0 ecosystem, Metaverse is characterized by a system architecture predicated on blockchain technology, emphasizing decentralization, self-governance, and intelligent systems. This paradigm shift facilitates real-time information sharing and collaboration among supply chain participants, fostering a high-trust environment.

Notably, this system possesses the capability to expeditiously transition the entire supply chain to a fully Metaverse-centric modality during exigent circumstances or crises. This feature ensures efficient resource reallocation and maintains the security and continuous operation of the comprehensive supply chain system. Furthermore, it engenders perfect synchronicity between virtual and tangible supply chains, significantly enhancing capabilities for pre-emptive risk identification.

The incorporation of DAO mechanisms cultivates mutual trust among participants and precludes the circulation of erroneous data within the supply chain. This elevation in the quality and reliability of information within the supply chain contributes to optimized decision-making processes. Consequently, it enables swift and efficacious responses to unforeseen circumstances, reinforcing operational continuity.

The implications of this research are multifaceted. Firstly, it suggests the feasibility of a truly decentralized supply chain model that does not rely on traditional centralized platforms. Secondly, it presents a new direction for supply chain management that extends beyond immersive experiences through Metaverse, offering practical and efficient solutions. Thirdly, it demonstrates the potential for dramatically improving supply chain resilience against global crises and technological failures.

In conclusion, the construction of Metaverse environments based on Web 3.0 ecosystems and DAO mechanisms shows significant potential in enhancing supply chain resilience. Future research should address the technical challenges of implementing this concept, as well as legal and ethical considerations. Additionally, further investigation into its applicability across different industry sectors and analysis of long-term economic impacts are warranted.

References

Abbate, S., Centobelli, P., Cerchione, R., Oropallo, E., & Riccio, E. (2022). A first bibliometric literature review on Metaverse, *2022 IEEE technology and engineering management conference* (TEMSCON EUROPE), pp. 254-260.

Ali, N., Ahmed, A., Anum, L., Ghazal, T., Abbas, S., Khan, M., Alzoubi, H., & Ahmad, M. (2021). Modelling supply chain information collaboration empowered with machine learning technique. *Intelligent Automation & Soft Computing, 29*, 243–257. https://doi.org/10.32604/iasc.2021.018983

Alnajim, A., Salehi, S., Shen, C., & Smith, M. (2023). *Traffic characteristics of extended reality.* ArXiv, abs/2304.07908. https://doi.org/10.48550/arXiv.2304.07908.

Aloqaily, M., Bouachir, O., Karray, F., Ridhawi, I., & Saddik, A. (2022). Integrating digital twin and advanced intelligent technologies to realize the Metaverse. *IEEE Consumer Electronics Magazine, 12*, 47–55.

Alpala, L., Quiroga-Parra, D., Torres, J., & Peluffo-Ordóñez, D. (2022). Smart factory using virtual reality and online multi-user: Towards a Metaverse for experimental frameworks. *Applied Sciences., 12*. https://doi.org/10.3390/app12126258

Alzahrani, B. (2020). An information-centric networking based registry for decentralized identifiers and verifiable credentials. *IEEE Access, 8*, 137198–137208.

Anand, M., Mithun, S., Shree, L., & Ranjith, M. (2023). Survey on connecting to the decentralized storage using IPFS protocol with web 3 technology. *2023 international conference for advancement in technology* (ICONAT), pp. 1-4.

Asif, R., & Hassan, S. (2023). Exploring the confluence of IoT and Metaverse: Future opportunities and challenges. *IoT., 4*. https://doi.org/10.3390/iot4030018

Aslam, A., Chaudhary, R., Bhardwaj, A., Budhiraja, I., Kumar, N., & Zeadally, S. (2023). Metaverse for 6G and beyond: The next revolution and deployment challenges. *IEEE Internet of Things Magazine, 6*, 32–39.

Bag, S., Rahman, M., Srivastava, G., & Shrivastav, S. (2023). Unveiling metaverse potential in supply chain management and overcoming implementation challenges: An empirical study. *Benchmarking: An International Journal., 32*. https://doi.org/10.1108/bij-05-2023-0314

Balakrishnan, S., Hameed, M., Venkatesan, K., & Aswin, G. (2021). Interaction of spatial computing in augmented reality. *2021 7th international conference on advanced computing and communication systems* (ICACCS), 1, pp. 1900-1904.

Balance. (2023). *Case studies of 3d utilization on automotive industry websites: What is rich design for branding?* (in Japanese), BALANCe Magazine. https://balance.bz/magazine/category/web/3d-automobile-website/

Bambacht, J., & Pouwelse, J. (2022). Web3: A decentralized societal infrastructure for identity, trust, money, and data. ArXiv, abs/2203.00398. https://doi.org/10.48550/arXiv.2203.00398.

Bhattacharya, P., Saraswat, D., Savaliya, D., Sanghavi, S., Verma, A., Sakariya, V., Tanwar, S., Sharma, R., Răboacă, M., & Manea, D. (2023). Towards future internet: The Metaverse perspective for diverse industrial applications. *Mathematics., 11*. https://doi.org/10.3390/math11040941

Boulos, M., & Wheeler, S. (2007). The emerging web 2.0 social software: An enabling suite of sociable technologies in health and health care education. *Health Information and Libraries Journal, 24 1*, 2–23.

Cheng, R., Wu, N., Chen, S., & Han, B. (2022). Will Metaverse be NextG internet? Vision, hype, and reality. *IEEE Network, 36*, 197–204.

Choi, T. M., Wen, X., Sun, X., & Chung, S. H. (2019). The mean-variance approach for global supply chain risk analysis with air logistics in the blockchain technology era. *Transportation Research Part E Logistics and Transportation Review., 127*, 178–191. https://doi.org/10.1016/j.tre.2019.05.007

Chotkan, R., Decouchant, J., & Pouwelse, J. (2022). Unstoppable DAOs for web3 disruption. *Proceedings of the 3rd international workshop on distributed infrastructure for the common good*. https://doi.org/10.1145/3565383.3566112.

Cui, Y., Idota, H., & Ota, M. (2019). Improving supply chain resilience with implementation of new system architecture. *2019 IEEE social implications of technology (SIT) and information management* (SITIM), pp. 1-6. https://doi.org/10.1109/SITIM.2019.8910226.

Dimitri, N. (2023). *"Voting in DAOs", Distributed ledger technologies: Research and practice*. https://doi.org/10.1145/3624574.

Dolgui, A., & Ivanov, D. (2023). Metaverse supply chain and operations management. *International Journal of Production Research, 61*, 8179–8191.

Dwivedi, Y. K., Hughes, L., Baabdullah, A. M., Ribeiro-Navarrete, S., Giannakis, M., Al-Debei, M. M., Dennehy, D., Metri, B., Buhalis, D., Cheung, C. M. K., Conboy, K., Doyle, R., Dubey, R., Dutot, V., Felix, R., Goyal, D. P., Gustafsson, A., Hinsch, C., Jebabli, I., & Wamba, S. F. (2022). Metaverse beyond the hype: Multidisciplinary perspectives on emerging challenges, opportunities, and agenda for research, practice and policy. *International Journal of Information Management, 66*, 1–55. https://doi.org/10.1016/j.ijinfomgt.2022.102542

Fadzli, F., Kamson, M., Ismail, A., & Aladin, M. (2020). 3D telepresence for remote collaboration in extended reality (xR) application. *IOP Conference Series: Materials Science and Engineering, 979*. https://doi.org/10.1088/1757-899X/979/1/012005

Faqir-Rhazoui, Y., Arroyo, J., & Hassan, S. (2020). An overview of decentralized autonomous organizations on the blockchain. *Proceedings of the 16th international symposium on open collaboration*. https://doi.org/10.1145/3412569.3412579.

Feichtinger, R., Fritsch, R., Vonlanthen, Y., & Wattenhofer, R. (2023). *The hidden shortcomings of (D)AOs–an empirical study of on-chain governance*. ArXiv, abs/2302.12125. https://doi.org/10.48550/arXiv.2302.12125.

French, A., Shim, J., Risius, M., Larsen, K., & Jain, H. (2021). The 4th industrial revolution powered by the integration of AI, Blockchain, and 5G. *Communications of the Association for Information Systems, 49*, 6–286.

Fuchs, C. (2011). New media, web 2.0 and surveillance. *Sociology Compass, 5*, 134–147.

Gent, E. (2022). Lessons from a second life > before meta, Philip Rosedale created an online universe. *IEEE Spectrum, 59*, 19.

Golan, M., Jernegan, L., & Linkov, I. (2020). Trends and applications of resilience analytics in supply chain modeling: Systematic literature review in the context of the COVID-19 pandemic. *Environment Systems & Decisions, 40*, 222–243. https://doi.org/10.1007/s10669-020-09777-w

Guan, J., Morris, A., & Irizawa, J. (2023). Extending the Metaverse: Hyper-connected smart environments with mixed reality and the internet of things. *2023 IEEE conference on virtual reality and 3D user interfaces abstracts and workshops* (VRW), pp. 817-818.

Guidi, B., Michienzi, A., & Ricci, L. (2022). Evaluating the decentralisation of filecoin. *Proceedings of the 3rd international workshop on distributed infrastructure for the common good*. https://doi.org/10.1145/3565383.3566108.

Hafid, A., Hafid, A., & Samih, M. (2020). Scaling Blockchains: A comprehensive survey. *IEEE Access, 8*, 125244–125262.

Hahn, G. (2020). Industry 4.0: A supply chain innovation perspective. *International Journal of Production Research, 58*, 1425–1441.

Hamidouche, W., Bariah, L., & Debbah, M. (2023). *Immersive media and massive twinning: Advancing towards the Metaverse*. ArXiv, abs/2307.01522. https://doi.org/10.48550/arXiv.2307.01522.

Haque, A., & Rahman, M. (2020). *Blockchain technology: Methodology, application and security issues. ArXiv*, abs/2012.13366.

Huynh, D., Khosravirad, S., Masaracchia, A., Dobre, O., & Duong, T. (2022). Edge intelligence-based ultra-reliable and low-latency Communications for Digital Twin-Enabled Metaverse. *IEEE Wireless Communications Letters, 11*, 1733–1737.

Ivanov, D. (2021). Supply chain viability and the COVID-19 pandemic: A conceptual and formal generalisation of four major adaptation strategies. *International Journal of Production Research, 59*(12), 3535–3552. https://doi.org/10.1080/00207543.2021.1890852

Jain, S., Erichsen, L., & Weyl, G. (2022). *A plural decentralized identity frontier: Abstraction v. Composability Tradeoffs in Web3. ArXiv,* abs/2208.11443. https://doi.org/10.48550/arXiv.2208.11443.

Jha, M., Yogeshwari, A., Rubini, P., & Singh, M. (2023). Converge of IoT and AI in Metaverse: Challenges and opportunities. *2023 5th international conference on inventive research in computing applications* (ICIRCA), pp. 1462-1467.

Kim, J. (2021). Advertising in the Metaverse: Research agenda. *Journal of Interactive Advertising, 21*, 141–144.

Kim, R., Song, H., & Park, G. (2023). Moving real-time services to web 3.0: Challenges and opportunities. *IEEE Transactions on Services Computing, 16*, 4041–4059.

Kouhizadeh, M., Saberi, S., & Sarkis, J. (2021). Blockchain technology and the sustainable supply chain: Theoretically exploring adoption barriers. *International Journal of Production Economics., 231*. https://doi.org/10.1016/j.ijpe.2020.107831

Krebs on Security. (2024). *Global Microsoft meltdown tied to Bad Crowdstrike update, University of Maryland Global Campus*. Accessed Aug 10, 2024, from https://krebsonsecurity.com/2024/07/global-microsoft-meltdown-tied-to-bad-crowstrike-update/

Lawson, G., Salanitri, D., & Waterfield, B. (2016). *The future of virtual reality in the automotive industry*. https://www.sciencedirect.com/science/article/abs/pii/S0003687015300260.

Lee, L., Braud, T., Zhou, P., Wang, L., Xu, D., Lin, Z., Kumar, A., Bermejo, C., & Hui, P. (2021). *All one needs to know about Metaverse: A complete survey on technological singularity*. Virtual Ecosystem, and Research Agenda. *ArXiv*, abs/2110.05352. https://doi.org/10.13140/RG.2.2.11200.05124/1.

Lehner, R., Misyura, I., Schneider, B., & Asprion, P. (2023). *Potentials of decentralized autonomous organizations: A conceptual model to address agency conflicts in traditional corporations*. https://ceur-ws.org/Vol-3645/bes4.pdf.

Masip-Bruin, X., Marín-Tordera, E., Ruiz, J., Jukan, A., Trakadas, P., Cernivec, A., Lioy, A., López, D., Santos, H., Gonos, A., Silva, A., Soriano, J., & Kalogiannis, G. (2021). Cybersecurity in ICT supply chains: Key challenges and a relevant architecture. *Sensors (Basel, Switzerland), 21*. https://doi.org/10.3390/s21186057

METI. (2022). *Approach to improving Web3.0 business environment, Ministry of Economy, Trade and Industry (METI)*. https://www.meti.go.jp/policy/economy/keiei_innovation/sangyokinyu/Web3/web3.pdf

Murugesan, S. (2007). Understanding Web 2.0. *IT Professional, 9*. https://doi.org/10.1109/MITP.2007.78

Okwu, M., Tartibu, L., Maware, C., Enarevba, D., Afenogho, J., & Essien, A. (2022). Emerging Technologies of Industry 4.0: Challenges and Opportunities. *2022 International conference on artificial intelligence, big data, computing and data communication systems* (icABCD), pp. 1–13.

Petcu, A., Pahontu, B., Frunzete, M., & Stoichescu, D. (2023). A secure and decentralized authentication mechanism based on web 3.0 and Ethereum Blockchain technology. *Applied Sciences., 13*. https://doi.org/10.3390/app13042231

Pierro, G., & Tonelli, R. (2022). Can Solana be the solution to the Blockchain scalability problem?. *2022 IEEE international conference on software analysis, evolution and reengineering* (SANER), pp. 1219-1226.

Queiroz, M., Wamba, S., Pereira, S., & Jabbour, C. (2023). The metaverse as a breakthrough for operations and supply chain management: Implications and call for action. *International Journal of Operations & Production Management*. https://doi.org/10.1108/ijopm-01-2023-0006

Ramanathan, U. (2014). Performance of supply chain collaboration–a simulation study. *Expert Systems With Applications, 41*, 210–220.

Rathor, S., Zhang, M., & Im, T. (2023). Web 3.0 and sustainability challenges and research opportunities. *Sustainability., 15*. https://doi.org/10.3390/su152015126

Roy, S., Ganguli, R., & Goswami, S. (2019). Transformation of supply chain provenance using Blockchain—A short review. *Adv. Intell. Syst. Comput.* https://doi.org/10.1007/978-981-13-7403-6_51

Saberi, S., Kouhizadeh, M., Sarkis, J., & Shen, L. (2018). Blockchain technology and its relationships to sustainable supply chain management. *International Journal of Production Research, 57*, 2117–2135.

Salvetti, F., Gardner, R., Minehart, R., & Bert agni, B. (2023). Intelligent avatars, holographic tools, digitized objects: An extended reality simulation demonstration. *International Journal of Advanced Corporate Learning (iJAC).* https://doi.org/10.3991/ijac.v16i3.35739

Saraswat, D., & Tripathi, D. (2020). Cloud computing: Comparison and analysis of cloud service providers-AWs, Microsoft and Google. *2020 9th international conference system modeling and advancement in research trends* (SMART), pp. 281-285.

Sguanci, C., Spatafora, R., & Vergani, A. (2021). *Layer 2 Blockchain scaling: A survey*. ArXiv, abs/2107.10881.

Sharma, T., Kwon, Y., Pongmala, K., Wang, H., Miller, A., Song, D., & Wang, Y. (2023). Unpacking how decentralized autonomous organizations (DAOs) work in practice. *ArXiv,* abs/2304.09822. https://doi.org/10.48550/arXiv.2304.09822.

Sharma, T., Potter, Y., Pongmala, K., Wang, H., Miller, A., Song, D., & Wang, Y. (2024). *Future of algorithmic organization: Large-scale analysis of decentralized autonomous organizations (DAOs).* https://arxiv.org/pdf/2410.13095

Siaw, C., & Sarpong, D. (2021). Dynamic exchange capabilities for value co-creation in ecosystems. *Journal of Business Research, 134*, 493–506.

Siddiqui, M., Khurshid, K., Rashid, I., & Khan, A. (2020). Artificial Intelligence based 6G Intelligent IOT: Unfolding an Analytical Concept for Future Hybrid Communication Systems. *Proceedings of 2020 the 10th International Workshop on Computer Science and Engineering.* https://doi.org/10.18178/wcse.2020.02.021.

Sims, A. (2019). Blockchain and decentralised autonomous Organisations (DAOs): The evolution of companies? *Economics of Networks eJournal.* https://doi.org/10.2139/ssrn.3524674

Suryavanshi, A., Apoorva, G., Mohan Babu, T.N., Rishika, M., Haq, A. (2023). The integration of Blockchain and AI for Web 3.0: A security Perspective. *2023 4th International Conference on Innovative Trends in Information Technology* (ICITIIT), pp. 1–8.

Tana, G., & Chai, J. (2023). Digital transformation: Moderating supply chain concentration and competitive advantage in the service-oriented manufacturing industry. *Systems., 11.* https://doi.org/10.3390/systems11100486

Tennakoon, D., & Gramoli, V. (2022). *Smart red belly Blockchain: Enhanced transaction management for decentralized applications.* ArXiv, abs/2207.05971. https://doi.org/10.48550/arXiv.2207.05971

Tukur, M., Schneider, J., Househ, M., Dokoro, A., Ismail, U., Dawaki, M., & Agus, M. (2023). The metaverse digital environments: A scoping review of the challenges, privacy and security issues. *Frontiers in Big Data, 6.* https://doi.org/10.3389/fdata.2023.1301812

Turner, F. (2005). Where the counterculture met the new economy: The WELL and the origins of virtual community. *Technology and Culture, 46*, 485–512.

Wang, Y. (2023). *Digital transformation and resilience in global supply chains: A comprehensive review and future directions in the post-COVID-19 era.* Advances in Economics, Management and Political Sciences. https://doi.org/10.54254/2754-1169/51/20230649.

Warin, C., & Reinhardt, D. (2022). Vision: Usable privacy for XR in the era of the Metaverse. *Proceedings of the 2022 European symposium on usable security.* https://doi.org/10.1145/3549015.3554212.

Xevgenis, M., Kogias, D., Karkazis, P., & Leligou, H. (2023). Addressing ZSM security issues with Blockchain technology. *Future Internet, 15*, 129.

Zack, J. (2007). The ultimate company town: Wading in the digital marsh of second life. *University of Pennsylvania Journal of Constitutional Law, 10*, 225.

Zaczyk, M. (2023). *The concept of a decentralized autonomous organization as an innovative organizational structure.* Scientific Papers of Silesian University of Technology. Organization and Management Series. https://doi.org/10.29119/1641-3466.2023.174.30.

Zarick, R., Pellegrino, B., & Banister, C. (2021). *LayerZero: Trustless Omnichain interoperability protocol.* ArXiv, abs/2110.13871.

Part IV
Artificial Intelligence in Supply Chain Operations

Chapter 13
Causal Machine Learning: An Empirical Approach to Supply Chain Management

Alfredo Roa-Henriquez, Juan Fung, Ruhaimatu Abudu, Jennifer Helgeson, and Douglas Thomas

Abstract Over the past two decades, artificial intelligence (AI) has revolutionized industries, with machine learning (ML) at its core. While ML has enhanced supply chain management (SCM) in efficiency and resilience, it often relies on correlations, risking decisions based on spurious relationships. Causal machine learning (CML) offers a solution by focusing on cause-and-effect relationships, promising accuracy and better decision-making. Leading companies like Amazon and Walmart are beginning to harness CML for predictive modeling and AI-driven pricing strategies. Despite its potential, CML in SCM remains nascent, with limited empirical research. This chapter delves into the integration of causal inference in SCM, emphasizing methods that avoid pre-selecting covariates. It reviews foundational causal estimation techniques, surveys CML algorithms, and showcases their application through a simulation. By connecting theoretical advancements with practical implementations, the chapter underscores CML's transformative potential for SCM research and practice, and outlines future directions for its development.

Keywords Supply chain management · Causal machine learning · Predictive analytics · Decision-making

---

A. Roa-Henriquez (✉) · R. Abudu
College of Business, North Dakota State University (NDSU), Fargo, ND, USA
e-mail: alfredo.roa@ndsu.edu

J. Fung · J. Helgeson · D. Thomas
Applied Economics Office (AEO), Engineering Lab, National Institute of Standards and Technology (NIST), Gaithersburg, MD, USA

Y. Khojasteh, S. M. M. Jahan Abadi (eds.), *Emerging Technologies in Supply Chains*, International Series in Operations Research & Management Science 373, https://doi.org/10.1007/978-3-032-01218-0_13

13.1 Introduction

Over the past two decades, artificial intelligence (AI) has experienced substantial developments and entailed transformative changes for businesses globally (Dutta et al., 2024). Its core technology involves algorithms broadly defined as machine learning (ML) methods. These computationally intensive techniques learn from experience (i.e., past data) to make predictions and improve accuracy, performance, and decision-making (Mohri et al., 2018). The evolution of AI has progressed from ML to more elaborate forms like deep learning (DL), natural language processing (NLP), and generative AIs (GenAI) (Goodfellow et al., 2016; Simonite, 2023; Burtell & Toner, 2024). Although organizations initially employed AI systems for basic tasks, recent theoretical and practical implications for firms and their supply chains have received significant attention, particularly after large investments in AI and ML adoption took place in the late 2010s (Dutta et al., 2024). Today, essential advancements propelled by large language models (LLMs) and GenAI technology have created capabilities that aim to improve the management of supply chains (Dutta et al., 2024; Liddell, 2023).

Some scholars suggest that firms employing AI have opportunities to enhance their supply chain resilience by increasing visibility and improving sourcing, distribution, and last-mile delivery (Modgil et al., 2021). Others indicate that firms implementing AI can gain efficiencies in supply chain processes by taking advantage of digitalization and unphysicalization (Loske & Klumpp, 2022). Another perspective suggests that supply chain management (SCM), in theory and practice, still needs to prepare for this capability and technological transformation because it entails risks and challenges associated with AI-induced bias (Hendriksen, 2023).

The large-scale implementation of ML methods in SCM began in the late 1990s, initially focusing on logistics, inventory management, and demand forecasting (e.g., Carbonneau et al., 2008). As computational power increased and access to big data grew in the 2010s, the adoption of ML in SCM expanded significantly (Ni et al., 2020; Tirkolaee et al., 2021). Most current applications rely on learning algorithms designed to maximize an objective or minimize a loss function central to the models' training and fine-tuning, ensuring they perform as accurately as possible (Mohri et al., 2018). In ML, these "black box" models are created directly from the data provided to the learning algorithms. Often even the designers of these models cannot explain how variables are selected and used to make predictions due to the complexity and opacity involved in the process (Rudin & Radin, 2019; Wolf & March, 2024). Sometimes, though, using ML methods to address specific problems in different fields of social sciences (e.g., Economics, SCM, and Marketing) requires careful tuning and adaptation to exploit the structure of those problems. This structure refers, among others, to the causal nature of many variables, endogeneity, panel structure, optimization criteria, and selection of confounders (Athey & Imbens, 2019).

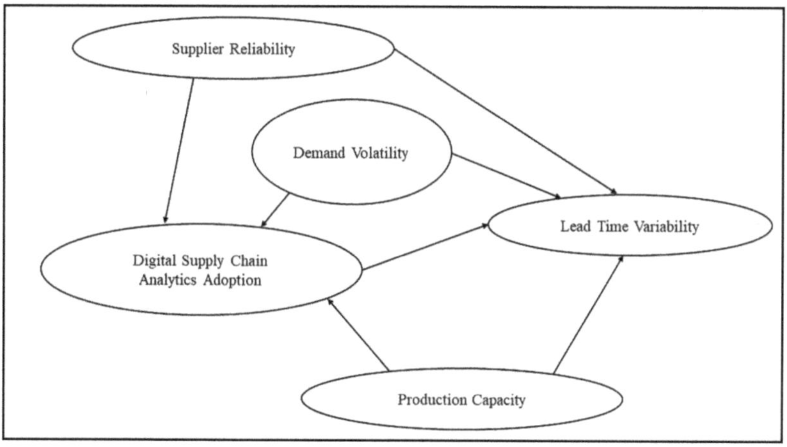

Fig. 13.1 Directed acyclic graph (DAG)

In SCM, understanding the structure or underlying causality mechanism that informs decision-making is essential for businesses to improve their competitive position (Helgeson & Roa-Henriquez, 2022). By identifying which actions actually cause improvements, firms can allocate resources more effectively and invest in initiatives that will certainly improve performance rather than relying on intuition or misleading correlations. For example, if a firm learns that adopting an advanced digital analytics tool can causally reduce lead time variability within a supply chain, whereas other factors such as demand volatility, production capacity, and supplier reliability only correlate with lead time variability, the firm can confidently invest in the tool and gain a competitive edge (see Fig. 13.1). The figure is a Directed Acyclic Graph (DAG) depicting a hypothetical causal relationship between the causal factor (Digital Supply Chain Analytics Adoption), and the outcome (Lead Time Variability), including potential confounders.

Under this scenario, it is possible to combine the advantages provided by ML models to answer prediction questions with the framework provided by causal inference to respond "what if" type of questions (Facure, 2023). Causal machine learning (CML)—or causal AI, more generally—is a recent scientific innovation that goes beyond traditional predictive models by distinguishing between genuine drivers of improvement and spurious relationships of factors that only appear related. In this sense, it addresses the limitations of conventional data or ML models by integrating them with the principles governing causal inference (Athey & Imbens, 2019; Wolf & March, 2024). This new approach has received particular attention from firms in different sectors with capabilities to store, process, and analyze big data, as summarized in Table 13.1 below.

The applications of CML across various industries highlight its transformative potential to derive actionable insights from complex datasets. Each application

Table 13.1 Applications of CML in SCM

Application	Industry	Details	CML Approach
Predicting Supply Chain Disruptions (Bodendorf et al., 2023).	Automotive	Combines deep learning with causal inference to assess supply chain reliability based on supplier data.	Integration of deep learning and causal inference to identify causal disruptions.
Optimal Rework Policies (Schwarz et al., 2024).	Manufacturing	Utilizes data from semiconductor manufacturing to estimate rework policies.	Data-driven causal inference model for policy optimization.
Visual component impact (Engin et al., 2023).	Entertainment (Netflix)	Analyzes the causal impact of visual components on viewer engagement.	Causal machine learning to identify key drivers of engagement.
Economic decision-making (ALICE, n.d.).	Technology (Microsoft)	Focuses on demand estimation, price optimization, and advertising effectiveness.	Uses causal inference to guide economic strategies.
Logistics and Inventory Efficiency (Götz & Budhathoki, 2022).	E-commerce (Amazon)	Causal analysis of factors affecting delivery times and inventory levels.	Causal inference to enhance logistics operations.
Causal Vector Analysis (Cheikhrouhou, 2023; Chen & Kasiviswanathan, 2020).	E-commerce (Amazon)	Investigates permutations of contextual factors for hypothesis testing.	Contextual vector analysis for causal hypothesis testing.
AI-driven Pricing Models (Pearson, 2021)	Retail (Walmart, Kroger)	Explores the causal effect of AI-powered pricing on sales.	Causal analysis to identify pricing strategies' impact on sales.

leverages causal inference to address specific industry challenges, ranging from supply chain disruptions to customer engagement and pricing strategies:

1. **Automotive Industry**: The integration of deep learning with causal inference in predicting supply chain disruptions (Bodendorf et al., 2023) underscores the need for robust models that can simultaneously handle large datasets and identify causative factors. This approach enhances reliability in predicting disruptions, offering a significant advantage in a sector where supply chain continuity is critical.
2. **Manufacturing**: The use of causal models for optimizing rework policies in semiconductor manufacturing (Schwarz et al., 2024) illustrates the potential of CML to improve operational efficiency. By focusing on causal relationships rather than correlations, manufacturers can make informed decisions that directly impact production efficiency and cost-effectiveness.
3. **Entertainment**: Netflix's application of CML to evaluate the impact of visual components (Engin et al., 2023) demonstrates the importance of causality in user engagement strategies. Understanding what visual cues drive engagement

enables more personalized content delivery, enhancing user satisfaction and retention.

4. **Technology**: Microsoft's ALICE project highlights the role of causal inference in economic decision-making. By understanding causal factors in demand estimation and pricing, technology firms can develop more effective strategies that align with market dynamics.
5. **E-commerce**: Amazon's exploration of causal factors in logistics and inventory efficiency (Götz & Budhathoki, 2022) showcases CML's ability to optimize supply chain operations. By identifying causative elements, firms can streamline processes, reduce delivery times, and increase customer satisfaction.
6. **Retail**: The use of AI-driven pricing models by Walmart and Kroger (Pearson, 2021) to explore causal effects on sales exemplifies how CML can refine pricing strategies. By focusing on causation, retailers can better understand the impact of pricing decisions on consumer behavior and sales performance.

As Table 13.1 presents, CML is a recent approach employed in the field of SCM to obtain empirical evidence. Although traditional empirical methods applied to observational studies have been more broadly implemented, the application of these methods is still challenging if endogenous shocks are present, simultaneously affecting inputs and firms' outcomes (e.g., Dormady et al., 2022). In this regard, because supply chains are deeply ingrained into the broader economy, the empirical approach needed for causal identification of effects on firms' outcomes becomes much more complex. Additionally, as the application of these advanced methods mature, they may help respond important policy questions associated with incentive misalignments along supply and value chains when the bearers of disaster impacts in one location are different from those managing much of the infrastructure of supply chains (Thomas & Helgeson, 2021, 2022) or when disruptions such as power outages affect delivery downstream (Thomas & Fung, 2022).

In this chapter, we discuss methods not often employed in SCM applications but that supply chain practitioners can use for assessing causal effects without the common practice of pre-selecting covariates. These CML approaches aim to identify treatment effects while providing valid inferences under a data-driven selection of covariates (Huber, 2023). However, before presenting this approach with a simulation example, we provide a review of the causal inference methods under the potential outcome framework.

The chapter is organized as follows. Section 13.2 presents the classical approach to causal effect estimation employed in Randomized Control Trials (RCT) and from observational data, the potential outcome framework—also known as the Rubin causal model. Section 13.3 presents a survey of different causal machine learning algorithms that are suitable to be employed in the context of SCM applications. Lastly, section 13.4 presents a simulation example taken from the literature (Yılmaz et al., 2024) that was formerly implemented through matching methods (i.e., techniques that require the pre-selection of covariates by creating comparable groups for treated and untreated subjects). We employ the same data and implement it by using a causal machine learning approach. In sect. 13.5, we summarize future research and provide a conclusion.

13.2 Causal Inference in SCM

SCM integrates *operations processes* needed to transform resources into products and *logistics processes* needed to supply inputs and services to transformation centers (e.g., manufacturing plants) and distribute products and services to customers (Jacobs & Chase, 2024). In terms of research, the field has experienced a surge in empirical analysis over the last two decades with most of empirical papers in operations management employing observational data to study causal effects derived from, for instance, the implementation of programs or the adoption of policies (Yılmaz et al., 2024), the analysis of the relationship between visibility and performance (Swift et al., 2019), and the implication of supply chain disruptions on prices of shipping containers (Mechai & Wicaksono, 2024). Given the challenges posed by observational data for the estimation of causal effects, we begin the discussion of how causal inference within the context of randomized experiments can be implemented in the study of supply chains. Many of the foundational concepts used for causal inference with observational data were developed for randomized experiments in the early twentieth century (Imbens & Rubin, 2015) and remain relevant in this context.

Many people are familiar with randomized control trials; for instance, medical trials for new fever-reducing medication are discussed in the context of randomized control trial efficacy. The classic setup is that volunteers sign up for a medical trial and are randomly assigned to either the treatment group (i.e., receive the medication) or the control group (i.e., receive the placebo drug). Data is collected on all the volunteers to draw inferences on the effect of the medication for the reduction of fevers. Assuming random assignment, any observed fever reductions *on average* can be attributed to the treatment. In particular, the causal effect of the medication can be estimated as the average difference in fevers for those in the treatment group and those in the control group. Moreover, this causal effect is unbiased (Imbens, 2024). Random assignment is key because it provides confidence that the observed response to the medication (i.e., fever reduction) is purely due to the treatment and not due other factors (i.e., confounders).

In the context of SCM, similar methodologies are employed. As mentioned in the introduction, retailers often utilize AI to experiment with price adjustments (Pearson, 2021). These price adjustments can be part of strategies to optimize resource allocation, such as inventory management or order fulfillment. For instance, A/B testing, or split testing, serves as a practical example of randomized control trials in this domain (EasyBA.Co, 2023).

13.2.1 Potential Outcomes Framework

This concept is generally intuitive to those without any domain-specific expertise, but there is a mathematical framework that provides the foundation for identifying such causal effects. The potential outcomes framework is based on the assumption that causation is tied to an action (i.e., exposure to treatment) applied to a unit (e.g., individual, household, firm) (Imbens & Rubin, 2015). The key observation is that while the individual was exposed to a particular condition (say the medication), they could just as easily have been exposed to another condition (the placebo). In that sense, each individual has a set of *potential* outcomes for each of the possible experimental conditions.

As with the preceding example of a medical trial for a new fever-reducing medication, assume we have a binary exposure (i.e., the treatment variable), $W = \{0, 1\}$. Let i be denote an individual volunteer in the medical trial. For each of the potential exposures in $w \in W$, we assume there is a corresponding potential outcome, $Y_i(0)$ and $Y_i(1)$, where $Y_i(w)$ would denote the individual's fever when their exposure is set to $W = w$.

By comparing the two potential outcomes, for instance the measured difference in fever with and without the treatment, $Y_i(1) - Y_i(0)$, we obtain an individual's causal effect from the medication. In other words, the causal effect compares an individuals' actual outcome with their counterfactual outcome: what would plausibly have happened if the individual's exposure had been changed.

However, in a randomized experiment, it is impossible to observe all potential outcomes for each individual, as we can only observe one; this is known as the "fundamental problem of causal inference.[1] Instead, we assign many individuals to one of the exposures to observe outcomes across both conditions. We can then estimate causal effects by comparing observed outcomes across multiple units.

Suppose we have N individuals, indexed $i = 1, \ldots, N$ and let W_i denote the exposure indicator. That is, $W_i = 1$ if individual i receives the treatment (medication) and $W_i = 0$ if individual i receives the control (placebo). Using this notation, the observed potential outcome for each individual is expressed as:

$$Y_i^{\text{obs}} = Y_i(W_i) = \begin{cases} Y_i(0), & \text{if } W_i = 0 \\ Y_i(1), & \text{if } W_i = 1 \end{cases} \#(1) \tag{13.1}$$

Without loss of generality, we write $Y_i = Y_i^{\text{obs}}$ to denote the observed outcome. Moreover, we can express the unobserved ("missing") potential outcome as:

$$Y_i^{\text{mis}} = Y_i(W_i) = \begin{cases} Y_i(1), & \text{if } W_i = 0 \\ Y_i(0), & \text{if } W_i = 1 \end{cases} \#(2) \tag{13.2}$$

[1] Note that some experiments can be designed to observe the same units over time (Imbens & Rubin, 2015).

Under a randomized assignment mechanism, the potential outcomes are independent of exposure:

$$W_i \perp\!\!\!\perp \left(Y_i(0), Y_i(1)\right) \tag{13.3}$$

We can then estimate the causal effect of interest, such as the sample average difference in outcomes:

$$\tau \equiv \frac{1}{N}\sum_{N}^{i=1}\left(Y_i(1) - Y_i(0)\right)\#(4) \tag{13.4}$$

This causal effect, τ, is called the **Average Treatment Effect** (ATE) and is one of many potential *estimands* of interest, which we will discuss further in Sect. 13.2.5.

The potential outcomes framework allows us to model the counterfactual (the "what if") as the unobserved potential outcome. Importantly, this relies on a critical assumption: that individual 's observed outcome is not influenced by the exposure condition of any other individual. This assumption is known as the **Stable Unit Treatment Value Assumption** (SUTVA), defined by Imbens and Rubin (2015) as follows:

- **No interference**: The outcome of unit i is unaffected by treatment exposure of another unit, i.e., $Y_i(W_i, W_{j \neq i}) = Y_i(W_i)$;
- **Counterfactual consistency**: Each unit's observed outcome corresponds to the potential outcome, given exposure: that is, if $W_i = w$ then $Y_i(\mathrm{w}) = Y_i$.

While SUTVA is a foundational assumption in causal inference, it is recognized as a strong assumption that does not always hold in practice. There is a body of literature addressing the estimation of causal effects in the presence of spillovers, where the treatment of one unit may affect the outcomes of another (Imbens, 2024). However, causal inference in the presence of such spillovers is beyond the scope of this chapter.

Identifying causal effects typically relies on SUTVA and other critical assumptions. Among these are *exclusion restrictions*, which assert that any effect on the outcome is solely through the treatment, with no other pathways. While some exclusion restrictions are testable, others remain untestable and depend on the context and domain knowledge.

Exchangeability implies that conditional on observed covariates, the distribution of potential outcomes is the same for treated and untreated units. **Positivity** requires that there is a non-zero probability of receiving each treatment level for all values of the covariates. Closely related is the assumption of **Conditional Independence** (or selection on observables), which posits that, given a set of observed covariates, the treatment assignment is independent of the potential outcomes. These assumptions are crucial when observing the same or different units over time and underpin many causal inference methodologies. In the subsequent sections, we will delve deeper into the roles of Positivity and Conditional Independence in causal inference.

13.2.2 Observational Studies

The concept of the counterfactual, or unobserved potential outcome, is fundamental to the potential outcomes framework. It formally asks the question, "What if we had done B instead of A?" In the social sciences, conducting idealized randomized experiments is often challenging, if not impossible. Practical or ethical constraints frequently make it infeasible to randomly assign households or businesses to different treatments solely to observe their responses. For instance, when a small business receives a grant for workforce development, we cannot observe the business outcomes had the grant not been awarded. Similarly, it would be unethical to provide disaster assistance "at random" to small businesses affected by infrastructure damage simply for experimental purposes.

Consequently, many causal inference studies rely on observational data. However, the real world presents complexities, and in most cases, potential outcomes derived from observational data are not independent of exposure (Imbens, 2024). While observational data is not generated through an experimental process, we can sometimes make certain assumptions about the data-generating process so that the observed data appears "as if" random.

The literature on causal inference with observational data is rich and diverse, spanning various fields and providing insights into numerous complex relationships. Classical examples include Angrist and Krueger's (1991) study on the returns to schooling, which utilized the quarter of birth as proxy (or instrumental variable, to be defined below) to address endogeneity in educational attainment. In public health, observational studies often explore the effects of lifestyle factors on health outcomes, such as diet and exercise, on cardiovascular health, assuming that the selection of observables can adequately control for confounding variables.

In the context of SCM, Bodendorf et al. (2023) investigate the impacts of plausibly exogenous factors, such as changes in political stability and variations in upstream and downstream precipitation and wind speeds, on delivery reliability. This study highlights the use of observational data to infer causal relationships by focusing on external factors that are likely independent of the supply chain's internal dynamics. Another example in SCM might involve assessing the impact of supplier diversity on procurement costs, assuming selection on observables to control for firm size and industry characteristics that influence both supplier diversity and costs. These examples illustrate the breadth of applications and the potential of observational data to yield valuable causal insights across different domains.

A major challenge in observational studies is confounding, which involves disentangling the causal effect of a treatment from other factors that influence the outcome. Addressing these challenges requires key identification assumptions to condition observed data, allowing us to interpret comparisons between treated and control units as causal effects. We discuss these assumptions below, noting that identifying a causal effect depends on a combination of both testable and untestable assumptions. In some instances, interpreting observed outcomes as causal effects may be impossible due to unobserved confounding.

13.2.3 Selection on Observables

In controlled experiments, treatment is assigned randomly to the population under a random assignment mechanism. In this setting, if potential outcomes are independent of the exposure variable and the two groups (treated and control) are similar regarding outcomes, the causal interpretation is justified (Imbens & Rubin, 2015).

With observational data, units such as individuals, households, or firms have agency over their exposure: they can select the treatment. For example, treatments in Bodendorf et al. (2023) include shipping modes and changes in trading-partner economic conditions, which may not be plausibly exogenous. Although we may not have a controlled experiment, we can compare outcomes when the treated and non-treated groups differ only by observable characteristics.

A key assumption that justifies a causal interpretation in such cases is the **Conditional Independence Assumption**, also known as **Selection on Observables**:

$$W_i \perp\!\!\!\perp (Y_i(0), Y_i(1)) | X_i \tag{13.5}$$

where X_i denotes a vector of pretreatment variables observed by the analyst. This assumption posits that treatment assignment is independent of potential outcomes, conditional on observables. It is also referred to as Unconfounded Assignment, as opposed to randomized or individualized assignment (Imbens & Rubin, 2015).[2]

This assumption is typically paired with the assumption of **positivity**, sometimes called overlap:

$$e(x) = pr(W_i = w \mid X_i = x) \in (c, 1-c), \quad \text{for some } c > 0\#(6) \tag{13.6}$$

Here, $e(\cdot)$ is known as the **propensity score** and models the exposure, conditional on observables. For many problems, the propensity score is central to causal inference, as we show in the discussion of estimators. Together, the conditional independence assumption and positivity are known as the strong ignorability (or ignorable treatment) assumption.

In practice, a common mistake when selecting variables for estimating the Average Treatment ATE under Conditional Independence is to use variables that are causally influenced by the treatment or the outcome (Imbens, 2024). For example, in SCM, consider a scenario where a company implements a new inventory management system (the treatment) to improve delivery times (the outcome). If analysts mistakenly include delivery times as a variable to condition on when estimating the ATE of the new system, they risk biasing the results because delivery times are

[2] Individualized assignment refers to a treatment assignment mechanism where the decision to assign a treatment is based on individual characteristics and not randomly determined. This contrasts with randomized assignment, where treatment is assigned randomly, and unconfounded assignment, treatment is assigned randomly conditional on observed variables (Imbens & Rubin, 2015).

directly affected by the treatment itself. Instead, they should focus on variables that are not influenced by the treatment or outcome, such as historical demand patterns.

13.2.4 Causal Inference Without Unconfounded Assignment

In observational studies, the conditional independence assumption that treatment assignment is independent of potential outcomes often does not hold, leading to endogeneity and bias. Endogeneity is a pervasive issue, particularly in economics and other social sciences, arising from omitted variables, measurement error, and simultaneity. These issues can lead to biased estimators, complicating causal inference. For example, omitted variables correlated with both the treatment and the outcome can confound the estimated treatment effect.

To address selection on unobservables, researchers employ several methods to infer causality in natural or quasi-experimental settings. These methods include Instrumental Variables, Regression Discontinuity, Difference-in-Differences, and Synthetic Control Designs. Each method has specific strengths and weaknesses, depending on the context and available data. We refer the reader to Imbens and Rubin (2015) for more details.

1. **Instrumental Variables (IV):** IV methods are used to address endogeneity by identifying variables (instruments) that are correlated with the treatment but not directly with the outcome. The key conditions for a valid IV are relevance (the instrument is correlated with the treatment) and exogeneity (the instrument affects the outcome only through the treatment).
2. **Regression Discontinuity (RD):** RD designs exploit situations where treatment assignment is based on a cutoff score, such as a test score or income level. By comparing individuals just above and below the cutoff, researchers can estimate the causal effect of the treatment in a local neighborhood around the cutoff.
3. **Difference-in-Differences (DiD):** DiD methods compare the changes in outcomes over time between a treatment group and a control group. This method assumes that, in the absence of treatment, the average change in the outcome would have been the same for both groups.
4. **Synthetic Control Designs:** Synthetic control methods construct a weighted combination of control units to serve as a counterfactual for the treated unit. This approach is particularly useful in comparative case studies where a single treated unit is compared to a synthetic control group.

13.2.4.1 Instrumental Variables

For illustrative purposes, we present the IV approach to selection on unobservables. While other methods for unconfounded assignment (including RD, DiD, and synthetic control) are also pivotal in causal inference, our emphasis is motivated in part by its natural extension from the estimators case study we present in this chapter. Understanding the IV approach also complements the CML techniques we explore, as there is research on ML for IV (Chen & Hsiang, 2019; Wang et al., 2022). For detailed discussions on other methods, we refer the reader to Imbens (2024).

IV methods address endogeneity by identifying instruments that are correlated with the treatment but not directly with the outcome. The key conditions for a valid IV are:

- **Relevance**: The instrument must be correlated with the treatment.
- **Exogeneity**: The instrument affects the outcome only through the treatment.

Despite their potential, IV methods pose theoretical and empirical challenges. One challenge in using IVs is finding instruments that satisfy both conditions. Theoretically, researchers must prove instrument validity, while empirically, they must find suitable instruments. Conducting robustness checks and sensitivity analyses is crucial to validate findings and ensure reliable causal estimates. Weak instruments, which show a weak correlation with the treatment, can introduce bias. Diagnostic tests, such as the F-statistic, help identify weak instruments. Researchers can use multiple instruments or the Limited Information Maximum Likelihood (LIML) estimator to address this issue. Angrist and Pischke (2009) cover these and other challenges in depth.

For researchers considering IVs or other methods to address selection on unobservables, careful instrument selection, diagnostic testing, and cautious interpretation are paramount. Robustness checks and sensitivity analyses are essential to ensure the reliability of causal estimates. Moreover, recent advances have integrated machine learning techniques with IV estimation to improve robustness and accuracy. Additionally, combining IV with other methods, like RD, enhances causal effect identification.

13.2.5 Estimands and Estimators

In the potential outcomes framework, we define a range of potential estimands of interest. An *estimand* is the target quantity we aim to estimate, such as the ATE introduced earlier. An estimand may be estimated using various *estimators*, which are methods for quantifying the estimand.

Estimands are defined with respect to specific populations of interest.

- ATE measures the average effect of a treatment across the entire population:

$$\tau_{ATE} = E\left[Y(1) - Y(0)\right].$$

- The Average Treatment Effect on the Treated (ATT) measures the average effect of a treatment on those who have received it:

$$\tau_{ATT} = E\left[Y(1) - Y(0) \mid W = 1\right].$$

- The Average Treatment effect on the Untreated (ATU) can be defined analogously, focusing on those that did not receive the treatment:

$$\tau_{ATU} = E\left[Y(1) - Y(0) \mid W = 0\right].$$

- More generally, the Conditional Average Treatment Effect (CATE) targets heterogeneous treatment effects and is defined as the average effect within a subpopulation characterized by specific pretreatment observables $X = x$

$$\tau_{CATE}(x) = E\left[Y(1) - Y(0) \mid X = x\right].$$

Estimators are methods used to estimate these estimands and can be parametric, semi-parametric, or non-parametric. In CML, the focus is often on developing robust estimators that minimize the need for stringent parametric assumptions, allowing for more flexible modeling of complex relationships.

- Parametric Estimators: Rely on specific functional forms and distributional assumptions.
- Semi-parametric Estimators: Combine parametric and non-parametric elements, offering a balance between flexibility and interpretability.
- Non-parametric Estimators: Make minimal assumptions about the functional form, allowing for greater flexibility in capturing complex patterns.

CML is particularly concerned with developing and applying estimators that can accurately capture causal effects while reducing reliance on parametric assumptions.

In practice, estimands are obtained from a sample of the population of interest, calculated analogously to the sample ATE presented in Sect. 13.2.1. Understanding and selecting appropriate estimators is crucial for accurate causal inference, especially in complex data environments.

13.2.6 Mediation Analysis

In this section, we focus on mediation analysis. We refer the reader to Hernan and Robins (2024) for a detailed discussion on causal mediation. In the context of supply chain management, by identifying mediators (i.e., variables that transmit the effect of a predictor, such as a supply chain disruption, to an outcome, such as supply chain performance), this analysis helps uncover the indirect pathways through which resilience is achieved. This is the case of supply chain resilience, for instance. The resilience of a supply chain might be influenced not only directly by the robustness of its suppliers but also indirectly through factors such as inventory management practices or the flexibility of logistics operations. Understanding these mediating effects can provide deeper insights into the mechanisms by which supply chains absorb and adapt to disruptions. This understanding allows supply chain managers to identify factors and design more effective interventions that enhance resilience by strengthening direct and indirect pathways.

Causal mediation analysis also facilitates targeted improvements in supply chain resilience by distinguishing between direct and indirect effects of risk management strategies. For example, a strategy aimed at diversifying suppliers might directly benefit from reducing dependency on any single supplier. However, its full impact might be mediated through improvements in lead time variability and overall flexibility of the supply chain. By quantifying these mediating effects, organizations can prioritize investments in areas that yield the most significant overall resilience gains, making the supply chain more resistant to disruptions and more adaptive to changing market conditions.

13.3 Causal Machine Learning in Supply Chain Management

If firms are solely interested in "good" predictions (relative to the decision maker's loss function), machine learning might be sufficient, as it focuses on finding patterns in data to make accurate forecasts without explicitly understanding why those patterns exist. However, CML becomes crucial when firms need to understand the underlying structure of cause-and-effect relationships that drive these patterns. Causal insights allow firms to make informed, strategic decisions that go beyond correlation (Athey & Imbens, 2019; Brand et al., 2023). This can be critical when economic conditions change and spurious correlations are broken. For instance, spurious correlations may no longer apply as a result of an economic downturn, a natural hazard, or other supply chain disruption. When spurious correlations no longer apply and there are limited data observations to correct specification, a model can make suboptimal predictions. Additionally, it can provide insight into what factors can be leveraged to change outcomes and the magnitude of their effects.

Machine learning is typically described as a discipline focused on prediction rather than inference (Breiman, 2001; Athey & Imbens, 2019). In causal inference, the goal is to understand how an exposure (a feature or input into the model) affects the outcome of interest, providing confidence in the magnitude of the estimand, such as through standard errors or confidence intervals. In contrast, the goal of machine learning is to predict the outcome of interest "really well," which typically means optimizing metrics such as accuracy or precision based on a loss function. The causal relationship is interesting only to the extent that it minimizes prediction error, with less focus on the magnitude or confidence of estimated parameters like treatment effects.

13.3.1 Why Causal Machine Learning?

In the last decade, causal inference has increasingly incorporated machine learning methods while maintaining the primary goal of inference. As highlighted in the introduction, notable applications including Amazon's delivery-time analysis and Microsoft's model focusing on demand estimation and price optimization demonstrate the value of CML.

While causal inference fundamentally relies on identification assumptions rather than specific estimators, machine learning can offer estimators with desirable properties, such as efficient first-stage estimators, that can be seamlessly integrated into causal inference frameworks. Despite these advances, inference remains a challenge, with ongoing differences between the econometrics- statistics focus on inference and the machine learning focus on prediction.

Machine learning facilitates "model free" causal inference. Typically, estimating an estimand requires unnecessary parametric assumptions on the effect of X on the outcome. For instance, in a model such as $Y_i = \tau W_i + \beta X_i$, under reasonable conditions, we can relax the parametric assumption and use machine learning methods to learn the functional form of the relationship between X and the outcome: $Y_i = \tau W_i + f(X_i)$. Under certain conditions, we can achieve *consistent* estimators and *asymptotic normality*, providing confidence intervals for inference.

Parametric assumptions often lead to modeling choices that constrain the functional form of f. In a linear model, it can be challenging to include many interaction effects or to estimate heterogeneous treatment effects. Relaxing parametric assumptions allows for more flexible modeling of the outcome and estimation of a broader range of estimands. However, semi- or non-parametric model, or more complex machine learning models, may not always improve inference. As we relax parametric assumptions, more observations (N) are needed to estimate f. More flexibility also allows us to include more features, increasing the dimension K of X. Both factors can introduce the curse of dimensionality, resulting in biased or inefficient (i.e., not consistent) estimators as N or K (or both) increase.

In the next subsection, we elaborate on a few key semi- and non-parametric estimators that are suitable in settings with selection on observables. We then discuss a few key machine learning techniques, including regularization and ensemble methods, which can be used to implement a broad range of estimators.

Figure 13.2 illustrates the intersection of machine learning and causal inference and some of the topics we cover (as well as topics beyond the current chapter scope). As depicted in this figure, the causal inference identifies causal links, while machine learning excels in prediction, and causal machine learning leverages both to provide deeper causal insights.

To set the stage, Table 13.2 below provides an overview of the machine learning models we discuss, categorizing them into traditional and advanced models. For each model, we provide an explanation, use cases, strengths, and weaknesses. This table is intended to serve as a reference point, aiding readers in comparing and contrasting the various models discussed.

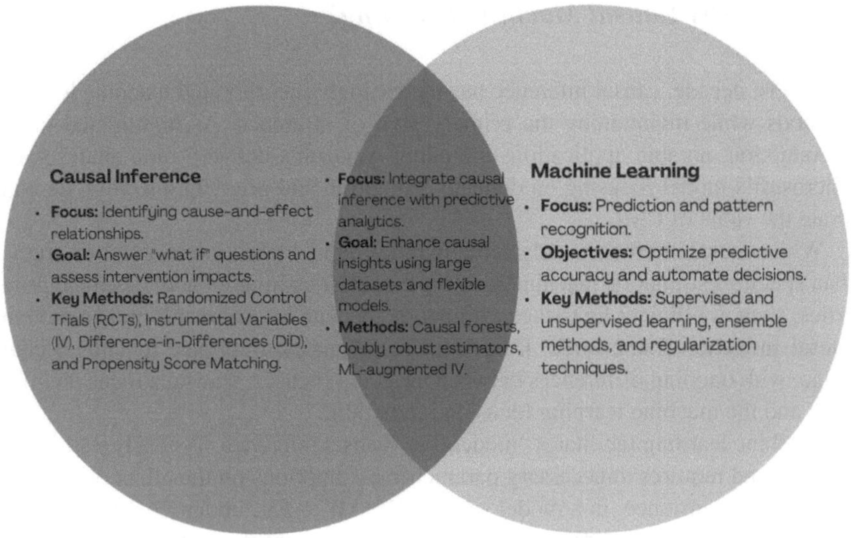

Fig. 13.2 Intersection of Causal Inference and Machine Learning

13.3.2 Key Estimators in Causal Inference

Inverse-Propensity Weighting (IPW) adjusts for confounding by estimating treatment probabilities based on observable covariates, assuming selection on observables and SUTVA (Rosenbaum & Rubin, 1984). Let $\hat{e}(x) = P(W = 1 \mid x)$ denote the estimator of the exposure model (or propensity score), or the probability of treatment given observable covariates. The IPW estimator of the ATE is

$$\hat{\tau}_{IPW} = \frac{1}{n} \sum_{n}^{i=1} \left(\frac{W_i Y_i}{\hat{e}(X_i)} - \frac{(1 - W_i) Y_i}{1 - \hat{e}(X_i)} \right) \# (7) \tag{13.7}$$

Under correct specification of the propensity score model and the assumption of selection on observables, the IPW estimator is consistent for the ATE (Imbens, 2004). The efficiency of the IPW estimator is contingent upon accurately estimating the propensity scores. Misspecification can lead to inefficient estimates. Finally, while IPW can handle high-dimensional covariates, it is sensitive to extreme propensity score values, which can inflate the variance.

The Augmented Inverse-Propensity Weighting (AIPW) estimator, introduced in Robins et al. (1994), additionally incorporates outcome estimation for each of the potential outcomes $\hat{Y}_1(x)$ and $\hat{Y}_0(x)$. Like IPW, AIPW assumes selection on observables and SUTVA. Combining both the propensity score model and the outcome model to estimate the ATE is intended to mitigate potential biases from misspecification. The AIPW estimator of the ATE is

Table 13.2 Overview of machine learning methods, including use cases, strengths, and weaknesses

Model	Explanation	Use Cases	Strengths	Weaknesses
Linear regression	A basic regression model that assumes a linear relationship between variables.	Predicting sales, demand forecasting	Simplicity, easy interpretability	Assumes linearity, sensitive to outliers
Decision trees	A model that splits data into subsets based on feature value decisions.	Classification, risk assessment	Interpretability, handles both numerical and categorical data	Prone to overfitting, can be unstable
Random forests	An ensemble of decision trees to improve prediction accuracy.	Demand forecasting, anomaly detection	Robust to overfitting, handles high dimensionality	Less interpretable, computationally intensive
Gradient boosting	An ensemble model that iteratively improves weak models.	Price optimization, customer behavior	High accuracy, handles complex relationships	Prone to overfitting, requires careful tuning
Neural networks	Deep learning models capable of capturing complex patterns in data.	Image recognition, demand prediction	High flexibility, can model non-linear relationships	Requires large datasets, opaque model structure
Ridge regression	Linear regression with L2 regularization to prevent overfitting by penalizing large coefficients.	Handling multicollinearity in supply chain data; improving model robustness.	Reduces model variance; stabilizes estimates in presence of multicollinearity.	Does not perform variable selection; all variables are retained in the model.
Lasso regression	Linear regression with L1 regularization to enforce sparsity in features by shrinking some coefficients to zero.	Feature selection, predictive modeling	Reduces overfitting, performs automatic feature selection	May exclude relevant features, sensitive to choice of regularization
Elastic net	Combines L1 and L2 regularization for feature selection and model stability.	Predictive modeling	Balances feature selection and model stability, handles correlated features	Requires careful tuning, more complex than lasso or ridge

$$\hat{\tau}_{AIPW} = \frac{1}{n}\sum_{n}^{i=1}\left(W_i \frac{Y_i - \hat{Y}_1(X_i)}{\hat{e}(X_i)} - (1-W_i)\frac{Y_i - \hat{Y}_0(X_i)}{1-\hat{e}(X_i)} + \left(\hat{Y}_1(X_i)\_\hat{Y}_0(X_i)\right)\right)\#(8) \tag{13.8}$$

The AIPW estimator often achieves higher efficiency compared to IPW alone by leveraging information from both models. Moreover, by incorporating outcome models, the AIPW estimator can reduce bias when the propensity score model is misspecified, offering more reliable causal estimates. Finally, the AIPW estimator is **doubly robust**.[3]

Doubly robust estimators, including AIPW and Targeted Maximum Likelihood Estimation (TMLE),[4] combine the exposure model ($W = g(Z)$) with the outcome model ($Y = f(X)$) to estimate causal effects. A notable feature of these estimators is that the models can take any functional form and the sets X and Z do not need to coincide. This flexibility allows for more tailored modeling approaches.

One advantage of doubly robust estimators is the property of **weak double robustness**, which refers to their ability to provide consistent estimates if either the exposure model, g, or the outcome model, f, is correctly specified, though not necessarily both (Funk et al., 2011).[5] This property provides a safeguard against misspecification, making these estimators particularly useful in complex observational settings. More importantly, when both the exposure and outcome models are correctly specified, the estimator achieves **strong double robustness**, allowing the estimator to exhibit optimal statistical properties, such as consistency and asymptotic normality, ensuring minimal variance.

Doubly robust estimators are desirable in settings where there is uncertainty about which model (exposure or outcome) is correctly specified. They are particularly valuable in observational studies with high-dimensional data or when there is potential for complex interactions and non-linear relationships among covariates, allowing for the use of semi- and non-parametric machine learning methods for the exposure or the outcome. Moreover, inference properties hold even when the estimator uses cross-validation for model selection. The ability to achieve consistent estimation even with partial model specification knowledge makes these estimators a powerful tool in causal inference.

However, despite their advantages, doubly robust estimators also have several potential disadvantages. Implementing these estimators can be computationally intensive, particularly in large datasets or when using complex models, such as those involving machine learning techniques. While they can be consistent if either model is correct, the choice of models and their specifications can be challenging. Poorly chosen models can lead to inefficiencies and increased variance. In practice, doubly robust estimators may exhibit poor performance in small samples, where the theoretical properties are less likely to hold. Although designed to be robust, the performance of doubly robust estimators can degrade if both models are severely misspecified, leading to biased estimates.

---

[3] We refer the reader to the excellent course notes by Stefan Wager (2022).

[4] Targeted Maximum Likelihood Estimation (TMLE) is an approach that iteratively refines initial estimates of the outcome model using the propensity score, ensuring that the final estimator is doubly robust. TMLE optimizes both the bias and variance trade-offs, leading to efficient and consistent estimates under correct model specifications (van der Laan and Rose, 2011).

[5] More precisely, if one of the models is *consistent*.

In settings with selection on unobservables, doubly robust estimators are typically not suitable. These estimators rely on the assumption of selection on observables, meaning all relevant confounding variables must be observed and accounted for in the models. When there is selection on unobservables, the underlying assumption that either the treatment or outcome model sufficiently adjusts for confounding is violated. In such cases, alternative methods that account for unobserved confounders, such as instrumental variable approaches or structural modeling techniques, may be more appropriate.

13.3.3 Advanced Machine Learning Techniques

Before diving into the specific techniques, it is important to understand why techniques such as regularization and ensemble methods hold significance in causal inference, particularly within SCM applications. Regularization methods, such as ridge and lasso, are powerful tools that help prevent overfitting by penalizing overly complex models, thus improving the generalizability of causal estimations. These techniques are especially useful in high-dimensional settings where the number of potential covariates is large, providing a disciplined approach to model selection and variable shrinkage. Similarly, ensemble methods like random forests offer robustness against overfitting by combining multiple models to produce more stable and reliable estimates. Their ability to capture complex interactions and non-linear relationships makes them valuable in exploring causal effects, particularly when traditional parametric models may fall short.

13.3.3.1 Regularization Techniques

Regularization is a critical technique in machine learning that addresses the challenge of overfitting by adding a penalty term to the loss function. This penalty discourages complex models that fit the training data too closely, enhancing the model's generalization to new data (Hastie et al., 2009). Regularization techniques help in feature selection and can handle multicollinearity among predictors (Tibshirani, 1996).

For simplicity, we discuss regularization in the context of ordinary least squares (OLS) regression, where $f(X) = x_i^T \beta + \varepsilon$ and the objective is to minimize the sum of squared residuals, $\left(Y_i - X_i^T \beta\right)^2$. Note that these techniques have been extended to the full family of generalized linear models (GLMs) that include logistic (binomial), Poisson, and gamma regression and the objective is to minimize the (negative) log-likelihood function.[6]

[6] For further details on GLMs and elastic net regularization for GLMs, see Tay et al. (2023).

Ridge regression (also known as L2 regularization) adds a penalty equal to the square of the magnitude of coefficients:

$$L_2 = \sum_{n}^{i=1} \left(Y_i - X_i^T \beta\right)^2 + \lambda \sum_{K}^{j=1} \beta^2 \#(9) \tag{13.9}$$

Lasso regression (L1 regularization) adds a penalty equal to the absolute value of the magnitude of coefficients:

$$L_1 = \sum_{n}^{i=1} \left(Y_i - X_i^T \beta\right)^2 + \lambda \sum_{K}^{j=1} |\beta| \#(10) \tag{13.10}$$

Elastic net is a regularization technique that combines the penalties of both ridge and lasso, providing a more flexible approach that can handle correlated variables and performs well in high-dimensional settings (Zou & Hastie, 2005).

$$L_\alpha = \sum_{n}^{i=1} \left(Y_i - X_i^T \beta\right)^2 + \lambda \left(\alpha \sum_{K}^{j=1} |\beta| + \frac{1-\alpha}{2} \sum_{K}^{j=1} \beta^2 \right) \#(11) \tag{13.11}$$

The regularization term in the elastic net is thus a mixture of L1 and L2, with ≥0 serving as a tuning parameter and $0 \leq \alpha \leq 1$ as the hyperparameter that determines the mixture between L1 and L2.

For causal inference, regularization can be particularly useful in selecting covariates and ensuring that the estimation models are not overly complex, which could lead to biased estimators (Belloni et al., 2014).

13.3.3.2 Ensemble Methods

Ensemble methods are a class of machine learning techniques that combine multiple models to improve predictive performance and robustness. By aggregating the outputs of various models, ensemble methods can mitigate the variance and bias typically associated with individual models, leading to more reliable estimations. Common ensemble techniques include bagging, boosting, and stacking. Bagging, or bootstrap aggregating, enhances model stability by training multiple versions of a model on different subsets of the data and averaging their predictions (Breiman, 1996). Boosting, on the other hand, sequentially applies models to the data, focusing on correcting the errors made by previous models, thus improving overall accuracy (Friedman, 2001). Stacking involves training a meta-model to combine the predictions of several base models, potentially capturing complex patterns that single models might miss (Wolpert, 1992).

In the realm of causal inference, ensemble methods like causal forests—a variant of random forests—can be particularly valuable. They estimate heterogeneous treatment effects by leveraging the ensemble's ability to handle complex interactions and non-linear relationships (Athey et al., 2019). This is especially useful in SCM, where diverse and intricate causal factors often influence outcomes. Ensemble methods thus provide a robust framework for exploring causal relationships, allowing researchers and practitioners to derive actionable insights from complex data landscapes.

13.3.3.3 Random Forests

Random forests are an ensemble learning method that constructs multiple decision trees during training and outputs the mode of their classes (for classification) or mean prediction (for regression) (Breiman, 2001). This method is known for its robustness and ability to handle large datasets with higher dimensionality, making it suitable for causal inference applications where complex interactions and nonlinearities are present.

In the context of CML, random forests can be adapted for causal inference through approaches such as causal forests, which are specifically designed to estimate heterogeneous treatment effects (Athey et al., 2019). These methods allow for flexible modeling of the treatment effect heterogeneity without the need for prespecifying interaction terms.

13.3.4 Summary of Strengths and Weaknesses

Machine learning techniques are increasingly applied in causal inference to address complex supply chain challenges. Each technique has its unique strengths and limitations, making it crucial for practitioners to understand when and how to use them effectively. Table 13.3 provides a comprehensive overview of these aspects, alongside recommendations for practitioners.

Integrating machine learning techniques such as regularization and ensemble methods with causal inference tools like IPW, AIPW, and doubly robust estimators offers substantial benefits but also presents challenges. The guidance for application of these methods in SCM can be summarized as follows:

- **IPW and AIPW**: These methods are foundational in adjusting for confounding variables, especially in high-dimensional settings. IPW is relatively straightforward but can be sensitive to extreme propensity scores, which destabilizes the variance of estimates. AIPW enhances this approach by incorporating outcome models, providing robustness against misspecification of either the exposure or outcome model. In SCM applications, where complex interactions are common, AIPW is particularly beneficial, although computationally intensive.
- **Doubly Robust Estimators**: These estimators are valuable in settings with uncertainty about model specification. By achieving consistency if either the treatment or outcome model is correctly specified, they offer a safeguard against potential misspecification errors. In practice, this requires careful consideration

Table 13.3 Summary of ML models and methods for causal inference

Model	Strengths	Weaknesses	Recommendations
IPW	Straightforward adjustment for confounding; suitable for high-dimensional data	Sensitive to extreme propensity scores; requires correct model specification	Use when selection on observables is plausible and data is well-controlled; avoid with extreme propensity values
AIPW	Combines IPW with outcome modeling for bias reduction; doubly robust	Computationally intensive; sensitive to model misspecification	Employ in settings with complex interactions; leverage machine learning for outcome modeling
Doubly robust estimators	Protection against model misspecification; consistent if one model is correct	High computational cost; challenging model specification	Ideal for high-dimensional data and potential interactions; ensure robust model validation
Regularization	Prevents overfitting; aids in feature selection	May oversimplify models if not carefully tuned	Suitable for high-dimensional settings; use cross-validation to select hyperparameters
Ensemble methods	Robustness against overfitting; captures complex interactions	Lack of interpretability; computationally demanding	Best for complex causal structures; use stacking to mitigate reliance on specific models
Random forests	Handles non-linear relationships; estimates heterogeneous treatment effects	Interpretability challenges; requires large datasets	Effective for exploring heterogeneous effects; ensure sufficient data to avoid overfitting

of model choices and validation processes to ensure reliability, especially in high-dimensional data environments.

- **Regularization Techniques**: Regularization methods like ridge, lasso, and elastic net are essential for handling multicollinearity and preventing overfitting, particularly in high-dimensional SCM data. They are useful for feature selection and can be tuned using cross-validation to optimize performance. However, practitioners must be cautious not to oversimplify models, especially when capturing complex causal relationships.
- **Ensemble Methods**: Techniques such as bagging, boosting, and stacking provide robustness by aggregating model outputs. They are particularly effective in scenarios with complex causal structures, offering enhanced predictive performance and stability. However, their lack of interpretability and computational demand should be carefully managed. Stacking, which combines multiple models, is recommended for reducing reliance on any single model specification.
- **Random Forests and Causal Forests**: These methods are adept at modeling non-linear relationships and estimating heterogeneous treatment effects. In SCM, where diverse and intricate causal factors influence outcomes, these approaches are powerful tools. Nonetheless, they require large datasets to avoid

overfitting and maintain interpretability. Practitioners should ensure data sufficiency and consider interpretability trade-offs when employing these models.

By understanding these strengths and limitations, practitioners can make informed decisions about the appropriate use of these methods in their causal inference efforts within SCM, ensuring both the reliability and relevance of their findings.

13.4 Application with Synthetic Data

With the purpose of highlighting the usefulness for the field of supply chain and operations management and guiding scholars on how to approximate randomized experiments by using synthetic control and matching methods, Yılmaz et al., 2024 presents implementation details in a simulation study comparing matching algorithms. The hypothetical scenario describes a brick-and-mortar store that is expanding to online sales, offering exclusive SKUs unavailable in-store. The observations are customer transactions. The outcome of interest is customer spending. The exposure is a promotion (special deal) that the retailer is offering to incentivize online sales. Thus, we are interested in the causal impact of the deal on spending. The data-generating process (DGP) follows a selection-on-observables model (satisfying SUTVA and the other key identification assumptions), as per Heckman and Robb Jr. (1985).

Table 13.4 presents a summary of the variables, including details on the outcome and exposure models. The outcome model is:

$$OnlineSpending = \tau Discount + f(X) + u\#(12) \tag{13.12}$$

The DGP specifies a linear relationship between *OnlineSpending* and *X*, where *X* includes all the covariates in Table 13.4 (with continuous covariates in log form).

The exposure model is

$$Discount^{*} = g(X') + v\#(13) \tag{13.13}$$

The DGP specifies a linear relationship between *X*' and the latent variable, *Discount**, where *X*' includes *Age*, log(*Distance*), and log(*PastVisits*). The observed binary exposure, $W = Discount$, is determined by a cutoff value, κ, such that $Discount = 1$ if $Discount^{*} > \kappa$.

We make a minor modification to the exposure model by adding a squared Age term. This simple modification allows us to explore the impacts of misspecification of our empirical models. In a sense, we have a semi-parametric exposure model: $Discount^{*} = \theta Z + g(Age) + v$, where $Z = (X' \setminus Age)$ and *g* is the linear combination of *Age* and Age^2. Given the new latent exposure, we use an imbalanced treatment rule such that κ is a random number between the eighth and ninth decile of the latent

Table 13.4 Descriptive statistics for synthetic data, based on Yılmaz et al. (2024)

		All (N = 10,000)				Control (N = 8637)	Treated (N = 1363)
Variable	Definition	Mean	SD	Min	Max	Mean	Mean
OnlineSpending	Spending at the first online transaction	50.59	18.68	18.46	239.01	48.59	63.25
Male	1 if customer is male, 0 otherwise	0.45	0.50	0	1	0.45	0.44
Age	Customer age	38.33	8.55	18	80	34.67	49.57
Income	Customer household income	59.92	35.03	10	250	55.70	73.90
PastStoreVisits	Number of store visits last year	5.96	3.15	1	23	5.94	6.32
PastStoreSpent	Amount spent in the store last year	208.22	314.51	60	4000	194.62	315.57
Midterm	Medium-term customer (24–47 months)	0.20	0.40	0	1	0.20	0.21
Longterm	Long-term customer (48+ months)	0.40	0.49	0	1	0.40	0.40
Distance	Distance to the store	20.00	2.99	9.01	30.95	20.05	19.64

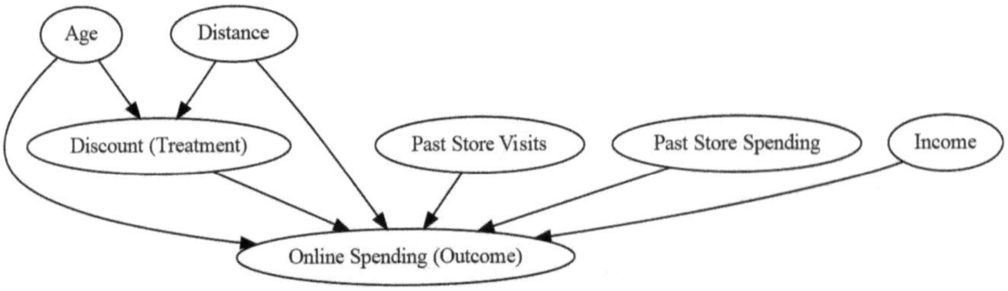

Fig. 13.3 Directed Acyclic Graph (DAG) depicting causal relationships between the covariates, the treatment (Discount), and the outcome (Online Spending), including potential confounders. Source: Authors

exposure. More details on the data generating process, including model equations and code, we refer to Yılmaz et al. (2024).

To provide a clearer understanding of the causal relationships inherent in our synthetic scenario, we present a causal graph in Fig. 13.3. This graph visually represents the assumptions about the relationships between the treatment (Discount), the outcome (Online Spending), and relevant covariates. The causal graph, constructed using a Directed Acyclic Graph (DAG), helps to illustrate how we believe the promotional offer influences online spending, accounting for potential confounding variables such as Age and Distance.

The causal graph in Fig. 13.3 depicts the hypothesized relationships among the treatment, outcome, and covariates. The directed edges (arrows) represent causal assumptions, where the direction indicates the presumed causal influence. For example, the treatment Discount is expected to directly affect the Online Spending outcome. Additionally, covariates such as Age and Distance are assumed to have causal effects on both the treatment and the outcome, suggesting their potential role as confounders. Understanding these relationships is crucial for identifying causal effects and ensuring that our analyses adjust appropriately for confounding.

13.4.1 Application Setup

Our estimand of interest is the ATE. In the true DGP, we know the ATE is $\tau = 10$. We begin our case study by illustrating naïve estimation of the ATE: a GLM for the outcome model that is used to predict both potential outcomes. The ATE is then the difference between the predicted potential outcomes. We then explore the impact of misspecification (of the outcome model) on the results from naïve estimation. To incorporate an exposure model, we then explore IPW estimates of the ATE. This time, we can explore the impact of misspecification of the exposure model as well as the outcome model. Finally, we explore doubly robust estimation with AIPW, where again we examine the impacts of misspecification. Moreover, we illustrate a few of the key machine learning techniques (i.e., regularization, random forests, and ensembles with stacking).

Our analysis uses the R programming language (version 4.4.0, build aarch64-apple-darwin20 for Mac OS Sequioa), including the following packages:

- boot and rsample: These packages were used to construct bootstrap standard errors and confidence intervals. The boot package allows for resampling with replacement, while rsample facilitates the creation of resampling objects, supporting various types of bootstrap strategies.
- WeightIt: This package was employed for implementing IPW in our analysis. It provides a flexible interface for creating weights that adjust for confounding, helping in estimating causal effects from observational data.
- AIPW and SuperLearner: For the AIPW analysis, we utilized the AIPW package. Coupled with SuperLearner, a framework for ensemble learning, we enhanced the robustness of our causal estimates by stacking multiple models and optimizing both bias and variance.

The authors did not generate the synthetic data used in this analysis. We adhered to the exact Stata code provided in the cited paper by Yılmaz et al. (2024) to replicate the data generation process, except for modifying the exposure model with the addition of the Age^2 term.@@@

13.4.2 Naive ATE

First, we "naively" fit a linear outcome model (in particular, a GLM, with Gaussian link, which is equivalent to classical linear regression, OLS). We use the linear model fit to predict the outcome:

- Y_1, when we assume all units are treated ($W_i = 1, \forall\, i$)
- Y_0, when we assume all units are untreated ($W_i = 0, \forall\, i$)

We then compute the ATE as $Y_1 - Y_0$. To obtain valid standard errors (and confidence intervals), we bootstrap the process (based on $n_{boot} = 4000$ replications).

The results are summarized in Table 13.5. The correctly specified model yields an ATE of 10.55 with a standard error of 0.103, and a 95% confidence interval of (10.34, 10.74). However, omitting a key variable, *Age*, results in a misspecified model, producing a biased ATE of 8.67 and increased standard errors, highlighting the sensitivity to model specification.

13.4.3 Inverse Probability Weighted Estimates

Using IPW, we estimate separate exposure and outcome models, allowing for flexible modeling. Results are shown in Table 13.6. With both models correctly specified, the estimated ATE is 10.08, but misspecification of either model leads to biased results. When the outcome model is misspecified, the ATE is underestimated (9.18) with a lower standard error. Misspecifying the exposure model also results in bias, though less severe, with a reduced standard error.

Table 13.5 Comparison between the correctly specified and missepecified models for ATE.

Model Specification	ATE	Standard error	95% Confidence interval		Bootstrap replicates	Calculation scale
			Lower interval	Upper interval		
Naive ATE	10.55	0.103	10.34	10.74	4000	Original scale
Naive ATE, misspecified	8.67	0.154	8.36	8.97	4000	Original scale

Table 13.6 IPW-estimated ATE, standard error, and 95% confidence interval with correct and misspecified outcome and exposure models

Model Specification	Estimated ATE	Standard error	95% Confidence interval	
			Lower interval	Upper interval
IPW with correct outcome and exposure model	10.08	0.501	9.10	11.07
IPW with misspecified outcome model	9.18	0.358	8.48	9.88
IPW with misspecified exposure model	9.99	0.235	9.53	10.45

13.4.4 Doubly Robust Estimates with AIPW

The doubly robust AIPW estimator is applied to further address model misspecification. We explore stacking—a form of ensemble learning combining multiple models (e.g., linear models, and random forests)—to mitigate restrictive assumptions. Stacking provides flexibility by reducing reliance on precise model specification.

The results are summarized in Table 13.7. When both models are correctly specified and stacked with differences in means and GLM, the estimated ATE is 10.09. Stacking with random forests and regularization demonstrate stable ATE estimates, though slightly larger standard errors.

With misspecification, the estimated ATE is biased downward, with a misspecified outcome model causing larger bias than a misspecified exposure model. However, stacking shows resilience against misspecification, maintaining fairly stable estimates despite increased standard errors.

13.4.5 Key Takeaways

Both naive and IPW methods demonstrate significant sensitivity to model specification, underscoring the importance of including relevant covariates. Machine learning methods, especially AIPW with regularization and ensemble techniques, exhibit enhanced robustness to model misspecification. Stacking, in particular, reduces dependency on precise parametric forms, providing more reliable estimates.

While the case study uses synthetic data, the findings illustrate the potential value of CML methods, like regularization and ensemble techniques, in SCM. These methods offer robust tools for estimating causal effects, even amidst complex interactions and high-dimensional data. By applying these approaches, practitioners can achieve more accurate causal inferences, aiding strategic decision-making in dynamic environments like supply chain management.

Table 13.7 AIPW-estimated ATE, standard error, and 95% confidence interval, with stacking and correct and misspecified outcome and exposure models

Model Specification	Estimated ATE	Standard error	95% Confidence interval	
			Lower interval	Upper interval
Correct (stacked with stacked differences in means and GLM)	10.09	0.069	9.953	10.223
Correct (stacked with stacked difference in means, GLM, and RF)	10.13	0.114	9.903	10.349
Correct (stacked with stacked difference in means, GLM, RF, and GLMnet)	10.00	0.114	9.78	10.226
Misspecified outcome (stacked with differences in mean and GLM)	8.77	0.069	8.637	8.909
Misspecified outcome (stacked with differences in mean, GLM, and RF)	8.89	0.111	8.67	9.106
Misspecified outcome (stacked with differences in mean, GLM, RF, and GLMnet)	8.87	0.108	8.658	9.080
Misspecified exposure (stacked with differences in mean and GLM)	9.95	0.149	9.653	10.238
Misspecified exposure (stacked with differences in mean, GLM, and RF)	9.94	0.149	9.649	10.234
Misspecified exposure (stacked with differences in mean and GLM, RF, and GLMnet)	9.90	0.149	9.61	10.193

Note: *GLM* Generalized Linear Model; *GLMnet* Generalized Linear Model with elastic net regularization; *RF* random forests

13.5 Conclusions and Future Directions

CML is a recent scientific innovation that has emerged as a transformative approach for supply chain management, enabling more accurate predictions and deeper understanding into the underlying mechanism and drivers of supply chain performance. Traditional machine learning methods often excel at pattern recognition but struggle to establish causation, a crucial aspect for supply chains where decisions often depend on understanding the causal impact of various strategies. By integrating causal inference techniques into ML, CML provides a more robust framework for predicting how interventions (e.g., changes in inventory policies, supplier adjustments, or transportation optimizations) will affect outcomes, including cost, lead time, and service levels. SCM has experienced and benefited from the application of recent advancements of causal inference techniques, such as propensity score matching, synthetic controls, and instrumental variable approaches, which have been adapted to SCM settings. These methods allow managers to evaluate the impact of interventions across the supply chain, even in the absence of randomized experiments. For example, CML has been applied to assess the effects of quality assurance programs (e.g., Giamattei et al., 2024), analyze demand forecasting methods' impacts on inventory levels (e.g., Wibowo, 2024), determine how transportation mode and order size influence delivery performance (Bo & Xiao, 2024), and analyze the causal mechanism driving supplier delivery delays (Wyrembek et al., 2025).

However, the adoption of CML in SCM is still in its early stages. The dynamic, interdependent nature of supply chains poses challenges regarding data availability, the complexity of confounding factors, and the contextual uniqueness of supply chain interventions. To enhance the robustness and applicability of CML, future research should prioritize the development of models that can handle the high-dimensional and often non-linear nature of SCM data. Advanced methods could improve the precision of causal estimates and better account for the complexities in supply chains. A crucial aspect is that CML holds significant potential to revolutionize the field of SCM by shifting the focus from descriptive and predictive analytics to truly prescriptive analytics based on causal understanding. This approach may help link disparate supply chains to help identify weak points of intersection across spatial and temporal dimensions. Continued research into more adaptable, scalable, and interpretable causal models will likely drive broader adoption in industry, ultimately leading to more resilient, efficient, and responsive supply chains.

References

Angrist, J. D., & Krueger, A. B. (1991). Does compulsory school attendance affect schooling and earnings? *The Quarterly Journal of Economics, 106*(4), 979–1014. https://doi.org/10.2307/2937954

Angrist, J. D., & Pischke, J.-S. (2009). *Mostly harmless econometrics: An empiricist's companion*. Princeton University Press.

ALICE. (n.d.). Microsoft research. Retrieved August 12, 2024, from https://www.microsoft.com/en-us/research/group/alice/

Athey, S., & Imbens, G. W. (2019). Machine learning methods that economists should know about. *Annual Review of Economics, 11*, 685–725. https://doi.org/10.1146/annurev-economics-080217-053433

Athey, S., Tibshirani, J., & Wager, S. (2019). Generalized random forests. *The Annals of Statistics, 47*(2), 1148–1178. https://doi.org/10.1214/18-AOS1709

Belloni, A., Chernozhukov, V., & Hansen, C. (2014). High-dimensional methods and inference on structural and treatment effects. *Journal of Economic Perspectives, 28*(2), 29–50. https://doi.org/10.1257/jep.28.2.29

Bo, S., & Xiao, M. (2024). Root cause attribution of delivery risks via causal discovery with reinforcement learning. *Algorithms, 17*(11), 498.

Breiman, L. (2001). Statistical modeling: The two cultures (with comments and a rejoinder by the author). *Statistical Science, 16*(3), 199–231. https://doi.org/10.1214/ss/1009213726

Breiman, L. (1996). Bagging predictors. *Machine Learning, 24*(2), 123–140. https://doi.org/10.1023/A:1018054314350

Bodendorf, F., Sauter, M., & Franke, J. (2023). A mixed methods approach to analyze and predict supply disruptions by combining causal inference and deep learning. *International Journal of Production Economics, 256*, 108708. https://doi.org/10.1016/j.ijpe.2022.108708

Brand, J. E., Zhou, X., & Xie, Y. (2023). Recent developments in causal inference and machine learning. *Annual Review of Sociology, 49*, 81–110. https://doi.org/10.1146/annurev-soc-030420-015345

Burtell, M., & Toner, H. (2024). The surprising power of next word prediction: Large language models explained, part 1. *Georgetown Center for Security and Emerging Technology.*. https://cset.georgetown.edu/article/the-surprising-power-of-next-word-prediction-large-languagemodels-explained-part-1/

Carbonneau, R., Laframboise, K., & Vahidov, R. (2008). Application of machine learning techniques for supply chain demand forecasting. *European Journal of Operational Research, 184*(3), 1140–1154. https://doi.org/10.1016/j.ejor.2006.12.004

Cheikhrouhou, H. (2023, September 15). How big tech companies use ML and causal inference to make data-driven decisions. *Medium.* https://medium.com/@haytham99cheikhrouhou/how-amazon-facebook-netflix-and-spotify-use-causal-inference-to-make-data-driven-decisions-6290cd6d6dec

Chen, J. E., & Hsiang, C. W. (2019). Causal random forests model using instrumental variable quantile regression. *Econometrics, 7*(4), 49. https://doi.org/10.3390/econometrics7040049

Chen, S., & Kasiviswanathan, S. (2020, September 9). Machine learning tools increase power of hypothesis testing. *Amazon Science.* https://www.amazon.science/blog/machine-learning-tools-increase-power-of-hypothesis-testing

Dormady, N. C., Rose, A., Roa-Henriquez, A., & Morin, C. B. (2022). The cost-effectiveness of economic resilience. *International Journal of Production Economics, 244*, 108371. https://doi.org/10.1016/j.ijpe.2021.108371

Dutta, S., Abielmona, A., Rana, P., Wasan, S., Chowdhary, K., & Anurag, R. (2024). Implications of AI and generative AI for the future of supply chains. *EY Center for Executive Leadership.* https://assets.ey.com/content/dam/ey-sites/ey-com/en_us/topics/coo/ey-implications-of-ai-and-generative-ai-for-the-future-of-supply-chains.pdf

EasyBA.Co. (2023). Elevate your supply chain management strategy with a/B testing results insights. *EasyBA.Co.* https://easyba.co/blog/business-analysis/elevate-your-supply-chain-management-strategy-with-a-b-testing-results-insights/.

Engin, B., Lan, Y., Tang, G., Segalin, C., Griggs, K., & Iyengar, V. (2023, January 14). Causal machine learning for creative insights. *Medium.* https://netflixtechblog.com/causal-machine-learning-for-creative-insights-4b0ce22a8a96

Facure, M. (2023). *Causal inference in python.* O'Reilly Media, Inc.

Friedman, J. H. (2001). Greedy function approximation: A gradient boosting machine. *The Annals of Statistics, 29*(5), 1189–1232. https://doi.org/10.1214/aos/1013203451

Funk, M. J., Westreich, D., Wiesen, C., Stürmer, T., Brookhart, M. A., & Davidian, M. (2011). Doubly robust estimation of causal effects. *American Journal of Epidemiology, 173*(7), 761–767. https://doi.org/10.1093/aje/kwq439

Giamattei, L., Guerriero, A., Pietrantuono, R., & Russo, S. (2024). Causal reasoning in software quality assurance: A systematic review. *Information and Software Technology, 107599.*

Goodfellow, I., Bengio, Y., & Courville, A. (2016). *Deep learning.* MIT Press.

Götz, P., & Budhathoki, K. (2022, May 31). AWS contributes novel causal machine learning algorithms to DoWhy Python library. *Amazon Science.* https://www.amazon.science/blog/aws-contributes-novel-causal-machine-learning-algorithms-to-dowhy

Hastie, T., Tibshirani, R., & Friedman, J. (2009). *The elements of statistical learning: Data mining, inference, and prediction* (2nd ed.). Springer. https://doi.org/10.1007/978-0-387-84858-7

Helgeson, J. F., & Roa-Henriquez, A. (2022). A theoretical framework for supply chain resilience planning. In Y. Khojasteh, H. Xu, & S. Zolfaghari (Eds.), *Supply chain risk mitigation: Strategies, methods and applications* (pp. 159–189). Springer International Publishing. https://doi.org/10.1007/978-3-031-09183-4_8

Hendriksen, C. (2023). Artificial intelligence for supply chain management: Disruptive innovation or innovative disruption? *Journal of Supply Chain Management, 59*(3), 65–76. https://doi.org/10.1111/jscm.12304

Hernan, M. A., & Robins, J. M. (2024). *Causal inference: What if* (1st ed.). Taylor and Francis.

Huber, M. (2023). *Causal analysis: Impact evaluation and causal machine learning with applications in R.* The MIT Press.

Imbens, G. W. (2004). Nonparametric estimation of average treatment effects under exogeneity: A review. *Review of Economics and Statistics, 86*(1), 4–29.

Imbens, G. W. (2024). Causal inference in the social sciences. *Annual Review of Statistics and Its Application, 11*, 123–152. https://doi.org/10.1146/annurev-statistics-033121-114601

Imbens, G. W., & Rubin, D. B. (2015). Causal inference for statistics, social, and biomedical sciences: An introduction. *Cambridge University Press.* https://doi.org/10.1017/CBO9781139025751

Jacobs, F. R., & Chase, R. B. (2024). *Operations and supply chain management (seventeenth edition).* McGraw Hill.

Liddell, P. (2023, December 11). *Supply chain trends 2024: The digital shake-up–KPMG global.* KPMG. https://kpmg.com/xx/en/home/insights/2023/12/supply-chain-trends-2024.html

Loske, D., & Klumpp, M. (2022). Verifying the effects of digitalisation in retail logistics: An efficiency-centred approach. *International Journal of Logistics Research and Applications, 25*(2), 203–227. https://doi.org/10.1080/13675567.2020.1815681

Mechai, N., & Wicaksono, H. (2024). Causal inference in supply chain management: How does ever given accident at the Suez Canal affect the prices of shipping containers? *Procedia Computer Science, 232*, 3173–3182. https://doi.org/10.1016/j.procs.2024.02.133

Modgil, S., Singh, R. K., & Hannibal, C. (2021). Artificial intelligence for supply chain resilience: Learning from Covid-19. *The International Journal of Logistics Management, 33*(4), 1246–1268. https://doi.org/10.1108/IJLM-02-2021-0094

Mohri, M., Rostamizadeh, A., & Talwalkar, A. (2018). *Foundations of machine learning* (2nd ed.). MIT Press. https://books.google.com?id=dWB9DwAAQBAJ

Ni, D., Xiao, Z., & Lim, M. K. (2020). A systematic review of the research trends of machine learning in supply chain management. *International Journal of Machine Learning and Cybernetics, 11*(7), 1463–1482. https://doi.org/10.1007/s13042-019-01050-0

Pearson, B. (2021, September 7). *Personalizing price with AI: How Walmart, Kroger Do It.* Forbes. https://www.forbes.com/sites/bryanpearson/2021/09/07/personalizing-price-with-ai-how-walmart-kroger-do-it/

Robins, J. M., Rotnitzky, A., & Zhao, L. P. (1994). Estimation of regression coefficients when some regressors are not always observed. *Journal of the American Statistical Association, 89*(427), 846–866.

Rosenbaum, P. R., & Rubin, D. B. (1984). Reducing bias in observational studies using subclassification on the propensity score. *Journal of the American Statistical Association, 79*(387), 516–524.

Rudin, C., & Radin, J. (2019). Why are we using black box models in AI when we Don't need to? A lesson from an explainable AI competition. *Harvard Data Science Review, 1*(2). https://doi.org/10.1162/99608f92.5a8a3a3d

Schwarz, P., Schacht, O., Klaassen, S., Grünbaum, D., Imhof, S., & Spindler, M. (2024, June 17). *Management decisions in manufacturing using causal machine learning–to rework, or not to rework?* https://doi.org/10.48550/arXiv.2406.11308.

Simonite, T. (2023). The WIRED Guide to artificial intelligence. *Wired* https://www.wired.com/story/guide-artificial-intelligence/

Swift, C., Guide, V. D. R., & Muthulingam, S. (2019). Does supply chain visibility affect operating performance? Evidence from conflict minerals disclosures. *Journal of Operations Management, 65*(5), 406–429. https://doi.org/10.1002/joom.1021

Tay, J. K., Narasimhan, B., & Hastie, T. (2023). Elastic net regularization paths for all generalized linear models. *Journal of Statistical Software, 106*(1), 1–31. https://doi.org/10.18637/jss.v106.i01

Thomas, D., & Fung, J. F. (2022). Measuring downstream supply chain losses due to power disturbances. *Energy Economics., 114*, 106314. https://doi.org/10.1016/j.eneco.2022.106314

Thomas, D., & Helgeson, J. (2021). The effect of natural/human-made hazards on business establishments and their supply chains. *International Journal of Disaster Risk Reduction., 51*(1), 102257. https://doi.org/10.1016/j.ijdrr.2021.102257

Thomas, D., & Helgeson, J. (2022). The effect of natural Hazard damage on manufacturing value added and the impact of spatiotemporal data variations on the results. *International Journal of Disaster Risk Science., 13*, 691–704. https://doi.org/10.1007/s13753-022-00438-x

Tibshirani, R. (1996). Regression shrinkage and selection via the lasso. *Journal of the Royal Statistical Society: Series B (Methodological), 58*(1), 267–288. https://doi.org/10.1111/j.2517-6161.1996.tb02080.x

Tirkolaee, E. B., Sadeghi, S., Mooseloo, F. M., Vandchali, H. R., & Aeini, S. (2021). Application of machine learning in supply chain management: A comprehensive overview of the Main areas. *Mathematical Problems in Engineering, 2021*, 1–14. https://doi.org/10.1155/2021/1476043

Van der Laan, M. J., & Rose, S. (2011). *Targeted learning: causal inference for observational and experimental data* (Vol. 4). New York: Springer.

Wager, S. (2022). *STATS 361: Causal Inference*. Stanford University Course Notes. https://web.stanford.edu/~swager/stats361.pdf

Wang, G., Li, J., & Hopp, W. J. (2022). An instrumental variable forest approach for detecting heterogeneous treatment effects in observational studies. *Management Science, 68*(5), 3399–3418. https://doi.org/10.1287/mnsc.2021.4084

Wibowo, B. S. (2024). When and how to adjust statistical forecasts in supply chains? Insight from causal machine learning. *Journal of Business Analytics, 7*(1), 25–41.

Wolf, T., and March, C. (2024, June 24). *Decision Intelligence and Causal AI are transforming the future of manufacturing*. New SCMR. https://www.scmr.com/article/decision-intelligence-and-causal-ai-are-transforming-the-future-of-manufacturing

Wolpert, D. H. (1992). Stacked generalization. *Neural Networks, 5*(2), 241–259. https://doi.org/10.1016/S0893-6080(05)80023-1

Wyrembek, M., Baryannis, G., & Brintrup, A. (2025). Causal machine learning for supply chain risk prediction and intervention planning. *International Journal of Production Research*, 1–20.

Yılmaz, Ö., Son, Y., Shang, G., & Arslan, H. A. (2024). Causal inference under selection on observables in operations management research: Matching methods and synthetic controls. *Journal of Operations Management, 70*(5), 831–859. https://doi.org/10.1002/joom.1318

Zou, H., & Hastie, T. (2005). Regularization and variable selection via the elastic net. *Journal of the Royal Statistical Society Series B: Statistical Methodology, 67*(2), 301–320. https://doi.org/10.1111/j.1467-9868.2005.00503.x

Chapter 14
Digital Supply Chain Capabilities and Machine Learning Models for Demand Forecasting

Eric Lambourdière and Elsa Corbin

Abstract Supply chain digitalization correlates positively with organizations' performance. To address the complexities and demands of digital supply chains (DSCs), create value, and remain competitive, supply chain stakeholders must develop and adopt new, specific digital and analytical capabilities and DSC management practices. Organizations are aiming to create more efficient, connected, and responsive supply chains through essential practices such as real-time data exchange, interoperability, and collaborative decision making. Based on these considerations and in response to recent calls to evaluate the use of machine learning (ML) models for forecasting, we investigated whether ML-based demand forecasting outperforms classical time-series forecasting methods such as autoregressive integrated moving average calculation, linear regression, and naïve forecasting. The core contribution of this chapter is to demonstrate the performance of ML models with structured datasets in the improvement of the operations management activity of forecasting.

Keywords Digital supply chain capabilities · Machine learning · Ensemble learning · Demand forecasting

14.1 Introduction

In today's business environment, digital supply chain (DSC) transformation and thereby the evolution of organizations' DSC initiatives have become imperative. Digital technologies have progressed swiftly, contributing to the emergence of advanced digital and analytical capabilities that enhance automation, data-driven decision making, real-time visibility, and seamless integration across DSCs (Park,

---

E. Lambourdière · E. Corbin (✉)
Department of Transport and Logistics Management, Institute of Technology of Martinique, University of French West Indies, Schoelcher, Martinique
e-mail: elsa.corbin@univ-antilles.fr

Y. Khojasteh, S. M. M. Jahan Abadi (eds.), *Emerging Technologies in Supply Chains*, International Series in Operations Research & Management Science 373, https://doi.org/10.1007/978-3-032-01218-0_14

2016; Ricárdez-Estrada et al., 2024; Saberi et al., 2019; Shi et al., 2020). Supply chain leaders are proficient in technology integration, data-driven decision making, and collaboration (Ricárdez-Estrada et al., 2024). Supply chain members involved in digitalization initiatives can take full advantage of the momentum of DSC strategies in changing their operations.

The field of supply chain management has expanded, and information technology and automation have enabled the alteration of the operations and demand sides of supply chains. Accurate planning systems and appropriate decision-making processes are crucial to help businesses deal with fluctuating product and resource demands (Ahamed et al., 2023), and forecasting accuracy is a crucial contributor to their development (Hsieh et al., 2020). Sales forecasting is an integral process of matching demand and supply via operations planning (Mentzer & Moon, 2005). Demand is the responsibility of sales and marketing teams, and supply is the responsibility of supply chain members undertaking procurement, logistics, and distribution activities.

Supply chain digitalization correlates positively with organizations' performance (Wang & Prajogo, 2024). To address the complexities and demands of DSCs, create value, and remain competitive, supply chain stakeholders must develop and adopt new, specific digital and analytical capabilities and DSC management practices. Organizations are aiming to create more efficient, connected, and responsive supply chains through essential practices such as real-time data exchange, interoperability, and collaborative decision making (Ricárdez-Estrada et al., 2024).

Based on these considerations and in response to recent calls to evaluate the use of machine learning (ML) models for forecasting (Ahamed et al., 2023), we investigated whether ML-based forecasting outperforms classical time-series forecasting methods such as autoregressive integrated moving average (ARIMA) calculation, linear regression, and naïve forecasting. The core contribution of this chapter is to paint a clear picture of the use of ML for forecasting and its performance with structured datasets. The chapter is structured as follows. Section 10.2 contains an overview of relevant primary concepts based on a literature review and descriptions of the analytical models used in this study. In sect. 10.3, the wholesaler case study, dataset, data preprocessing, and model evaluation are described. In sect. 10.4, the numerical results obtained with real-world data in the case study are presented. The implications of these results are discussed in sect. 10.5. Conclusions are presented and future research directions are proposed in sect. 10.6.

14.2 Literature Review

14.2.1 Supply Chain Digitalization

In the current business context, supply chains face numerous challenges that some organizations have addressed by building sophisticated technological capabilities and becoming heavily involved in supply chain digitalization initiatives. The digital transformation of organizations' supply chain operations is a crucial component of the shift from a traditional supply-driven model to a demand-driven system (Chen & Jing, 2023; Yuan & Pan, 2023). A new generation of DSCs with enhanced connectivity, collaboration, and efficiency has provided organizations such as Alibaba with a new source of competitive advantage.

A DSC is "an intelligent best-fit technological system based on the capability of massive data disposal and excellent cooperation and communication to support and synchronise interactions between organisations by making their services more valuable, accessible, and affordable with consistent, agile, and effective outcomes" (Büyüközkan & Göçer, 2018, p. 165). DSCs are also described as "interconnected systems that can integrate different processes, customers, and suppliers, internally and externally via enhancing communication, collection and transfer of data and information" (Weerabahu et al., 2023, p. 3047). DSCs are more networked and flattened than are traditional supply chains. Their members identify process deficiencies, make data-driven decisions, and improve supply chain efficiency, thereby reducing costs and enhancing supply chain resilience (Abdalla & Nakagawa, 2021; He et al., 2024; Rashid et al., 2024; Wang et al., 2022).

Digital transformation has a positive influence on supply chain management (Li & Zhao, 2024), elevating organizations' internal efficiency and bringing significant synergetic effects (Vu & Asongu, 2023; Yuan et al., 2024; Zhang et al., 2023). In fact, the capacity to meet customer demands and serve emerging markets—in other words, supply chain efficiency—is a source of competitive advantage (Ricárdez-Estrada et al., 2024), at the core of which lies the development of capabilities (Yu et al., 2018). These capabilities, also known as "ilities," define the attributes of successful supply chains (De Weck et al., 2012; Ricárdez-Estrada et al., 2024; Ross & Rhodes, 2015). Digital capabilities that can help organizations optimize DSCs (Büyüközkan & Göçer, 2018; Dolgui & Ivanov, 2022; Frederico et al., 2023; Ivanov, 2022; Muafi & Sulistio, 2022; Park, 2016; Roy, 2021) are depicted in Fig. 14.1.

14.2.2 Artificial Intelligence Capabilities and DSC Management

The integration of cutting-edge technical solutions into organizations' processes is the core component of digitization strategies for sustained competitive advantage based on trustworthy, effective, and efficient supply chains (Apte & Petrovsky, 2016; Mohsen, 2023). Organizations that are not using digital technologies such as big data, blockchains, the cloud, and the internet of things are unable to use the data at their disposal for supply chain decision making. Supply chain management involves the coordination of upstream to downstream (e.g., production to distribution) logistics operations for the synchronized circulation of physical, financial, and informational flows (Oettmeier & Hofmann, 2016). It provides a solid foundation

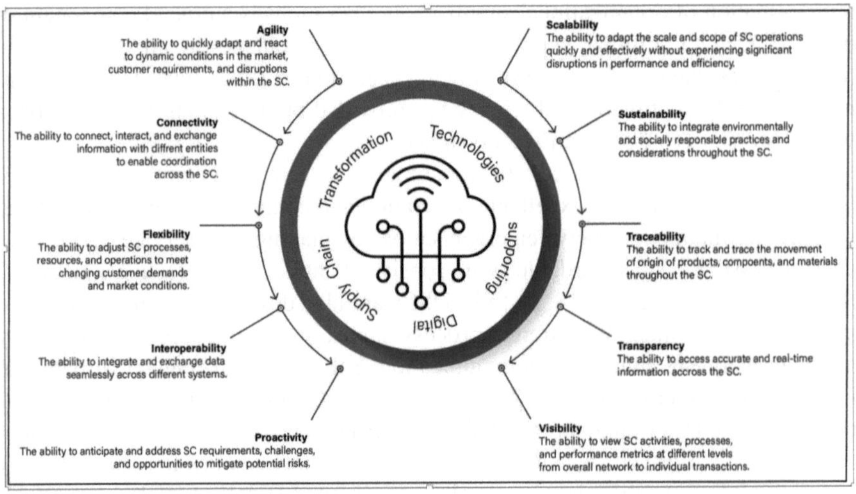

Fig. 14.1 Digital supply chain capabilities. Adapted from Ricárdez-Estrada et al. (2024)

for the smooth production and distribution of products within a given logistics system.

Organizations' digital technology adoption impacts supply chain information management (Mohsen, 2023). Artificial intelligence (AI) is among the most prominent digital technologies that can influence organizations' supply chain management practices (Raisch & Krakowski, 2021). This technology has been studied for a decade, and has only recently begun to provide value in these practices. The idea underlying the deployment of AI is to develop logistics systems that are capable of fulfilling tasks that formerly required humans, although the human factor remains key in supply chain and operations management (Brynjolfsson & Mcafee, 2017). Although computer-based decision making is not new, as is evident in the supply chain and operations management literature, AI applications for this purpose are novel due to their autonomous nature (Berente et al., 2021; Russell, 2022; Shrestha et al., 2019). These applications are capable of performing some tasks better than humans and do not have the limitations of human employees (Pournader et al., 2021). AI can reduce the cost and increase the efficiency of supply chain and operations management; it resolves problems more quickly and accurately, can enhance productivity, and is capable of managing large amounts of data (Toorajipour et al., 2021; Weisz et al., 2023). A comprehensively designed AI ecosystem allows seamless interactions between humans and machine and can form the core of the transformation from traditional to DSC management strategies (Galaz et al., 2021).

AI technology is based on three distinct but interrelated capabilities: the development of ML and deep learning, building of models to analyse datasets, and use of predictive analytics for accurate demand forecasting (Weisz et al., 2023). Among the mechanisms by which AI mimics human intelligence, ML is considered to be the most important (Adeyemi et al., 2024). ML can work 24 h a day and improve the efficiency and quality of supply chain management practices (Adeyemi et al., 2024), and ML and deep learning capabilities enable predictive forecasting (Weisz et al., 2023). The use of various ML models yields forecasting results that are more accurate than those obtained using popular statistical tools, such as exponential smoothing and the ARIMA (Riahi et al., 2021; Seyedan & Mafakheri, 2020).

The AI capabilities of learning, perception, prediction, interaction, adaptation, and reasoning have been applied strategically in supply chain and operations management (Jackson et al., 2024). Learning is the ability of machines to improve their performance through experience and databases. It may be classified as supervised, unsupervised, semi-supervised, and reinforcement learning (Berthelot et al., 2019; Hastie et al., 2009). In the supply chain management context, AI models learn from historical data and adapt to changing business environments, optimizing operations such as forecasting, inventory management, and route planning. Perception is the ability of a system to interpret sensory (visual, auditory, and tactile) data in the business environment. Computer vision and sensor-based technologies are the cornerstones of this AI capability. Perception provides real-time data capture and interpretation for supply chain actors (Chai et al., 2021; Purwins et al., 2019). Prediction is the ability to forecast outcomes based on past data, trends, and patterns, enabling more informed decision making in real time (Wu et al., 2022). Interaction encompasses communication between AI systems and humans or other systems via technologies such as natural language processing and chatbots (Panagou et al., 2024). Adaptation is a crucial feature for dynamic and complex systems, such as those in supply chains. It is the ability to change and improve behaviors and decision-making strategies while supply chain members consider new data, experiences, and/or business context changes. AI adapts using tools such as continuous learning and evolutionary models (Li et al., 2023; Xiao et al., 2014). Reasoning is the ability of machines to draw conclusions from available data and experience. It is crucial for decision making and planning in supply chain and operations management, as it involves the assessment of various possible outcomes and selection of optimal solutions (Leo Kumar, 2019; McDonnell et al., 2005) (Fig. 14.2).

Through the lens of RBV and Dynamic Capabilities, GAI is not merely an extension of traditional AI but a transformational force that unlocks new capabilities essential for competitive advantage. By enhancing learning, perception, prediction, interaction, adaptation, reasoning, and creativity, GAI equips firms with the tools to navigate complex, uncertain, and rapidly evolving business environments. In SCOM, its ability to generate, simulate, and optimize supply chain strategies in real-time marks a paradigm shift, paving the way for unprecedented operational resilience and innovation. Firms that effectively integrate GAI into their resource base and dynamic capabilities will be well-positioned to lead in the digital era of operations management.

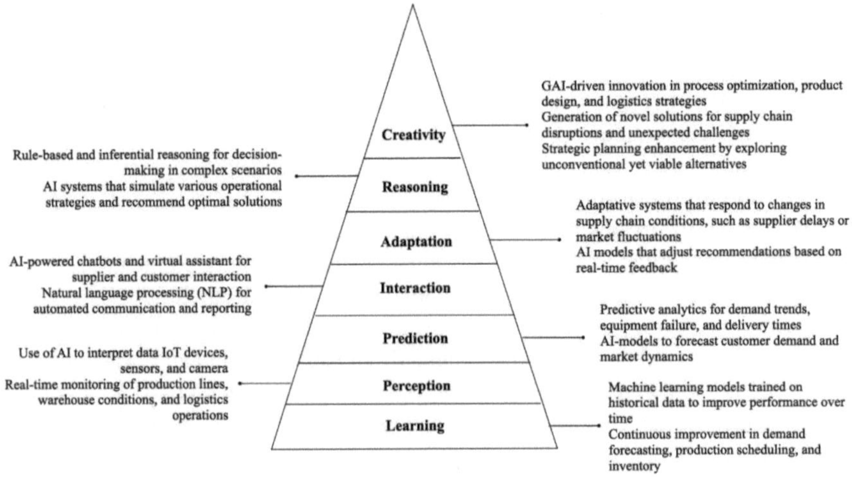

Fig. 14.2 AI Maturity pyramid in supply chain optimization—Source: Adapted from Jackson et al. (2024)

14.2.3 ML for Demand Forecasting

Demand forecasting plays an important, if not the most important, role in supply chain planning and decision-making processes (Ahamed et al., 2023). In a business environment in which organizations must deal with fluctuating product and resource demands, accurate planning systems for supply chain management cannot be neglected. Excess inventory necessitates a huge expenditure for storage and entails stock depreciation, and the likelihood that logistics systems contain some quantity of perishable items is very high. Organizations that must deal with such logistics issues lose money due to reductions in sales and customer satisfaction, and may have insufficient amounts of merchandise in their supply chains. Demand forecasting enables inventory management, with the aim of maintaining sufficient stock to satisfy customer demand while minimizing acquisitions and storage costs (Ahamed et al., 2023). Indeed, accurate demand forecasting can support inventory optimization (Graves & Willems, 2003) and the timely availability of products across supply chains, thereby playing a key role in the improvement of service levels and enhancement of customer satisfaction (Mentzer et al., 1999). It helps to align production schedules with anticipated demand, reducing lead times and avoiding overproduction (Jonsson & Mattsson, 2009). It also enables the efficient allocation of resources across supply chains.

Thus, demand forecasting directly affects the allocation of labor, machinery, and capital (Silver et al., 1998). Accordingly, demand management capability is a source of competitive advantage. Organizations that have been able to develop this ability plan and schedule asset acquisition accurately and in advance, estimating the amounts of raw materials, labor, and budget resources needed to produce specific items. Effective inventory management systems rely on accurate demand forecasts, which in turn increase profits, improve product design, and reduce losses and waste across supply chains (Goli et al., 2022). Demand forecasting helps organizations to make decisions and formulate plans. Thus, supply chain performance is related to its accuracy (Hsieh et al., 2020). As the odds that the future will be similar to the past are likely high (Srivastava et al., 2022), the analysis of past performance aids the prediction of product demand (Alhussein et al., 2020).

The design of nimble supply chains enables quick responses to changes in market demand. Accordingly, accurate real-time demand forecasting is needed for logistics systems to respond effectively to market volatility (Lee, 2004). However, the achievement of accurate demand forecasting in supply chain operations management entails numerous challenges, such as those posed by missing data, errors, and inconsistencies (Sanders & Manrodt, 2003). Customer demand is influenced by factors such as seasonality, trends, and the business environment, and such demand variability is difficult to capture with traditional forecasting models (Fildes et al., 2008). In addition, supply chains are complex flow systems in which multiple stakeholders manage unique forecasting requirements and lead times, and this complexity amplifies forecasting errors (Christopher, 2016). Moreover, external disruptions such as natural disasters and pandemics can disrupt supply chains and alter demand patterns (Ivanov & Dolgui, 2021). Small changes in demand at the customer level can cause important fluctuations in upstream supply chains (referred to as the bullwhip effect), making accurate forecasting difficult for supply chain actors (Lee et al., 2023). Finally, variations in suppliers' lead times are sources of additional complexity for demand planning and may cause forecasting errors (Chen & Lee, 2012). ML techniques effectively address these challenges, as they can handle complex patterns and high-dimensional data (Makridakis et al., 2018; Seeger et al., 2016), address data quality issues and improve accuracy (Guo et al., 2022; Xie et al., 2017), be used to simulate various scenarios and conduct "what-if" analyses (Bertsimas & Kallus, 2020), and provide real-time complex demand forecasting (Choi et al., 2020; Mutavhatsindi et al., 2020). ML can be used to analyse past data, identify trends and patterns, and make demand predictions accordingly. It can minimize the bullwhip effect (Bray & Mendelson, 2012), predict potential disruptions and adjust demand forecasting accordingly (Ivanov & Dolgui, 2021), capture seasonality and trends to manage demand variability (Smyl, 2020), and optimize inventory policies and lead time management (Oroojlooyjadid et al., 2017).

14.2.4 Demand Forecasting Models

14.2.4.1 Model M1: Naïve Prediction

M1 uses simple statistics (arithmetic means) to generate predictions and served as a baseline model in this study. In the time-series context, the mean of values covering a certain period is computed, and the predicted values are assumed to be equal to that mean. The naïve forecasting approach assumes that the demand for the next period will be the same as the current period.

14.2.4.2 Model M2: Linear Regression

M2 is a relatively simple statistical prediction method with which the linear relationships between input features and a target variable are modelled. Past sales data are used to establish a linear relationship between time and demand. The least squares equation is among the most commonly used in linear regression. This method enables the minimization of the residual sum of squares.

14.2.4.3 Model M3: Time-Series Modelling Family, ARMA (M3a) and ARIMA models (M3b)

Time-series forecasting problems can be solved using several techniques (Huang & Petukhina, 2022a, 2022b, 2022c; Ensafi et al., 2022). M3a have autoregressive (p) and moving average (q) components. Autoregressive outputs are obtained using the equation:

$$x_t = \vartheta_0 + \vartheta_1 x_{t-1} + \vartheta_2 x_{t-2} + \cdots + \vartheta_p x_{t-p} + \varepsilon_t = \vartheta_0 + \sum_{p}^{i=1} \vartheta_i x_{t-i} + \varepsilon_t, \tag{14.1}$$

where ϑ_0 is the constant, $\vartheta_{i\,(i\,=\,1,\,2,\,\ldots\ldots\ldots\ldots,\,k)}$ are the model parameters, p is the order of the time series model, ε_t represents random shocks at time t, and x_i are the observed values. Moving average outputs are obtained using the equation:

$$x_t = \mu + \varepsilon_t - \alpha\varepsilon_{t-2} - \cdots - \alpha_q \varepsilon_{t-q} = \mu + \varepsilon_t - \sum_{q}^{i=1} \alpha_{i\varepsilon_{t-i}}, \tag{14.2}$$

where μ is the mean of the series and $\alpha_{i\,(i\,=\,1,\,2,\,\ldots\ldots\ldots\ldots,\,q)}$ represents model parameters with order q, with random errors ε_t assumed as white noise. Combined M3a outputs are obtained using the equation:

$$y_t = \mu + \sum_{p}^{i=1} \vartheta_i x_{t-i} + \varepsilon_t - \sum_{q}^{i=1} \alpha_{1i\varepsilon_{t-i}}. \tag{14.3}$$

M3b (*p*, *d*, *q*), such as that tested in this study, incorporate a lag operator (*L*). Their outputs are obtained using the equation:

$$\vartheta = (L)(1-L)^d x_t = \mu + \alpha(L)_{\varepsilon_t}, \tag{14.4}$$

$$\left(1 - \sum_{p}^{i=1} \vartheta_i L^i\right)(1-L)^d x_t = \mu + \left(1 - \sum_{p}^{i=1} \alpha_i L^i\right)\varepsilon_t,$$

where d is the integrated term, or the difference required to achieve a stationary series.

14.2.4.4 Model M4: K–Nearest Neighbor Regression (K-NN)

M4 is another method that has been compared with ML forecasting techniques. M4 enables the creation of a function that employs Euclidian distance to describe the relationships between input and output variables. M4 can be used to analyse small numbers of observations. Its outputs are generated using the equation:

$$D(p,q) = \sqrt{(p_1 - q_1)^2 + (p_2 - q_2)^2 + \ldots\ldots + (p_n - q_n)^2}, \tag{14.5}$$

where p and q are variables to be compared with n characteristics.

14.2.4.5 Model M5: Random forest regression

M5 is an ensemble method based on decision trees. A decision tree is a flexible model that sequentially applies a series of hierarchical conditions from the "root" to the "leaves." M5 minimizes model bias and variance. Its outputs are generated using the equation:

$$\hat{f}_{\mathrm{RF}}^{C}(x) = \frac{1}{C} \sum_{C}^{i=1} T_i(x), \tag{14.6}$$

where $\mathcal{X}$ is a vectored input variable, the combination of the predictions of C decision trees produces an output, and $T_i(\mathcal{X})$ represents a single regression tree generated using a subset of input variables.

14.2.4.6 Model M6: Extra Tree Regression

M6 is a conventional top-down ensemble learning model developed from the random forest model. A set of unpruned regression trees is first generated, a split point is randomly selected for each feature, and the best split among these options is selected. This distinctive regression strategy balances unpredictability and accuracy. M6 outputs are generated using the equation:

$$Y = \frac{1}{N} \sum_{N}^{i=1} T_i\left(X\right), \tag{14.7}$$

where Y is the output that is determined by N number of trees in the ensemble, T_i is the ith tree in the ensemble, and X is the input feature vector.

14.2.4.7 Model M7: Gradient-Boosting Regression

With M7, a set of simple decision trees is combined to create a strong predictive model. This model uses gradient descent to update the weights of the decision trees on each repetition. The model iteratively gathers decision trees that are capable of correcting errors made by previous trees. On each repetition, it learns the difference between the previous model's predictions and the actual values for the target variable. M7 outputs are obtained using the equation:

$$f\left(x\right) = \sum_{M}^{m=1} \beta_m h_m\left(x\right), \tag{14.8}$$

where $f(x)$ is the predicted target variable, β_m is the weight assigned to the mth decision tree, $h_m(x)$ is the prediction of the mth decision tree for input x, and M is the number of trees in the model.

14.3 Methodology

14.3.1 Problem Context

The case investigated in this study is that of a US wholesaler that decided to maintain its supply chain relevance by adopting DSC practices for the movement of its products, which include shaving foam, razors, razor blades, and diapers. Omega (a fictitious name used for privacy protection) is a traditional wholesaler founded in 2017 that has 16 employees. The wholesaler recognized the need for accurate demand forecasting due to its significant impacts on inventory management and logistics, and thereby profitability. Historically, the company employed basic forecasting methods, such as M1 and M2, which yielded suboptimal results due to the dynamic nature of market demand. Although simple to implement, M1 did not meet Omega's needs because it fails to capture underlying patterns and seasonality in demand. The wholesaler thus introduced the use of M2, a slightly more sophisticated method. However, this method's assumption of a constant rate of change in demand over time does not align with the business environment in which Omega evolved. It produced inaccurate results, especially for periods with non-linear trends or external disruptions. Given these limitations of the traditional methods, the wholesaler adopted an ML-based approach to improve its forecasting accuracy.

14.3.2 Approach to Problem Analysis

For this analysis, we selected the time series for Omega's monthly shaving foam sales. The data preprocessing method used can have a large impact on the performance of univariate time-series forecasting models. We performed pre-processing to "clean" the datasets, for instance by removing outliers and addressing missing data, and to create additional features with the raw data to feed the models. In other words, as Ahamed et al. (2023) explained, we transformed the time series to make forecasting "easier" for the models. We performed feature engineering to transform the data into features (quantifiable variables) that served as model inputs. Feature selection depends on the problem being addressed, and the use of good-quality features can considerably improve overall model performance.

14.3.3 Model Types Tested

Several forecasting designs, whose characteristics are summarized in Table 14.1., were adopted. The ML design was used to split the data into training and testing sets (Shah, 2022) and to compare the forecasting ability of M5, M6, and M7 ensemble models. Ensemble learning involves the combined application of several models selected from a list of multiple regressors. Each regressor is trained and collectively contributes while making a prediction. Other classic and non-ensemble models used by wholesalers for forecasting, namely M1, M2, M3b, and M4, were also employed. Python libraries such as Scikit-learn 1.6.1. were used to train the models. Training involved hyperparameter optimization using grid search and cross-validation techniques.

Table 14.1 Characteristics of forecasting models used in supply chain and operations management

Code	Model	ML Model	Capabilities	Applications in Supply Chain & Ops	Pros	Cons
M1	Naïve forecasting	No	Basic forecasting based on simple parameters (e.g., past values or averages)	Demand forecasting, inventory level prediction	Simple, quick to implement, requires no training	Low accuracy, cannot capture trends or seasonality
M2	Linear regression	Yes	Predicts outcomes by fitting a linear relationship between input variables and output	Sales prediction, cost estimation, lead time prediction	Easy to interpret, fast, works well for simple linear problems	Assumes linearity, sensitive to outliers
M3 M3a M3b	Time-series modelling family ARMA ARIMA	No	Combines autoregression and moving average; ARIMA adds differentiation for non-stationary series	Demand forecasting, price forecasting, time series analysis	Good for univariate time series with trends/seasonality	Requires stationarity, parameter tuning is complex
M4	K-NN regression	Yes	Predicts output based on k-nearest neighbors in the feature space	Supply chain network optimization, demand prediction	Simple, non-parametric, works well with non-linear data	Sensitive to noise, computationally expensive for large datasets
M5	Random forest regression	Yes	Ensemble method using multiple decision trees	Demand forecasting, supplier risk assessment, inventory optimization	Handles high-dimensional data, reduces overfitting	Can be slow, less interpretable
M6	Extra tree regression	Yes	Similar to random forest regression but uses random splits for trees	Delivery time estimation, cost prediction, inventory optimization	Faster than random forest regression, reduces variance	Can overfit if not properly tuned
M7	Gradient-boosting regression	Yes	Ensemble method that builds trees sequentially to minimize errors	Demand forecasting, production schedule optimization, predictive maintenance	High accuracy, handles complex non-linear relationships	Computationally intensive, prone to overfitting without regularization

14.3.4 Model Evaluation

The trained models were evaluated by feeding them the test set, i.e., new restructured data with the same features as the training dataset (Lazzeri, 2020). We evaluated the models' ability to predict the target variable, or the prediction error, by comparing actual and predicted target values for a 5-month period (December 2023–April 2024). This quantitative measurement allowed us to fine tune the models and validate whether the error subsequently decreased. Prediction errors and time-series forecasting performance can be measured in various ways (Ensafi et al., 2022); in this study, the symmetric mean absolute percentage error (sMAPE), mean absolute percentage error (MAPE), root mean squared error (RMSE), and mean absolute scaled error (MASE) were used. The RMSE is the standard deviation of the predicted errors and is scale dependent. The MAPE is a function in which the error is expressed as a percentage (Hamzaçebi, 2008).

The formulas are:

$$sMAPE = \sum_{H}^{t=1} \frac{200|e_t|}{|y_t| + |\hat{y}_t|} \tag{14.9}$$

$$MAPE = \sum_{H}^{t=1} \frac{100|e_t|}{y_t} \tag{14.10}$$

$$RMSE = \sqrt{\frac{1}{H} \sum_{H}^{t=1} \left(e_t^2\right)} \tag{14.11}$$

$$MASE = \frac{1}{H} \sum_{H}^{t=1} |SE|, \tag{14.12}$$

where y_t and $\hat{y}_t$ are the actual observation and forecast for time t, respectively; H is the forecast horizon; M is the time series, which is indexed by m; and $e_t = y_t - \hat{y}_t$, giving the error at time t.

14.4 Results

Table 14.2. shows forecast values generated by the ML and statistical models tested compared to actual values (y) covering the study period. The actual values fluctuate notably over time, ranging from 1068 to 2496, reflecting important changes in trends. The forecasts differ widely among models, highlighting variations in the ability to capture temporal patterns and adapt to changing trends. The baseline M1 generated widely varying predictions, with the tendency to overestimate and underestimate relative to the other models. For instance, it significantly underestimated

Table 14.2. Forecasts generated by the models tested

Date	y	M1	M2	M4	M6	M5	M7	M3b
2023-12-01	1602	2039	2119	2509	2374	2362	2274	2498
2024-01-01	2304	1409	2655	2486	2183	2095	1898	2384
2024-02-01	1068	826	2139	1487	1629	1713	1553	1764
2024-03-01	2070	1626	1680	2307	2115	2150	1984	2516
2024-04-01	2496	1011	2343	1681	1786	1883	2386	1915

Note: y denotes the actual demand.

the January 2024 value of 2304, predicting a value of 1409.0. This performance reflects the simplicity of this model and its dependence on recent observations without trend extrapolation. The predictions generated with M2 are generally closer to the actual values, but exhibit notable variability. For instance, the predicted value of 2655.00 for January 2024 is an overestimation, and that of 2342.76 for April 2024 is also an overestimation ($y = 2496$) but is closer to the actual January 2024 value. M4 performed reasonably well in certain cases, such as for March 2024 (2306.90; $y = 2070$), but not in others, such as for April 2024 (1680.85). The predictions obtained with M6 largely remain within an acceptable range of the actual values, with some variation in performance. For example, the predicted value of 2115.37 for March 2024 is close to the actual value, whereas that of 1786.24 for April 2024 deviates significantly from the actual value. The predictions obtained with M5 are generally consistent, but this model sometimes underestimated slightly. For example, the predicted value of 2095.48 for January 2024 is an underestimation; that of 2149.63 for March 2024 is closer to the actual value. M7 shows strong performance in closely matching the actual values, with small deviations. The predicted value of 2385.80 for January 2024 nearly matches the actual value, and that of 1552.57 for February 2024 is also fairly close to the actual value of 1068. M3b also shows strong performance. For instance, the predicted value of 2515.70 for March 2024 is, a reasonable overestimation of the actual value. However, this model did not handle large deviations with trend changes well; the predicted value of 2497.97 for December 2023 is a considerable overestimation of the actual value of 1602.

Figure 14.3 presents a comparative analysis of forecasting models based on four evaluation metrics: Symmetric Mean Absolute Percentage Error (sMAPE), Mean Absolute Percentage Error (MAPE), Root Mean Squared Error (RMSE), and Mean Absolute Scaled Error (MASE). Each subplot compares the performance of various models including M7, M5, M6, M2, M4, M3b, and M1. Lower values across all four metrics indicate better forecasting performance, as they represent smaller deviations between predicted and actual values. As shown, M7 consistently achieves the lowest error values, indicating superior predictive accuracy compared to other models (sMAPE = 0.1835, MAPE = 0.2074, RMSE = 389.43, MASE = 0.6023).

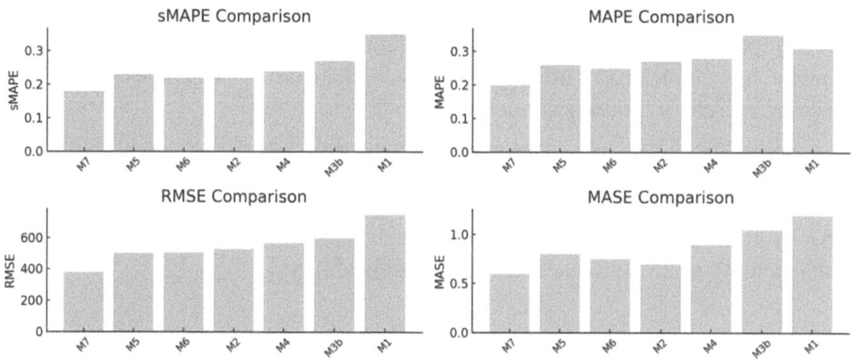

Fig. 14.3 Model performance graph

14.5 Discussion

The comparative analysis of forecasting model performance provides valuable insights into the models' suitability for application in supply chain and operations management, particularly for wholesaling activities. In this study, M7 performed best according to all evaluation metrics applied, achieving the most accurate demand forecasting. These results indicate that Omega should prioritize the adoption of M7 and M5 to improve the accuracy of its forecasting, particularly in the management of demand variability. They align with recent findings reflecting the ability of ensemble methods such as M7 to capture complex, non-linear data patterns (El Alaoui et al., 2022; Wu & Levinson, 2021). Unlike simpler models, such as those based on the M3b or M2, M7 can be used to effectively model interactions among features such as seasonality, promotional effects, and unexpected demand spikes. Although its output is often considered to be less interpretable than that of traditional models, modern implementations enable the understanding of feature importance. This capability enables the identification of key aspect of demand, such as products, categories of items, customer segments, and regional trends. The insights obtained from M7 can be incorporated into strategic decision making. By iteratively adding weak learners and fine tuning their weights, this technique minimizes overfitting relative to that occurring with other ensemble methods. This feature is particularly beneficial in forecasting applications for which historical demand data may include outliers or noise. M7 also allows for the customization of loss functions, enabling optimization for specific metrics, such as the sMAPE or RMSE. This adaptability makes it a versatile tool that can be used to meet diverse forecasting objectives, such as the minimization of absolute percentage errors for short-term planning or squared errors for the management of long-term trends. In wholesalers' operations, intermittent demand patterns pose significant challenges due to the irregular and unpredictable nature of orders. Highly precise M7 can be employed to reduce inventory-related costs. In addition, their robustness allows supply chain members to conduct scenario analyses with the adjustment of key input variables, such as the expected demand for a new product line or changes in market conditions. Planning based on such analysis can bolster decision support systems and capabilities. The scalability of M7 makes it suitable for integration with enterprise resource planning and supply chain management. Supply chain stakeholders can leverage this model to optimize inventory replenishment schedules and reduce stockout, thereby improving service.

The findings of this study also underscore the importance of using multiple error metrics to evaluate forecasting models. The observed differences in performance across metrics highlights the need to use a theoretical framework that accommodates multidimensional evaluation. Such a framework provides a more holistic understanding of model effectiveness and applicability.

14.6 Conclusions, Limitations, and Directions for Future Research

Demand forecasting is a crucial function within SCOM. Managers rely on this function to take strategic decision related to inventory, production, and distribution. The integration of AI into demand forecasting has enhance the effectiveness activities related to this function. Indeed, the AI-driven technology models can detect complex patterns one can find in a dataset and produce accurate predictions. Due to their adaptive learning capabilities, AI models can improve forecasting accuracy. The outcome of this case study offers a comprehensive insight in the functionalities and practical applications of AI in the domain of SCOM. The research highlights the potential and implications of AI through forecasting demand processes enhancements.

The case study discussed in this research provides a comprehensive overview of AI functionalities and their practical implementation in SCOM. By demonstrating the enhancements achievable through AI in demand forecasting processes, the research underscores the extensive potential and impactful implications of AI for SCOM.

Additionally, a positive case study serves as a valuable resource for both researchers and practitioners in the field. For researchers, it offers empirical evidence that validates theoretical assumptions and guides further investigation into innovative applications of AI. For managers, it provides practical examples and benchmarks that can inform strategic decisions, highlighting specific operational improvements and potential efficiency gains achievable through AI integration. Thus, a successful case study fosters greater confidence in adopting AI solutions, ultimately encouraging broader implementation and continuous innovation within supply chain and operations management.

This study provides evidence that ensemble ML models, particularly M7, significantly outperfom traditional demand forecasting approaches. The ability of these models to capture non-linear relationships, adapt to meet various forecasting objectives, and deliver highly accurate results make them indispensible tools for wholasalers aiming to optimize their supply chain operations.

The study has certain limitations. As it was performed with a dataset from a single wholesaler, the findings me be highly contextual and not generalizable to other industries or supply chain contexts. In addition, the analysis relied solely on historical demand data, with no consideration of external factors such as market trends, promotional activities, market disruptions, or macroeconomic indicators. These variables could significently influence the forecasting accuracy and should be included in future studies for the more comprehensive evaluation of forecasting model performance. Moreover, although a range of models was evaluated in this study, hybrid approches were not explored. Hybrid models may offer significantly improved performance via the leveraging of the strengths of both statistical and ML methods. Finally, the successful implementation of M7 requires careful consideration of computational resources, data quality and availability, and output interpretability, underscoring the importance of aligning forecasting solutions with organizational capabilities. Future research should focus on expanding the generalizability of the present findings by testing model performance across diverse industries, incorporating external variables, and exploring hybrid forecasting methods. By addressing the limitations of this study, researchers can provide more robust recommendations for model selection and implementation, ultimately advancing the theory and practice of supply chain demand forecasting.

References

Abdalla, S., & Nakagawa, K. (2021). The interplay of digital transformation and collaborative innovation on supply chain ambidexterity. *Technology Innovation Management Review, 11*(3).

Adeyemi, O. A., Pinto, P. M. G., Sunmola, F., Aibinu, A. M., Okesola, J. O., & Adeyemi, E. O. (2024). Towards the adoption of industry 4.0 technologies in the digitalization of manufacturing supply chain. *Procedia Computer Science, 232*, 337–347.

Ahamed, S. F., Vijayasankar, A., Thenmozhi, M., Rajendar, S., Bindu, P., & T. Subha Mastan Rao. (2023). Machine learning models for forecasting and estimation of business operations. *The Journal of High Technology Management Research, 34*(1), 100455.

Alhussein, M., Aurangzeb, K., & Haider, S. I. (2020). Hybrid Cnn-Lstm model for short-term individual household load forecasting. *IEEE Access, 8*, 180544–180557.

Apte, S., & Petrovsky, N. (2016). Will blockchain technology revolutionize excipient supply chain management? *Journal of Excipients and Food Chemicals, 7*(3), 76–78.

Berente, N., Bin, G., Recker, J., & Santhanam, R. (2021). Managing artificial intelligence. *MIS Quarterly, 45*, 1433–1450.

Berthelot, D., Carlini, N., Goodfellow, I., Papernot, N., Oliver, A., & Raffel, C. A. (2019). Mixmatch: A holistic approach to semi-supervised learning. *Advances in Neural Information Processing Systems, 32*.

Bertsimas, D., & Kallus, N. (2020). From predictive to prescriptive analytics. *Management Science, 66*(3), 1025–1044.

Bray, R. L., & Mendelson, H. (2012). Information transmission and the bullwhip effect: An empirical investigation. *Management Science, 58*(5), 860–875.

Brynjolfsson, E., & Mcafee, A. (2017). Artificial intelligence, for real. *Harvard Business Review, 1*, 1–31.

Büyüközkan, G., & Göçer, F. (2018). Digital supply chain: Literature review and a proposed framework for future research. *Computers in Industry, 97*, 157–177.

Chai, J., Zeng, H., Li, A., & Ngai, E. W. T. (2021). Deep learning in computer vision: A critical review of emerging techniques and application scenarios. *Machine Learning with Applications, 6*, 100134.

Chen, L., & Lee, H. L. (2012). Bullwhip effect measurement and its implications. *Operations Research, 60*(4), 771–784.

Chen, Y., & Jing, X. (2023). Digital transformation and firm cost stickiness: Evidence from China. *Finance Research Letters, 52*, 103510.

Choi, E., Cho, S., & Kim, D. K. (2020). Power demand forecasting using long short-term memory (Lstm) deep-learning model for monitoring energy sustainability. *Sustainability, 12*(3), 1109.

Christopher, M. (2016). *Logistics and supply chain management* (5th ed.). Pearson.

De Weck, O. L., Ross, A. M., & Rhodes, D. H. (2012). *Investigating relationships and semantic sets amongst system lifecycle properties (Ilities)*. Third International Engineering Systems Symposium, TU Delft, June 18-20.

Dolgui, A., & Ivanov, D. (2022). 5G in digital supply chain and operations management: Fostering flexibility, end-to-end connectivity and real-time visibility through internet-of-everything. *International Journal of Production Research, 60*(2), 442–451.

El Alaoui, O., Zerouaoui, H., & Idri, A. (2022). Deep stacked ensemble for breast cancer diagnosis. In *Information systems and technologies* (pp. 435–445). Springer International Publishing.

Ensafi, Y., Amin, S. H., Zhang, G., & Shah, B. (2022). Time-series forecasting of seasonal items sales using machine learning–a comparative analysis. *International Journal of Information Management Data Insights, 2*(1), 100058.

Fildes, R., Nikolopoulos, K., Crone, S. F., & Syntetos, A. A. (2008). Forecasting and operational research: A review. *Journal of the Operational Research Society, 59*(9), 1150–1172.

Frederico, G. F., Kumar, V., Garza-Reyes, J. A., Kumar, A., & Agrawal, R. (2023). Impact of I4. 0 technologies and their interoperability on performance: Future pathways for supply chain resilience post-covid-19. *The International Journal of Logistics Management, 34*(4), 1020–1049.

Galaz, V., Centeno, M. A., Callahan, P. W., Causevic, A., Patterson, T., Brass, I., Baum, S., Farber, D., Fischer, J., & Garcia, D. (2021). Artificial intelligence, systemic risks, and sustainability. *Technology in Society, 67*, 101741.

Goli, A., Golmohammadi, A.-M., & Edalatpanah, S. A. (2022). Application of artificial intelligence in forecasting the demand for supply chains considering industry 4.0. In J. M. Chatterjee, H. Garg, & R. N. Thakur (Eds.), *A roadmap for enabling industry 4.0 by artificial intelligence* (pp. 43–55).

Graves, S. C., & Willems, S. P. (2003). Supply chain design: Safety stock placement and supply chain configuration. In S. C. Graves & A. G. de Kok (Eds.), *Handbooks in operations research and management science* (pp. 95–132). Elsevier.

Guo, Z., Keping, Y., Li, Y., Srivastava, G., & Lin, J. C.-W. (2022). Deep learning-embedded social internet of things for ambiguity-aware social recommendations. *IEEE Transactions on Network Science and Engineering, 9*(3), 1067–1081.

Hamzaçebi, C. (2008). Improving artificial neural networks' performance in seasonal time series forecasting. *Information Sciences, 178*(23), 4550–4559.

Hastie, T., Tibshirani, R., & Friedman, J. (2009). Unsupervised learning. In T. Hastie, R. Tibshirani, & J. Friedman (Eds.), *The elements of statistical learning: Data mining, inference, and prediction* (pp. 485–585). Springer New York.

He, J., Fan, M., & Fan, Y. (2024). Digital transformation and supply chain efficiency improvement: An empirical study from a-share listed companies in China. *PLoS One, 19*(4), e0302133.

Hsieh, M.-C., Giloni, A., & Hurvich, C. (2020). The propagation and identification of arma demand under simple exponential smoothing: Forecasting expertise and information sharing. *IMA Journal of Management Mathematics, 31*(3), 307–344.

Huang, C., & Petukhina, A. (2022a). Arma and arima modeling and forecasting. In C. Huang & A. Petukhina (Eds.), *Applied time series analysis and forecasting with python* (pp. 107–142). Springer International Publishing.
Huang, C., & Petukhina, A. (2022b). Exploratory time series data analysis. In C. Huang & A. Petukhina (Eds.), *Applied time series analysis and forecasting with python* (pp. 37–69). Springer International Publishing.
Huang, C., & Petukhina, A. (2022c). Stationary time series models. In C. Huang & A. Petukhina (Eds.), *Applied time series analysis and forecasting with python* (pp. 71–106). Springer International Publishing.
Ivanov, D. (2022). Viable supply chain model: Integrating agility, resilience and sustainability perspectives—lessons from and thinking beyond the covid-19 pandemic. *Annals of Operations Research, 319*(1), 1411–1431.
Ivanov, D., & Dolgui, A. (2021). Or-methods for coping with the ripple effect in supply chains during Covid-19 pandemic: Managerial insights and research implications. *International Journal of Production Economics, 232*, 107921.
Jackson, I., Ivanov, D., Dolgui, A., & Namdar, J. (2024). Generative artificial intelligence in supply chain and operations management: A capability-based framework for analysis and implementation. *International Journal of Production Research, 62*(17), 6120–6145.
Jonsson, P., & Mattsson, S. A. (2009). *Manufacturing planning and control*. McGraw-Hill Higher Education.
Lazzeri, F. (2020). *Machine Learning for Time Series Forecasting with Python*. John Wiley & Sons.
Lee, H. L. (2004). The triple-a supply chain. *Harvard Business Review, 82*(10), 102–112.
Lee, S., Park, S. J., & Seshadri, S. (2023). Variations of the bullwhip effect across foreign subsidiaries. *Manufacturing & Service Operations Management, 25*(1), 1–18.
Leo Kumar, S. P. (2019). Knowledge-based expert system in manufacturing planning: State-of-the-art review. *International Journal of Production Research, 57*(15–16), 4766–4790.
Li, J., Jin, L., Wang, Z., Peng, Q., Wang, Y., Luo, J., Zhou, J., Cao, Y., Zhang, Y., Zhang, M., Qiu, Y., Qiang, H., Chen, L., Xiaoyu, Y., Zhou, X., Li, Q., Zhou, S., Huang, S., Luo, D., Mao, X., Yi, Y., Yang, X., Pan, C., Li, H., Wang, J., & Liao, J. (2023). Towards precision medicine based on a continuous deep learning optimization and ensemble approach. *npj Digital Medicine, 6*(1), 18.
Li, P., & Zhao, X. (2024). The impact of digital transformation on corporate supply chain management: Evidence from listed companies. *Finance Research Letters, 60*, 104890.
Makridakis, S., Spiliotis, E., & Assimakopoulos, V. (2018). Statistical and Machine Learning Forecasting Methods: Concerns and Ways Forward. *PLoS One, 13*(3), e0194889.
McDonnell, P., Joshi, S., & Qiu, R. G. (2005). A learning approach to enhancing machine reconfiguration decision-making games in a heterarchical manufacturing environment. *International Journal of Production Research, 43*(20), 4321–4334.
Mentzer, J. T., & Moon, M. A. (2005). *Sales forecasting management: A demand management approach*. Sage Publications.
Mentzer, J. T., Flint, D. J., & Kent, J. L. (1999). Developing a logistics service quality scale. *Journal of Business Logistics, 20*.
Mohsen, B. M. (2023). Developments of digital technologies related to supply chain management. *Procedia Computer Science, 220*, 788–795.
Muafi, M., & Sulistio, J. (2022). A nexus between green intellectual capital, supply chain integration, digital supply chain, supply chain agility, and business performance. *Journal of Industrial Engineering and Management, 15*(2), 275–295.
Mutavhatsindi, T., Sigauke, C., & Mbuvha, R. (2020). Forecasting hourly global horizontal solar irradiance in south africa using machine learning models. *IEEE Access, 8*, 198872–198885.
Oettmeier, K., & Hofmann, E. (2016). Impact of additive manufacturing technology adoption on supply chain management processes and components. *Journal of Manufacturing Technology Management, 27*(7), 944–968.

Oroojlooyjadid, A., Nazari, M., Snyder, L., & Takáč, M. (2017). A deep q-network for the beer game: A reinforcement learning model to solve inventory optimization problems. In *Neural Information Processing Systems (NIPS), Deep Reinforcement Learning Symposium 2017*.

Panagou, S., Patrick Neumann, W., & Fruggiero, F. (2024). A scoping review of human robot interaction research towards industry 5.0 human-centric workplaces. *International Journal of Production Research, 62*(3), 974–990.

Park, S. (2016). Development of innovative strategies for the korean manufacturing industry by use of the connected smart factory (Csf). *Procedia Computer Science, 91*, 744–750.

Pournader, M., Ghaderi, H., Hassanzadegan, A., & Fahimnia, B. (2021). Artificial intelligence applications in supply chain management. *International Journal of Production Economics, 241*, 108250.

Purwins, H., Li, B., Virtanen, T., Schlüter, J., Chang, S. Y., & Sainath, T. (2019). Deep learning for audio signal processing. *IEEE Journal of Selected Topics in Signal Processing, 13*(2), 206–219.

Raisch, S., & Krakowski, S. (2021). Artificial intelligence and management: The automation–augmentation paradox. *Academy of Management Review, 46*(1), 192–210.

Rashid, A., Rasheed, R., Ngah, A. H., Jayaratne, M. D. R. P., Rahi, S., & Tunio, M. N. (2024). Role of information processing and digital supply chain in supply chain resilience through supply chain risk management. *Journal of Global Operations and Strategic Sourcing, 17*(2), 429–447.

Riahi, Y., Saikouk, T., Gunasekaran, A., & Badraoui, I. (2021). Artificial intelligence applications in supply chain: A descriptive bibliometric analysis and future research directions. *Expert Systems with Applications, 173*, 114702.

Ricárdez-Estrada, J., Garay-Rondero, C. L., Romero, D., Wuest, T., & Pinto, R. (2024). Identifying digital supply chain capabilities. *Procedia Computer Science, 232*, 1182–1191.

Ross, A. M., & Rhodes, D. H. (2015). Towards a prescriptive semantic basis for change-type ilities. *Procedia Computer Science, 44*, 443–453.

Roy, V. (2021). Contrasting supply chain traceability and supply chain visibility: Are they interchangeable? *The International Journal of Logistics Management, 32*(3), 942–972.

Russell, S. (2022). Artificial intelligence and the problem of control. In H. Werthner, E. Prem, E. A. Lee, & C. Ghezzi (Eds.), *Perspectives on digital humanism* (pp. 19–24). Springer International Publishing.

Saberi, S., Kouhizadeh, M., Sarkis, J., & Shen, L. (2019). Blockchain technology and its relationships to sustainable supply chain management. *International Journal of Production Research, 57*(7), 2117–2135.

Sanders, N. R., & Manrodt, K. B. (2003). The efficacy of using judgmental versus quantitative forecasting methods in practice. *Omega, 31*(6), 511–522.

Seeger, M. W., Salinas, D., & Flunkert, V. (2016). Bayesian intermittent demand forecasting for large inventories. In *30th Conference on Neural Information Processing Systems (NIPS 2016)*.

Seyedan, M., & Mafakheri, F. (2020). Predictive big data analytics for supply chain demand forecasting: methods, applications, and research opportunities. *Journal of Big Data, 7*(1), 53.

Shah, C. (2022). *A hands-on introduction to machine learning*. Cambridge University Press.

Shi, Z., Xie, Y., Xue, W., Chen, Y., Liuliu, F., & Xiaobo, X. (2020). Smart factory in industry 4.0. *Systems Research and Behavioral Science, 37*(4), 607–617.

Shrestha, Y. R., Ben-Menahem, S. M., & von Krogh, G. (2019). Organizational decision-making structures in the age of artificial intelligence. *California Management Review, 61*(4), 66–83.

Silver, E., Pyke, D., & Peterson, R. (1998). *Inventory management and production scheduling* (Vol. 19, 3rd ed.). Wiley.

Smyl, S. (2020). A hybrid method of exponential smoothing and recurrent neural networks for time series forecasting. *International Journal of Forecasting, 36*(1), 75–85.

Srivastava, A. K., Tripathi, S. M., Kumar, S., Elavarasan, R. M., Gangatharan, S., Kumar, D., & Mihet-Popa, L. (2022). Machine learning approach for forecast analysis of novel covid-19 scenarios in India. *IEEE Access, 10*, 95106–95124.

Toorajipour, R., Sohrabpour, V., Nazarpour, A., Oghazi, P., & Fischl, M. (2021). Artificial intelligence in supply chain management: A systematic literature review. *Journal of Business Research, 122*, 502–517.

Vu, K., & Asongu, S. (2023). Patterns and drivers of financial sector growth in the digital age: insights from a study of industrialized economies. *Research in International Business and Finance, 66*, 102075.

Wang, M., & Prajogo, D. (2024). The effect of supply chain digitalisation on a firm's performance. *Industrial Management & Data Systems, 124*(5), 1725–1745.

Wang, X., Kumar, V., Kumari, A., & Kuzmin, E. (2022). Impact of digital technology on supply chain efficiency in manufacturing industry. In *Digital transformation in industry, 347-371*. Springer International Publishing.

Weerabahu, W. M. S., Kumari, P. S., Nakandala, D., & Hurriyet, H. (2023). Digital supply chain research trends: A systematic review and a maturity model for adoption. *Benchmarking: An International Journal, 30*(9), 3040–3066.

Weisz, E., Herold, D. M., & Kummer, S. (2023). Revisiting the Bullwhip Effect: How can AI Smoothen The bullwhip Phenomenon? *The International Journal of Logistics Management, 34*(7), 98–120.

Wu, H., & Levinson, D. (2021). The ensemble approach to forecasting: A review and synthesis. *Transportation Research Part C: Emerging Technologies, 132*, 103357.

Wu, J., Zhang, Z., & Zhou, S. X. (2022). Credit rating prediction through supply chains: A machine learning approach. *Production and Operations Management, 31*(4), 1613–1629.

Xiao, J.-h., Liu, B.-l., Huang, Y.-f., & Cheng, Z. (2014). An adaptive quantum swarm evolutionary model for partner selection in virtual enterprise. *International Journal of Production Research, 52*(6), 1607–1621.

Xie, K., Li, X., Wang, X., Xie, G., Wen, J., Cao, J., & Zhang, D. (2017). Fast tensor factorization for accurate internet anomaly detection. *IEEE/ACM Transactions on Networking, 25*(6), 3794–3807.

Yu, W., Chavez, R., Jacobs, M. A., & Feng, M. (2018). Data-driven supply chain capabilities and performance: A resource-based view. *Transportation Research Part E: Logistics and Transportation Review, 114*, 371–385.

Yuan, S., & Pan, X. (2023). The effects of digital technology application and supply chain management on corporate circular economy: A dynamic capability view. *Journal of Environmental Management, 341*, 118082.

Yuan, Y., Tan, H., & Liu, L. (2024). The effects of digital transformation on supply chain resilience: A moderated and mediated model. *Journal of Enterprise Information Management, 37*(2), 488–510.

Zhang, Z., Zhi, S., & Tong, F. (2023). Does digital transformation restrain corporate financialization? Evidence from China. *Finance Research Letters, 56*, 104152.

Chapter 15
Developing a Framework for Using Generative AI in Product Design

Tung Nhu Nguyen

Abstract Visually appealing product images are likely to positively impact online purchase behavior. This chapter presents a framework using Google Colab with Gemini and Vertex AI to create attractive product images, reducing design time and increasing personalization. Compared to models like Midjourney, it is more reliable by training on secondary and primary user data. Designers input simple text, and the AI generates a high-quality product image in seconds, which can be tailored iteratively. An example of bicycle design validated the output's aesthetics and personalization. Improvement features include gender considerations and ergonomics. The model is fast, simple, customizable, and suitable for non-technical users, but still requires human input to fix imperfections and address risks.

Keywords Product design · Kano's Model · House of Quality · Google Colab

15.1 Introduction

Artificial intelligence (AI) technology employs tools, methods, and applications that rapidly transform how businesses operate. For example, businesses can employ machine learning methods to predict sales, optimize prices, monitor employee performance, and predict attrition (Sharma & Gonaygunta, 2023). According to a global 2024 SAP report, 63% of the 2000 supply chain executives align AI strategies with business-level objectives, and 51% use AI for real-time business decisions.[1]

[1] https://www.sap.com/insights/research/ai-adoption-in-2024-sap-survey-insights.html

T. N. Nguyen (✉)
School of Business, International University–Vietnam National University Ho Chi Minh City, Ho Chi Minh City, Vietnam
e-mail: nntung@hcmiu.edu.vn

Y. Khojasteh, S. M. M. Jahan Abadi (eds.), *Emerging Technologies in Supply Chains*, International Series in Operations Research & Management Science 373, https://doi.org/10.1007/978-3-032-01218-0_15

15.1.1 Demand for Generative AI Applications in Product Design

Generative AI plays a crucial role in product design and development (Yehia, 2024). One of the most advanced AI applications in design is high-quality image creation, which can be used for product design and advertisement. This new technological application has more advantages than traditional tools do (Burnap et al., 2023). For example, traditional computer-aided design (CAD) uses computer graphics to assist in drawing products before they are produced. While a CAD designer can modify the monitor's design via communication devices such as a keyboard, a pen, or a joystick, AI-assisted design offers more benefits to designers. AI-assisted designers can use AI tools to quickly access large amounts of data from social media, emails, and the web to understand consumer preferences and behavior. It reduces product development time and creates products with specifications that best meet customers' requirements. For example, "machine learning" AI chatbots can be trained on datasets containing human conversational texts from messenger apps to understand what people want so that they can propose what product features can be added to meet their needs (Adulyasak et al., 2024). In the ideation phase of a product development project, AI-driven diffusion models can create visually pleasing product images from text descriptions (Chen et al., 2024; Yang et al., 2024).

Given those opportunities and needs, this chapter focuses on how to integrate two theoretical concepts in designing products: Kano's model for maximizing customer satisfaction and the quality function deployment (QFD) House of Quality, with the assistance of AI (Ji et al., 2014). AI tools are necessary to design new products. Product differentiation requires increasing "attractive" characteristics or aesthetics for a product. In aesthetic design, an "attractive" product is sporty, appealing, innovative, or luxurious. This product strategy results in economic benefits. Product aesthetics account for as many as 60% of consumers' purchasing decisions (Burnap et al., 2023). Research conducted on 93 companies spanning nine different product categories revealed that companies that made substantial investments in aesthetic design gained a 32% increase in earnings compared with industry averages (Hertenstein et al., 2005). Product aesthetics account for a significant proportion, up to 60%, of consumers' purchasing decisions (Kreuzbauer & Malter, 2005). The impact of automotive design on market performance is as follows. For example, the makeover of the 2005 Volkswagen Beetle resulted in a 54% market share increase within a single year (Burnap et al., 2023; Kreuzbauer & Malter, 2005).

Consumers consistently prioritize aesthetics as one of the top three influential factors in the decision-making process regarding product selection (Burnap et al., 2023). An illustration of this can be seen in the original iPod, where visual aesthetics achieved market acceptance. Particularly within sectors such as home appliances, aesthetics plays a significant role in enabling companies to create unique product feature (Burnap et al., 2023). Nike leverages AI to analyze extensive data collected from athletes, enabling the creation of customized shoe designs that meet individual needs. Considering unique biomechanics and personal preferences ensures that athletes enjoy optimal performance and comfort. Furthermore, Nike's smart sneakers feature AI-driven sensors that track athletic performance metrics, including speed, distance, and foot strike patterns.

However, for AI technology to be a competitive resource, companies must adopt it aggressively. As an emerging technology, AI technology has rarely been fully and deeply integrated into business processes, except for large and capable firms. Manufacturers must know how to use generative AI tools effectively and efficiently to create innovative products, especially for businesses that choose a product differentiation strategy for competitive advantage. In some cases, instead of hiring external designers at high contracting costs, their design team can learn how to use AI tools to create beautiful products at a minimal design cost.

Based on the aforementioned need, this chapter reviews theories related to the relationship between product characteristics and customer satisfaction and proposes a revised framework for using AI tools to create attractive product characteristics.

15.1.2 Objectives of the Study

This research addresses the challenges in designing a commercial product for online sales and presents the potential of AI as a tool to achieve the following objectives:

1. To develop a framework for using generative AI to support the design of aesthetic products for online sales.
2. To evaluate the performance and effectiveness of the AI platform Google Colab integrated with Gemini and Vertex AI in the case of bicycle design to generate user-based designs and styles.
3. To assess the risks of using text-to-image generating AI and to provide recommendations for designers to adopt this technology.

To achieve these objectives, this book chapter evaluates the current AI model to generate images, presents a new AI platform for high-quality image creation, and demonstrates an example of creating new bicycle designs with specific steps. As an illustrative example, we demonstrate how to use low-cost Google

Colab with Gemini and Vertex, an image-generating A.I. application, to design customized products to meet customers' quality requirements. This product strategy at the operational level is aligned with the business-level differentiation strategy. The following sections include a review of theories related to product design, including applications of generative AI for attractive characteristics in a new product, a methodology including the text-to-image process using a generative AI, an illustrative example following this method, and finally, a conclusion.

15.2 Literature Review

15.2.1 Related Theories

Theories related to visual effects include Exposure Theory, which suggests that frequent exposure to a stimulus increases a penchant for it (Ryu & Ryu, 2021). The Arousal Theory posits that individuals prefer stimuli with novel things (Berlyne, 1973), while the Fluency Theory emphasizes that individuals prefer visual displays that are easy to perceive (Mayer & Landwehr, 2018). These theories all provide a background for understanding how visual appeals influence consumer responses in e-commerce.

Studies related to the above theories have revealed a positive relationship between attractive product features and sales. For example, Hou and Tang (2023) discovered that appealing product color stimulated customer attention and emotion, increasing product engagement and purchase intention towards products in digital commerce platforms. This study also revealed that sales volume is positively associated with seasonal trending hues. Specifically, it showed that the red color, linked with warmth, vigor, and allure, creates great emotions and purchase inclinations (Hou & Tang, 2023). Additionally, Xia et al. (2020) revealed that product photo attributes and street scenes enhance sales in B2C e-commerce in the Chinese market. In addition, the study reveals that brand logos negatively impact men's clothing sales, which suggests that e-businesses should focus on adding attractive photos of their products on e-commerce platforms, avoiding brand logos. However, the study only examines the impact of product photos in general on sales volume, neglecting specific design characteristics like design attractiveness. Product details, including images, are key to customer purchase behavior. Bucko et al. (2018) found that product details, including photos, on an online store are among the key factors that affect consumers' willingness to purchase products.

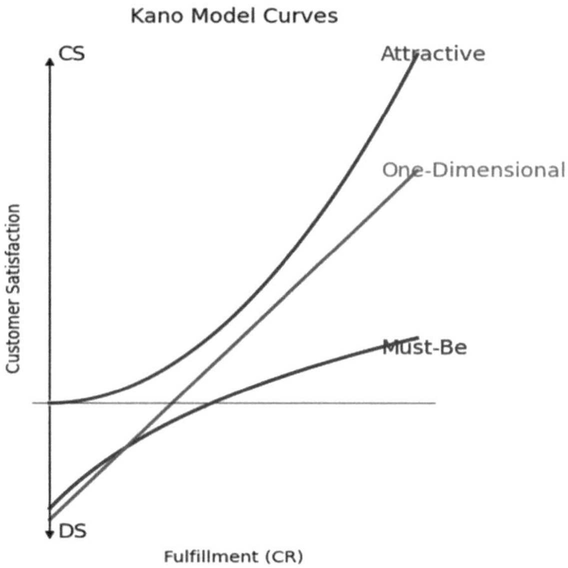

Fig. 15.1 Modified Kano's model (Dash, 2021; Ji et al., 2014)

15.2.2 Designs for Attractive Characteristics: Kano's Theory

Customer-focused products aim to maximize customer satisfaction. According to Kano's model, product characteristics are grouped into three categories: "must-be", "one-dimensional" and "attractive" characteristics (Ji et al., 2014). A person buying a product with "must-be" characteristics does not experience increased satisfaction beyond a certain peak point. A product with "one-dimensional" characteristics will result in a steady increase in customer satisfaction. However, a product with "attractive" characteristics will be exponentially satisfied if many of these characteristics are fulfilled. For example, a functioning tire has "must-be" characteristics but neither "one-dimensional" nor "attractive" characteristics. The greater durability of a tire is a "one-dimensional" characteristic. However, identifying "attractive" characteristics is difficult because customers do not express them. This is why customer surveys are important for understanding what customers want, identifying their desires, and determining which new features will be added to a product. A tool used for the whole product quality function deployment process is the House of Quality (C. C. Chen & Chuan, 2010; Liu et al., 2024). An extended Kano model contains five categories: reverse, indifferent, must-be, one-dimensional, and attractive characteristics (Dash, 2021).

Figure 15.1 visualizes the different quality attributes in **Kano's model**. Must-be quality requirements refer to the fundamental characteristics wanted by customers. Failure to meet these requirements will lead to high customer dissatisfaction. Nonetheless, fulfilling these needs does not enhance customer satisfaction (e.g., the effectiveness of a car's brakes). One-dimensional quality requirements are directly linked to how these needs are met. The greater the level of its fulfillment, the more content the customer will have (e.g., car fuel efficiency). Attractive quality requirements have the potential to surprise customers. These product characteristics differentiate a product from other competitors.

15.2.3 Applications of Generative AI to Identify and Fulfill "Attractive" Characteristics

Kano's model discussed above is necessary for customer-oriented organizations seeking new product characteristics. The more "attractive" product features are fulfilled, the higher customer satisfaction levels will increase exponentially.

One of the quality concepts companies can identify in product improvement front is to identify the gap in quality fulfillment between them and the best-practice organization using the benchmarking tool. Alternatively, understanding customers' quality requirements through customer surveys is also commonly used to determine the quality gap between the current design of a product and the expected or attractive features. This is the first stage in the process of quality function deployment (QFD). In this stage, the tool of the House of Quality is commonly used (Wei & Wang, 2023).

According to Ji et al. (2014), to optimize product design, we can quantify and integrate Kano's model parameters into the House of Quality. Formula (15.1) shows its objective function.

$$max \Sigma_{i=1}^{m} w_i s_i \tag{15.1}$$

where w_i is the weight of importance of each customer requirement i and where s_i is the degree of customer satisfaction with each customer requirement i. s_i is measured differently for Attractive (A), One-Dimensional (O) and Must-be Characteristics (C).

$$s_i = \begin{cases} \frac{\text{CS}_i - \text{DS}_i}{e-1} e^{y_i} - \frac{\text{CS}_i - e\text{DS}_i}{e-1}, & \text{for attractive attributes}(\text{A}) \\ (\text{CS}_i - \text{DS}_i) y_i + \text{DS}_i, & \text{for one-dimensional attributes}(\text{O}) \\ -\frac{e(\text{CS}_i - \text{DS}_i)}{e-1} e^{-y_i} + \frac{e\text{CS}_i - \text{DS}_i}{e-1}, & \text{for must-be attributes}(\text{M}) \end{cases}$$

where CS is customer satisfaction (fulfilled customer requirements), DS is customer dissatisfaction (unfulfilled customer requirements), and e is an exponential function.

AI-driven aesthetic diffusion models can create visually pleasing, flexible, and quick product designs from text descriptions (J. Chen et al., 2024; Yang et al., 2024). Traditionally, the design phase necessitates the creation of a two-dimensional or three-dimensional drawing, and the whole process is repeated until the drawing is satisfactory. Conversely, an AI-driven diffusion model relies on textual descriptions to generate initial design graphics, and it only requires adjustments to the textual prompts to regenerate a modified image (J. Chen et al., 2024). This significantly reduces the time and effort required compared to the lengthy traditional design process.

Figure 15.2 indicates the process of an aesthetic diffusion model for image creation. A traditional design process begins with technical drawings, and materials are applied to the model for a final design. In contrast, in this new model, AI translates descriptive texts into an initial design used to test potential users. The design improvement process repeats itself until total satisfaction is achieved, and only then is a final design accepted.

The most common AI models for text-to-image creations are Dall-E, Midjourney, Stable Diffusion, and Dreambooth (Alawadhi et al., 2023; J. Chen et al., 2024). Others are Adobe Firefly, ComfyUI, Automatic1111, Invoke, and Drawthings. These models have demonstrated remarkable performance in creating aesthetically pleasing designs for different products, such as interior designs, furniture, sunglasses, cosmetic products or flower vases with specified styles (Table 15.1).

Despite the advantages of AI-driven design models, their application has several limitations. First, to generate images, the models employ large datasets of images, such as the ImageNet image dataset with over 10 million images (Deng et al., 2009). However, there is a lack of datasets for indoor images with aesthetics and decorative styles, hampering the model's capability to create appealing product designs annotated with styles (Chen et al., 2024). It is time-consuming and costly for designers

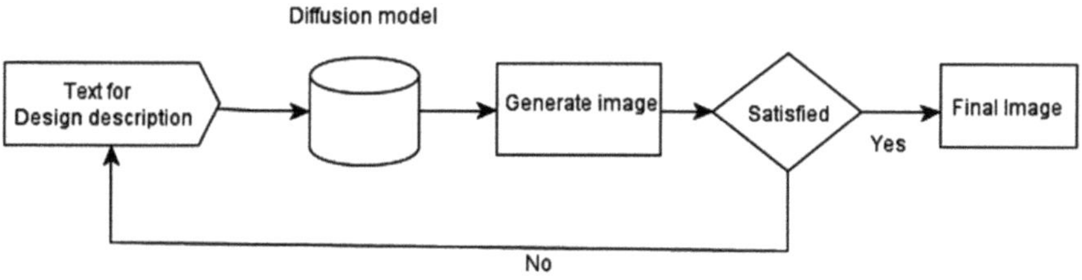

Fig. 15.2 Proposed text-to-image process (adapted from Chen et al., 2024)

Table 15.1 Comparison of common AI-driven models for text-to-image creation (adapted from Chen et al., 2024)

AI-Driven Model	Strengths	Weaknesses
Dall-E	• High-quality and diverse image generation.	• Requiring significant computational resources.
Midjourney	• Strong in creative and artistic image generation.	• Limited in generating highly realistic images.
Stable Diffusion	• Effective in generating images with diffusion techniques.	• Missing professional annotations leading to style and function confusion.
Dreambooth	• Fine-tunes neural network layers for improved domain-specific images.	• Specific to certain domains, may not generalize well across all areas.

to self-create image datasets. Another difficulty of using AI models for design is related to the prompt engineering capability of the designer, who needs to ask all AI questions related to product characteristics. For example, for visual effects on potential customers, product form requires shape color, texture, and decorative style considerations. Below is an example of using Midjourney to create a chair with prompts related to these features.

Figure 15.3 shows different dining chair images created via Midjourney application. Prompts were engineered following product attributes, including translucency, color, sculpture, form, styles, and texture, as suggested by Alawadhi et al. (2023).

Fig. 15.3 A collage of dining chair images generated by Midjourney application

However, the Midjourney application for product design aims to capture visual effects only (Jia & Song, 2024; Yin et al., 2023) In fact, product quality attributes cover not only aesthetics but also features, reliability, durability, and perceived value. Another pitfall of using Midjourney in the above illustrative example is that only a single material, plastic, is used for chair design. Will the application be able to create a design for a more sophisticated product composed of multiple materials or components?

To address the above pitfalls, this chapter introduces the use of another AI tool to create a more sophisticated product: a bicycle composed of many components and materials. The data used to train the model are from the secondary image dataset and primary data collected from users.

Due to the current limitations of common AI-driven models for text-to-image generation, as presented above, this chapter introduces a modified AI model that is easier to use, more affordable, and suitable for inexperienced designers. This proposed model utilizes Google Colab with Gemini and Vertex. It skips the requirements for prompting engineering skills and previous product design knowledge. Instead of using secondary data of images, this new model uses primary data by surveying a focus group of interest, whereby the product design is more specific to a group of users, which involves human-AI interaction for improved images.

15.3 A Proposed Framework for Using AI for Designing Products

Kano's model indicates that attractive product characteristics are necessary to dramatically increase customer satisfaction. Designers, therefore, focus on the attractiveness or aesthetics of a new product (Burnap et al., 2023). We combine the Quality Deployment Function—House of Quality with AI applications—for designing purposes to introduce a new framework for redesigning an attractive product (Fig. 15.4).

15.3.1 AI-Assisted Design Outcomes

15.3.1.1 "Attractive" Characteristics

The attractive characteristics of a product design increase customer satisfaction (Fig. 15.1). Attractive product designs are stylish, appealing, innovative, or sporty. They also draw the attention of impulsive customers, who are more likely to make abrupt purchases of products with nice-looking products (Burnap et al., 2023). This is where AI technology is applied to create aesthetically pleasing product images.

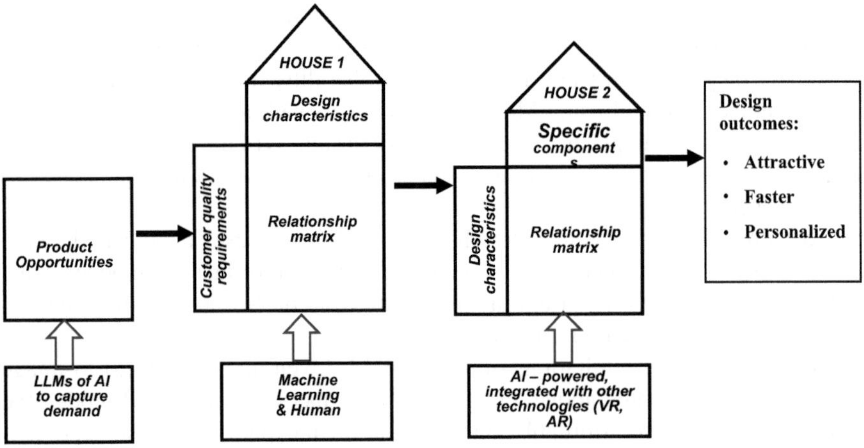

Fig. 15.4 Framework of using a machine learning tool to redesign a product

15.3.1.2 Faster Customer Responsiveness

Using machine learning algorithms allows the analysis of user requirements and market trends and quickly generates product prototypes and design solutions. As a result, the time allocated to the conventional manual design could be reduced. AI tools expedite the cycle of product ideation and development (Paliwal et al., 2024). Compared with the traditional process of product development, time is significantly reduced in the case of AI assistance. Combining AI with other technologies, such as virtual reality (VR) and augmented reality (AR), can improve the time needed to integrate customer wants into new products. Furthermore, trial production and physical market tests can be cut. For example, when AI technologies were used in vaccine development, pharmaceutical companies such as Moderna and Pfizer were able to reduce the time to market from normally 3 years down to just over 1 year.

AI tools allow designers to input specific parameters, which generate multiple design alternatives (Rane et al., 2023). This capability reduces the time spent on ideation and fosters innovation, resulting in highly optimized and visually appealing products.

15.3.1.3 Product Customization or Personalization

Through the collection and analysis of vast amounts of data, AI technology can understand user requirements and produce products that are more closely aligned with user expectations (Chen et al., 2024; Yang et al., 2024). AI enhances customer support and service by integrating AI-driven chatbots and virtual assistants, which can analyze user behavior and preferences to offer tailored recommendations for any product. By examining users' behavior data and input, user requirements can be identified, novel product attributes can be investigated, and a wider array of customized or personalized products can be presented.

15.3.2 AI-Assisted Product Design Process

Analysis of user-generated content (UGC) has been common in marketing and product development (Büschken & Allenby, 2016). This method is used to identify new product opportunities, understand customer desires, and specify product characteristics. It is depicted in Fig. 15.3 and explained below.

15.3.2.1 Identifying Product Opportunities

Before starting with a product development process, a business must identify whether the product idea is prospective. The factors affecting organizational decisions on a new product development project include external and internal factors. For example, for external factors, the forecast demand is favorable, technologies used for development are available, and the supply chain network is robust. Other external factors to consider are the regulatory framework and social preferences. For internal factors, organizational competencies to develop the product are available. Finally, the product type is still in the growth stage of its lifecycle.

AI can help marketers quickly scan external and internal data to identify improved or new product opportunities. For example, Boeing plans to delve deeply into its intensive internal data to understand customers' desires. For external environmental scanning, the LLM approach of AI can assist firms in rapidly capturing customer trends (Timoshenko & Hauser, 2019).

15.3.2.2 Capturing Customer Wants

Companies can utilize various tools and methods, including customer surveys, Google Trends, and real-time data, to better understand customer preferences (Timoshenko & Hauser, 2019). Another innovative method is conversational communication with customers through AI tools such as chatbots. These tools allow businesses to gather direct feedback and understand customer needs more deeply. Sentiment analysis of product reviews, tweets, and other user-generated data is also valuable. This analysis helps companies capture customer sentiment and identify areas needing improvement. Customers often express their frustrations to anonymous service agents, providing essential feedback for service enhancements. For example, Wendy's fast-food restaurant uses digital menus to update real-time offerings and improve the overall customer experience.

After that, large language models, as new approaches to AI, can be used in the ideation phase of the product development cycle. Figure 15.4 shows two houses in this phase. Below, we explain how AI tools are used in these two houses.

15.3.2.3 Constructing Product Specifications and Components

(i) **HOUSE 1: From Customer Quality Requirements to Design Specifications**

Designers can create aesthetic designs by augmenting traditional product development with machine learning tools. AI-embedded systems provide users with more comprehensive information and a wide array of features, enhancing the overall user experience. One notable AI-powered design tool is generative design. This tool allows designers to input specific parameters, enabling the AI to generate multiple design alternatives (Rane et al., **2023**). This capability reduces the time spent on ideation and fosters innovation, resulting in visually appealing products.

(ii) **HOUSE 2: From Design Specifications to Specific Components**

Integrating virtual reality (VR) and augmented reality (AR) allows customers to visualize products in virtual environments before making a purchase (Ibrahim et al., **2023**). This feature is particularly beneficial for furniture and home decor shopping, where customers can see how a piece fits and looks in their room via their smartphone or AR glasses. Apple's Vision Pro employs mixed reality technology for mixed reality applications to enhance the user's surroundings by overlaying interactive digital elements and information onto the real world. Vision Pro has high-resolution displays, spatial audio for an immersive sound experience, and intuitive controls. Vision Pro offers a premium mixed-reality experience by integrating these advanced features.

15.3.2.4 Human-AI Interaction for Product Design

AI is a tool that translates customer desires for a product into appealing product characteristics. Traditionally, companies must conduct customer surveys or focus groups to understand what customers want to improve their product design. According to the human-centered approach, product design should include the elements of design thinking, which involves a human-centered, iterative process for human feedback (Yehia, 2024). This approach focuses on empathy, creativity, and experimentation to understand user needs and create feasible and desirable solutions (Bongiovanni & Louis, 2021; Eisenbart et al., 2022). However, one major obstacle to applying the design thinking approach is that traditional engineering mindsets may prioritize technical specifications over user-centered design.

The proposed framework emphasizes AI machine learning and human interaction to provide satisfactory product features (Yehia, 2024). Customer engagement can be implemented through co-design strategies, leveraging customer knowledge, and utilizing technology-based interfaces. By engaging customers, companies can better satisfy self-determined needs and product innovation and foster a sense of ownership among consumers (Murthy, 2024). In summary, the traditional method for understanding customer quality requirements is the House of Quality, a matrix in quality function deployment (QFD). This chapter proposes a framework to integrate the House of Quality with Kano's Model with AI assistance and how to engage users in the product design process.

15.4 Methodology

This chapter introduces a text-to-image method revised from the aesthetic diffusion model by Chen et al. (2024) to suit the new integration of Gemini and Vertex AI in Google Colab. Suppose you want to sell a product on an e-commerce platform and create a visually appealing product image to stimulate sales. We propose using Google Colab with Gemini and Vertex for this purpose.

15.4.1 What is Google Colab with Gemini and Vertex?

Google Colaboratory (Colab) is a cloud-based platform that provides a Jupyter notebook environment, offering access to computing resources.[2] Gemini is Google's most advanced and capable AI model system. It is designed as a multimodal AI platform that can understand and process various types of information, including text, code, audio, images, and video.[3]

Vertex AI is Google Cloud's unified platform, integrating various AI and ML tools, including Gemini models, for developing and deploying machine learning solutions. It also provides a comprehensive suite of image generation for e-commerce applications.[4]

Google Colab, integrated with Gemini and Vertex AI, provides a Jupyter notebook environment for AI/ML projects. It involves generating product descriptions,

[2] https://codelabs.developers.google.com/; https://www.analyticsvidhya.com/blog/2020/03/google-colab-machine-learning-deep-learning/

[3] https://cloud.google.com/blog/products/ai-machine-learning/gemini-support-on-vertex-ai

[4] https://cloud.google.com/blog/products/ai-machine-learning/introducing-veo-and-imagen-3-on-vertex-ai

analyzing and creating content based on existing product images, and enhancing product images. This integrated system is designed for online sales.[5]

15.4.2 Difference between Gemini and Vertex and Midjourney Application

Gemini and Vertex AI on Google Colab have advantages and disadvantages over the well-known Midjourney in text-to-image creation. Table 15.2 shows the main differences between Gemini and Vertex AI and Midjourney.

15.4.3 Implementation Process

This new method for generating text-to-image uses Google Colab integrated with Gemini for prompt engineering and Vertex AI for image creation. Unlike the traditional House of Quality process, which starts with customer surveys for desired product features, the proposed design process using Google Colab integrated with Gemini and Vertex begins with an initial image generated by prompts created by Gemini.

The implementation process in Table 15.3 includes text-to-image steps according to the diffusion model by J. Chen et al. (2024):

We aim to generate a final product that satisfies users' aesthetic requirements. Traditional text-to-image AI models such as Midjourney and DALL-E require prompts for images, which many inexperienced designers struggle to provide. Knowing what to ask the AI for the above features requires knowledge of prompt engineering and product feature requirements. Suppose the designer has not experienced any prompt engineering or product design courses. In that case, he does not know what to ask the AI, and consequently, it takes him a long time to command the AI to create images that may not be satisfactory. Moreover, if their prompts are not accurate, AI-generated images are not as desirable, and the image-generating AI system provides unexpected image parts or components that consume the designer's time and effort to remove (Paananen et al., 2024).

The introduction of Gemini on the Google Colab platform makes it easy for novel designers to provide more accurate prompts for wanted images (Steps 1 and 2). Users enter simple text about product features, and Gemini will create a corresponding prompt.

---

[5] https://cloud.google.com/transform/101-real-world-generative-ai-use-cases-from-industry-leaders

Table 15.2 Main differences between Gemini and Vertex AI and Midjourney

Main differences	Gemini & Vertex AI	Midjourney
Data Structure	• Structured data preparation with explicit API integration • Supports multiple data formats (JSON, CSV, images, text) • Built-in data validation and preprocessing tools • Direct integration with Google Cloud Storage	• Discord-based interface with text prompts • Limited structured data support • Focus on natural language descriptions • No direct data preprocessing capabilities
Prompt Engineering	• Systematic prompt structuring • Context-aware prompt generation • Support for multi-turn conversations • Built-in prompt templates and galleries	• Parameter-based prompt structure • Specific syntax requirements • Limited conversation capability • Emphasis on artistic parameters
Data processing	• Automated data cleaning • Built-in quality checks • Scalable processing pipeline • Enterprise-grade security	• Manual data input • Limited preprocessing options • Fixed parameter structure • Community-based quality control
Integration	• Direct API access • Custom model training support • Batch processing capabilities • Advanced data transformation tools	• Discord-only interface • No direct API access • Limited data manipulation • Fixed processing pipeline
Development environment	• Full development environment • Custom code integration • Version control support • Collaborative features	• Command-line interface • No development environment • Limited customization • Fixed workflow
Customization	• Comprehensive, enterprise-focused data preparation environment with extensive customization options	• Streamlined, art-focused approach

Source: https://developers.google.com/machine-learning/resources/prompt-eng

Steps 1 to 3 are quickly implemented via Google Colab with Gemini and Vertex AI. In Step 4 (fine-tuning), we need to ask for user feedback. In the illustrative example of bicycle design, we surveyed a focus group of university students in Vietnam. The students aged 18 to 21 years were following the Operations Management course taught by the author at Vietnam National University—Ho Chi Minh City in Fall 2024. In one of the course sessions, given the bicycle image generated by Gemini and Vertex AI, the students were asked to complete a Google survey form with two questions:

Table 15.3 The procedures to use text-to-image Gemini and Vertex AI in Google Colab

Steps	Description	Resources
1. Text Input and Guidance	Inputting text descriptions that guide image generation.	Google Colab Designer—Gemini interaction
2. Diffusion and Denoising Modules	The diffusion module adds noise to the input image, while the denoising module learns to remove this noise, effectively generating the final image	Google Colab Designer—Vertex AI interaction
3. Learning from Data	The model is trained using a large dataset of paired text descriptions and images. This training enables the model to understand and generate images that match the input text descriptions	Google Colab Prompts generated by Gemini Vertex AI
4. Fine-Tuning	To improve performance in specific domains, the model can be fine-tuned	User feedback
5. Batch Generation and Flexibility	The model can generate multiple images in batches from the input text.	Google Colab Prompts generated by Gemini Vertex AI
6. Aesthetic Considerations	The AI-driven model integrates aesthetic considerations into the design process The final image must be visually appealing.	Google Colab Prompts generated Vertex AI

1. Rate the satisfaction level with the design of the new bicycle.
2. Why did you select your answer to the previous question?

Based on user feedback, Gemini and Vertex AI implement Step 5 (batch generation) and Step 6 (aesthetic consideration).

15.4.4 Evaluation

Our outcome-based evaluation is based on the framework built in Fig. 15.1. Specifically, the outcomes include image attractiveness, time of image creation, and personalization (Table 15.4).

Table 15.4 User feedback on AI-generated product images

Evaluation outcomes	What	When	How
Image attractiveness (Burnap et al., 2023).	User-perceived visual effects of product image	Step 4 (Fine-tuning) Step 6 (Aesthetic considerations)	Customer survey/ focus group discussion
Personalization (Chen et al., 2024; Yang et al., 2024)	User perception on product customization	Step 4 (Fine-tuning)	Customer survey/ focus group discussion
Fast responsiveness (Rane et al., 2023).	Time of image generation by AI	Step 3 (learning from data)	Recorded time in Google Colab platform

15.5 An Example

To demonstrate the advantages of using AI tools in designing a product, we use Google Colab with Gemini and Vertex, an AI application, to understand human requirements, create actionable prompts for AI software, and create appealing images. Its users can repeat the prompts until they are satisfied with the AI-generated product images. Fig. 15.5 analyzes the outputs and risks using the Google Colab with Gemini and Vertex AI.

15.6 Evaluation Results

15.6.1 Time to Generate a New Product Image

I provided a prompt to the Google Colab: "How long does it take Vertex to generate this image?". Figure 15.6 shows its answer:

The answer showed that Vertex AI created a bicycle image in only 7.73 seconds. I conducted five more repetitions and recorded the following times for this step: 7.59, 7.73, 7.54, 7.20, and 7.30 seconds per repetition. This reflects a substantial time-saving advantage of using Vertex AI compared to the traditional design process, where it takes many days for a designer to complete a 2D or 3D drawing.

Steps	Outputs	Risks

(i) The designer inputting text descriptions that guide image generation.	Use Gemini to generate an image prompt for your item
	`#@title Use Gemini to generate an image prompt for your item`
	`item_selling = 'bicycle' #@param {type: "string"}`
	`model = genai.GenerativeModel('gemini-pro')`
	`chat = model.start_chat(history=[])`
	`prompttext = f"""`
	`I'm selling {item_selling} online, and I need to generate an image of it.`
	`I need the image to be appealing, attractive, sporty, beautiful, innovative, colorful and luxurious to convince people to buy. The {item_selling} is mountain bicycle.`
	`The image has no rider. The bike must be foldable and light.`
	`Can you create a prompt I can use to generate an image of {item_selling} with Vertex?`
	`Respond with only the prompt, no other text. Be as verbose as possible.`
	`"""`
		Text descriptions provided by a human are vague, so AI creates unwanted images.
		
		Inconsistency in AI outputs. AI creates different prompts each time it reads text descriptions.

(ii) Gemini generating a full prompt and Vertex creating an image	Prompt generated by Gemini: *"Generate an image of an ultra-realistic foldable mountain bicycle positioned in 3/4 angle front view against an untextured plain white background. The bicycle should be a high-end model with a luxurious deep blue and gold color scheme and feature sleek, sharp angles, aerodynamic shapes, and innovative suspension and drivetrain systems. Include high-resolution details such as intricate spoke patterns, textured tire treads, plush saddles, ergonomic handlebars, and advanced electronic components. Ensure the lighting is dramatic and emphasizes the bike's dynamic lines and muscular stance. Showcase the bike's lightweight and compact foldable design in a way that highlights its convenience and portability. Render the image in a painterly style to enhance its visual appeal and artistic value."* Image created by Vertex: 	AI-generated prompts are not consistent, inaccurate. Inconsistency in AI outputs. AI creates different prompts each time of reading text descriptions. Inconsistency in AI outputs. AI creates different images each time of responding a prompt. Bias. AI uses a certain image database.
(iii) User feedback on the design	Key user comments: 1. The bike has a nice color, but the design is quite masculine, not very suitable for women 2. The frame could be lighter 3. The front saddle is so small; the bike does not have the back saddle 4. The body looks so thin compared to the thick wheels; this makes it look imbalanced	The respondents are users of the product. User comments/feedback are reliable and valid.

Fig. 15.5 Analysis of outputs and risks

Steps	Outputs	Risks
(iv) Designer training the model based on user feedback data	prompttext = f""" I'm selling {item_selling} online, and I need to generate an image of it. I need the image to be appealing, attractive, sporty, beautiful, innovative, colorful and luxurious to convince people to buy. The {item_selling} is mountain bicycle. The image has no rider. The bike must be foldable and light. The bike design must also be for women. The saddle should be bigger. The frame must be thinner. Can you create a prompt I can use to generate an image of {item_selling} with Vertex? Respond with only the prompt, no other text. Be as verbose as possible. """ response = chat.send_message(prompttext) response.text Fine-tuned prompt by Gemini: **Prompt:** *"Generate a stunning, high-quality image of a sleek, foldable, ultra-lightweight mountain bike designed for women, featuring:* ** A vibrant color scheme that exudes luxury and appeal* ** An eye-catching, innovative design that showcases the bike's advanced features* ** A spacious, comfortable saddle designed for female riders* ** A noticeably thin, sculpted frame that highlights the bike's lightweight construction* ** An overall composition that conveys a sense of speed, agility, and sophistication, inviting viewers to experience the joy of riding this exceptional bicycle"*	Inconsistency in AI outputs. AI creates different prompts each time it reads text descriptions.
(v) Designer may repeat the prompt for multiple images and (vi) accept the most appealing image option.	Some fine-tuned bike images created by Vertex: 	Inconsistency in AI outputs. AI creates different images each time it responds to a prompt.

Fig. 15.5 (continued)

15.6.2 User-Based Feedback on the AI-Generated Bicycle Design

15.6.2.1 Quantitative Results

When asked to rate their satisfaction level with the design of the new bicycle, the survey of 52 respondents revealed a moderately positive response. A total of 40.38% of the participants (21/52) expressed a neutral stance, while a significant 42.31% (22/52) showed positive feedback (28.85% satisfied and 13.46% extremely satisfied). Only 17.31% of the respondents (9/52) expressed dissatisfaction, with 11.54% being dissatisfied and 5.77% being extremely dissatisfied. The mean satisfaction score of 3.33 on a 5-point scale, which is slightly above the neutral point, suggests a generally favorable but cautious response to the design. The balanced distribution of the survey data, with a standard deviation of 1.04, indicates that while the bicycle design is appealing to many respondents, there is room for improvement (Fig. 15.7). The features that need to be improved are recognized from the users' feedback in the text.

```
Generated code may be subject to a license | vmmc2/Apollo
# prompt: How long does it take Vertex to generate this image

import vertexai
from vertexai.preview.vision_models import Image, ImageGenerationModel
from google.colab import auth
import google.generativeai as genai
from google.colab import userdata
from IPython.display import Image
import time

# ... (rest of your existing code)

#@title Use Vertex to generate an image

model = ImageGenerationModel.from_pretrained("imagegeneration@005")

start_time = time.time()  # Record the start time

images = model.generate_images(prompt=response.text)

end_time = time.time()  # Record the end time
elapsed_time = end_time - start_time

print(f"Image generation took {elapsed_time:.2f} seconds.")
```

Fig. 15.6 Speed to create a product image in Google Colab integrated with Vertex AI

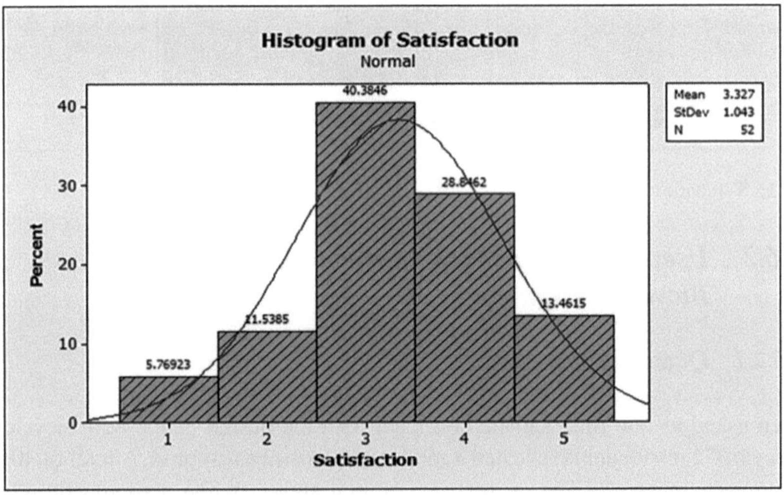

Fig. 15.7 Satisfaction rating of the AI-assisted design

15.6.2.2 Qualitative Results (Table 15.5)

Table 15.5 Summary of sentiment analysis

Expected Outcomes	Design Implications	Negative Sentiment	Risk Category
Attractiveness & Personalization	Aesthetics, gender considerations, ergonomic design	*"The bike has a nice color, but the design is quite masculine, not very suitable for women" (Quote 1)* *"I think this design is suitable for the sport bicycle. I think the color combination is attractive and makes user feel elegant. However, having more space for the water bottle will be more ok" (Quote 2)* *"…a mix of good aesthetics and some elements that could be refined., I rated the bicycle design 3 points because while I appreciate the unique color scheme and the premium-looking gold accents" (Quote 3)* *"The design is quite beautiful … it looks like a mounting bicycle, so it is hard to drive for me" (Quote 4)* *"Vintage color, looks sporty, the frame shape and geometry seem a bit generic and not particularly distinctive or innovative. It has a fairly standard mountain bike silhouette., not my favorite color," (Quote 5)*	Technical risk Demand risk
Responsiveness	Ergonomic design	*"It may be heavier and less efficient for daily road use. The frame could be lighter, or perhaps a bit more aerodynamic" (Quote 6)*	Operational risk Technical risk Demand risk
Personalization	Functionality, features, ease of use	*"The front saddle is so small., the bike does not have the back saddle., the body looks so thin compared to the thick wheels, this makes it look imbalanced" (Quote 8)*	Technical risk

Note: The respondents answered the question: "Why did you select your answer to the previous question?". Out of 52 responses, 48 responses were valid for further evaluation. Eliminated responses include unreadable characters. Fig 15.5 summarizes the sentiment analysis results using the software Orange

15.7 Risk Considerations

Despite the benefits of AI tools in creating attractive products, they pose technical, security, compliance, and operational risks. Technically, training and implementation of AI technology heavily relies on substantial data, but there is a technical risk of a paucity of relevant data and poor data quality (e.g., inaccessible product reviews and closed groups). Inadequate data samples or inaccurate data may also result in erroneous predictions of preferred product attributes. Data bias is another problem. For example, data are only collected from particular groups of users. Imagine that there is a bias toward only negative sentiment conversations; if the model undergoes training and testing solely within a biased dataset, it may fail to introduce optimal attributes for the associated products. Additionally, AI may generate some product features that are technically infeasible to manufacture or are mere hallucinations.

In terms of security risk, AI technology demands a substantial volume of data for both training and application, and this data may reveal confidential user information. Suppose a company slyly captures users' personal data, such as health history and sex orientation, and uses the data for commercial purposes, such as redesigning related devices targeting these users. Failure to safeguard the privacy and security of user data in the product design phase could lead to data breaches. Such violations damage a company's reputation and may violate privacy-related laws. Additionally, if a product has integrated AI capability into its product design, such as an AI-powered monitoring camera, it poses a significant risk to data security. Camera designers must ensure that captured videos are not used for unauthorized purposes. To understand this risk concern, companies deploying monitoring cameras have policies for data security. For example, organizers of the Paris 2024 Olympic Games guarantee to the public that their cameras, installed in and around stadiums and gymnasiums, aimed only to detect anomalies in crowds to prevent fights or locate unattended bags.

Furthermore, protecting individual privacy is not adequate. Compliance risk is another primary concern for AI-created visuals. Companies using AI tools to add more product features and functions should ensure these new features do not violate local laws and professional standards. For example, in new drug development, AI advice may differ from professional standards (Dai & Tayur, 2022). It is ethical for designers to comply with ethical standards when they use proper texts to ask AI to generate AI texts or images.[6] Good practices include separating the prompting and imaging phases from the review phase in a product deployment process. A review team, preferably from external organizations, will conduct objective and impartial audits of AI-created images to ensure they do not violate ethical standards.

Finally, operational risk refers to the inability to realize the design components suggested by AI tools. One of the big operational risks is the inconsistency in AI outputs. Specifically, with the same input text description, the AI model generates different versions of prompts. The same problem may happen with the image

[6] https://www.bcg.com/publications/2025/rethinking-medtech-marketing-gen-ai

output, meaning that with the same prompt, the system pops up different images. In addition, adding too many features to an advertised product but failing to materialize them in a real product may lead to customer dissatisfaction and probably legal action against fake marketing. AI tools can generate an aesthetic image of a product, but without human intervention in modifying the last design, it is technically challenging for producers to make. According to the theory of sustained competitive advantage, a company must develop its resources, including technology and human resources, into a competency or a bundle of valuable, invariable, rare, and organized (VRIO) resources (Barney, 1991; Porter, 1996). The ability to use AI effectively is important for it to pass the VRIO test. Using only AI tools to generate new design characteristics without human review may lead to an infeasible design. Humans still need to review the design's feasibility for manufacturing.

15.8 Conclusions and Future Work

Businesses can now use generative AI to create visually appealing product designs on e-commerce platforms to stimulate demand. The benefits of using Google Colab with Gemini and Vertex AI for product design include image attractiveness, speed, and personalization of a new product for online sales. It provides a process of automating the translation of product descriptions into a satisfactory product image. Despite some risks of using this model, technological opportunities related to this technology include more available larger image datasets in the future (Kate Knibbs, 2024). In terms of cost-effectiveness, a pay-as-you-go AI model such as Google Colab with Gemini and Vertex AI substantially saves design costs. Additionally, the time to market a new product using this technology is significantly reduced, thanks to shortened design time. By fine-tuning the initial image based on user feedback, the model has substantial potential to tailor design outputs, leading to customization. Compared with classical text-to-image AI models such as Midjourney, our model using Gemini and Vertex AI on the Google Colab platform is more reliable because it trains the model on secondary image datasets to create images and primary user-based data to fine-tune them.

An illustrative example of AI-generated bicycle design demonstrated that user-based design feedback validated its personalization, speed, and aesthetics. This new process can create a new product image as wanted in seconds. Design outputs are tailored in iterations to match user needs. The model, however, indicated several missing points: gender considerations and ergonomics in product design. The framework also implies that generative AI still needs humans (users, experts) to suggest any fixes to imperfections or absent elements in the AI-generated images.

In summary, the implementation process proved fast, simple, customizable, and suitable for non-technical and inexperienced designers who want to increase online sales in business-to-customer e-commerce platforms. Designers input only simple text for product descriptions, and the AI system generates a high-quality product image in seconds. Its ease of use and effectiveness suit non-technical or inexperienced designers. However, despite the benefits of AI tools in creating attractive products, they pose technical, security, and compliance risks. Human intervention in the design stage is important to manage those risks.

Limitations and Future Studies

First, the study scope is limited to product aesthetics, while product quality is associated with many dimensions, such as product performance, features, reliability, durability, standard compliance, etc. It does not present how to translate the desired product features into specific components for manufacturability. AI-generated images may be so complex and artistic that manufacturers find identifying the materials and methods to materialize them challenging. Second, the study does not discuss factors that affect product decisions, i.e., what to make. In the illustrative example of a bicycle, we assume that making bicycles is the best product decision. Product decisions should depend on product opportunities and the company's capabilities. External factors for product opportunities include the increasing demand for bicycles as an environmentally friendly means of transport, environmental behavior, socio-demographic differences, etc. Internal considerations for organizational capabilities include organizational resources and competencies, such as research and development activities. Third, this chapter focuses on fine-tuning images according to user feedback and ignores expert review. Standard-based quality also matters. Experts will implement the step of reviewing AI images to ensure that they comply with ethical and professional standards. This review step should be separated from text and image generation applications to ensure objectivity and impartiality.[7] Lastly, this study does not examine financial and non-financial justifications for an investment in generative AI for product development.

To address those limitations, future studies should cover the following topics: (i) AI capabilities to improve all product quality attributes, including product functionality, reliability, durability, compliance, and perceived values in all product development phases, not only in the design phase; (ii) the application of AI to dynamically capture product opportunities by analyzing external and internal factors; (iii) the design of a framework for expert-based evaluations of a generative AI model for product compliance; and (iv) multi-criteria evaluation model for design alternatives. For instance, businesses should consider the cost efficiency and non-financial scores of the different options for designing: (i) to purchase AI platforms such as Google Colab, with subscriptions to Gemini and Vertex to generate prompts and images automatically; (ii) to train in-house IT engineers on prompt engineering skills; and (iii) to outsource AI contractors to draw product images. Each option has advantages and disadvantages, and specialists must judge these options. Non-financial factors should include organizational capabilities for sustainable competitive advantages, such as digital capability (Tambe et al., 2021).

[7] https://www.bcg.com/publications/2025/rethinking-medtech-marketing-gen-ai

References

Adulyasak, Y., Benomar, O., Chaouachi, A., Cohen, M. C., & Khern-am-nuai, W. (2024). Using AI to detect panic buying and improve product distribution amid pandemic. *AI & SOCIETY, 39*(4), 2099–2128. https://doi.org/10.1007/s00146-023-01654-9

Alawadhi, A. M., Yousef, D. M., Mohammed, A., Diana, A., & Yousef, M. (2023). Revolutionizing plastic furniture design: Harnessing the power of artificial intelligence. *Arab International Journal of Information Technology & Data, 3*(4).

Barney, J. (1991). firm resources and sustained competitive advantage. *Journal of Management, 17*(1), 99–120. https://doi.org/10.1177/014920639101700108

Berlyne, D. E. (1973). Aesthetics and psychobiology. *The Journal of Aesthetics and Art Criticism, 31*(4), 553. https://doi.org/10.2307/429334

Bongiovanni, I., & Louis, C. P. (2021). Theory and practice of design thinking: perspectives of designers and business consultants. *International Journal of Design Creativity and Innovation, 9*(3), 174–191. https://doi.org/10.1080/21650349.2021.1929501

Bucko, J., Kakalejčík, L., & Ferencová, M. (2018). Online shopping: Factors that affect consumer purchasing behavior. *Cogent Business and Management, 5*(1), 1–15. https://doi.org/10.1080/23311975.2018.1535751

Burnap, A., Hauser, J. R., & Timoshenko, A. (2023). Product aesthetic design: A machine learning augmentation. *Marketing Science, 42*(6), 1029–1056. https://doi.org/10.1287/mksc.2022.1429

Büschken, J., & Allenby, G. M. (2016). Sentence-based text analysis for customer reviews. *Marketing Science, 35*(6), 953–975. https://doi.org/10.1287/mksc.2016.0993

Chen, C.-C., & Chuan, M.-C. (2010). An extended Kano model for deciding improvement priority of product attributes. In *The 40th International Conference on Computers & Industrial Engineering* (pp. 1–6). https://doi.org/10.1109/ICCIE.2010.5668252

Chen, J., Shao, Z., Zheng, X., Zhang, K., & Yin, J. (2024). Integrating aesthetics and efficiency: AI-driven diffusion models for visually pleasing interior design generation. *Scientific Reports, 14*(1), 1–14. https://search.ebscohost.com/login.aspx?direct=true&db=aph&AN=175832274&site=ehost-live

Dai, T., & Tayur, S. (2022). Designing AI-augmented healthcare delivery systems for physician buy-in and patient acceptance. *Production and Operations Management, 31*(12), 4443–4451. https://doi.org/10.1111/poms.13850

Dash, S. K. (2021). Identifying and classifying attributes of packaging for customer satisfaction-A Kano model approach. *International Journal of Production Management and Engineering, 9*(1), 57–64.

Deng, J., Dong, W., Socher, R., Li, K., & Fei-Fei, L. (2009). ImageNet: A large-scale hierarchical image database. *IEEE Conference on Computer Vision and Pattern Recognition, 2009*, 248–255. https://doi.org/10.1109/CVPR.2009.5206848

Eisenbart, B., Bouwman, S., Voorendt, J., McKillagan, S., Kuys, B., & Ranscombe, C. (2022). Implementing design thinking to drive innovation in technical design. *International Journal of Design Creativity and Innovation, 10*(3), 141–160. https://doi.org/10.1080/21650349.2022.2048698

Hertenstein, J. H., Platt, M. B., & Veryzer, R. W. (2005). The impact of industrial design effectiveness on corporate financial performance. *Journal of Product Innovation Management, 22*(1), 3–21.

Hou, M., & Tang, Y. (2023). The influence of visual features in product images on sales volume: A machine learning approach to extract color and deep learning super sampling features. *Traitement Du Signal, 40*(4), 1469–1477. https://doi.org/10.18280/ts.400415

Ibrahim, F. A., Boying, W., Abdul Rahim, N. S., Mohd Shafiei, M. W., & Zainol, N. Z. (2023). The influence of augmented reality on purchase intention in the Housing Industry. *Human Behavior and Emerging Technologies, 2023*(1), 5728088–5728011.

Jia, X.-H., & Song, J.-S. (2024). Research on design image creation through collaboration between designers and generative AI: Focusing on ChatGPT and Midjourney. *Korea Soc Pub Des, 13*(13), 30–42. https://doi.org/10.54545/kspd.2024.06.30

Ji, P., Jin, J., Wang, T., & Chen, Y. (2014). Quantification and integration of Kano's model into QFD for optimizing product design. *International Journal of Production Research, 52*(21), 6335–6348. https://doi.org/10.1080/00207543.2014.939777

Kate Knibbs. (2024, December 12). *Harvard is releasing a massive free AI training dataset funded by OpenAI and Microsoft.* https://www.wired.com/story/harvard-ai-training-dataset-openai-microsoft/

Kreuzbauer, R., & Malter, A. J. (2005). Embodied cognition and new product design: Changing product form to influence brand categorization. *Journal of Product Innovation Management, 22*(2), 165–176.

Liu, P., Zhang, K., Dong, X., & Wang, P. (2024). A big Data-Kano and SNA-CRP based QFD model: Application to product design under chinese new E-commerce model. *IEEE Transactions on Engineering Management, 71*. https://doi.org/10.1109/TEM.2022.3227094

Mayer, S., & Landwehr, J. R. (2018). Quantifying visual aesthetics based on processing fluency theory: Four algorithmic measures for antecedents of aesthetic preferences. *Psychology of Aesthetics, Creativity, and the Arts, 12*(4), 399–431. https://doi.org/10.1037/aca0000187

Murthy, D. N. (2024). Role of design thinking and customer engagement in product development. *Economic Sciences, 20*(2) https://economic-sciences.com

Paananen, V., Oppenlaender, J., & Visuri, A. (2024). Using text-to-image generation for architectural design ideation. *International Journal of Architectural Computing, 22*(3), 458–474. https://doi.org/10.1177/14780771231222783

Paliwal, G., Donvir, A., Gujar, P., & Panyam, S. (2024). Accelerating time-to-market: The role of generative AI in product development. *IEEE Colombian Conference on Communications and Computing (COLCOM), 2024*, 1–9. https://doi.org/10.1109/COLCOM62950.2024.10720255

Porter, M. E. (1996). What is strategy? *Harvard Business Review, 74*(6), 61–78. https://doi.org/10.1098/rspb.2008.0355

Rane, N., Choudhary, S., & Rane, J. (2023). Enhanced product design and development using Artificial intelligence (AI), Virtual reality (VR), Augmented reality (AR), 4D/5D/6D Printing, internet of things (IoT), and blockchain: A review. *SSRN Electronic Journal.* https://doi.org/10.2139/ssrn.4644059

Ryu, S., & Ryu, S. (2021). Feeling excited and fluent: how consumers respond to the visual appeals of products in an online shopping environment. *Behavior and Information Technology, 40*(11), 1219–1233. https://doi.org/10.1080/0144929X.2021.1942989

Sharma, P., & Gonaygunta, H. (2023). Role of AI in product management automation and effectiveness. *SSRN Electronic Journal.* https://doi.org/10.2139/ssrn.4637857

Tambe, P., Hitt, L. M., Rock, D., & Brynjolfsson, E. (2021). *Digital capital and superstar firms.* https://www.brookings.edu/research/digital-capital-

Timoshenko, A., & Hauser, J. R. (2019). Identifying customer needs from user-generated content. *Marketing Science, 38*(1). https://doi.org/10.1287/mksc.2018.1123

Wei, W., & Wang, Z. (2023). An improved QFD method for rapid response to customer requirements in product optimization design. *Procedia CIRP, 119.* https://doi.org/10.1016/j.procir.2023.01.010

Xia, H., Pan, X., Zhou, Y., & Zhang, Z. (2020). Creating the best first impression: Designing online product photos to increase sales. *Decision Support Systems, 131*, 113235. https://doi.org/10.1016/j.dss.2019.113235

Yang, L., Zhang, Z., Song, Y., Hong, S., Xu, R., Zhao, Y., Zhang, W., Cui, B., & Yang, M.-H. (2024). Diffusion models: A comprehensive survey of methods and applications. *ACM Computing Surveys, 56*(4), 1–39. https://doi.org/10.1145/3626235

Yehia, E. (2024). Developments on generative AI. In *AI and emerging technologies* (1st ed., pp. 139–160). CRC Press. https://doi.org/10.1201/9781003501152-9

Yin, H., Zhang, Z., & Liu, Y. (2023). The exploration of integrating the midjourney artificial intelligence generated content tool into design systems to direct designers toward future-oriented innovation. *Systems, 11*(12), 566. https://doi.org/10.3390/systems11120566

MIX
Papier aus verantwortungsvollen Quellen
Paper from responsible sources
FSC® C105338

If you have any concerns about our products,
you can contact us on
ProductSafety@springernature.com

In case Publisher is established outside the EU,
the EU authorized representative is:
Springer Nature Customer Service Center GmbH
Europaplatz 3, 69115 Heidelberg, Germany

Printed by Libri Plureos GmbH
in Hamburg, Germany